Working with

Molecular Cell Biology
Fifth Edition

Student Companion and Solutions Manual

Brian Storrie, Eric A. Wong, Richard A. Walker,
Glenda Gillaspy, Jill Sible, and Muriel Lederman

Virginia Polytechnic Institute and State University

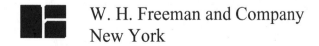
W. H. Freeman and Company
New York

To our students at Virginia Tech
(Virginia Polytechnic Institute and State University)
and elsewhere, and to our teachers

Copyright © 2004 by W. H. Freeman and Company

All rights reserved.

Printed in the United States of America

ISBN: 0-7167-5993-4

First Printing 2003

Table of Contents

2 Chemical Foundations

PART A: *Chapter Summary*

Molecular cell biology aims at understanding biological organization, that is, the structure and function of organisms and cells, in terms of the properties of individual molecules, such as proteins and nucleic acids. Complex cellular processes follow the rules of chemistry and physics. Here, the important chemical concepts required to comprehend biological organization are reviewed. The topics discussed are fundamental to the concepts and experiments presented in later chapters.

Atomic bonds, the strongest of which are covalent bonds, connect individual atoms in a molecule. Noncovalent bonds are weaker but are important stabilizing forces between groups of atoms within larger molecules and between different molecules. Important noncovalent interactions within and among cellular molecules include ionic bonds, hydrogen bonds, Van der Waals interactions and the hydrophobic effect.

Proteins, nucleic acids and polysaccharides are biological macromolecules composed of polymers of covalently linked subunits. Proteins consist of amino acids joined by peptide bonds. Distinguishing molecular properties of amino acids—the building blocks of proteins—result from the molecular composition of their side chains. Ribonucleic acids and deoxyribonucleic acids are polymers of nucleotides joined by phosphodiester bonds. Polysaccharides are linear or branched polymers of pentose or hexose monosaccharides joined by glycosidic bonds. Biological membranes are primarily composed of phospholipids, which form from the esterification of fatty acid chains with hydrophilic heads. Phosopholipids associate by noncovalent interactions to form membranes.

Chemical reactions involve the breaking and reforming of covalent bonds. Chemical equilibrium occurs when the forward and reverse rates of a reaction are equal, although reactions in a cell exist in a steady state rather than equilibrium. The ratio of products to reactants at equilibrium, is described by the equilibrium rate constant, K_{eq}. Another important chemical reaction is the dissociation of ions in water, described by the pH value. The directionality of a chemical reaction depends upon the release of free energy as measured by the ΔG value. In a cell, unfavorable chemical reactions can be coupled to favorable reactions, often the hydrolysis of ATP. Oxidation-reduction reactions involve the transfer of electrons and often involve coenzymes NAD+ and FAD as electron donors and acceptors.

PART B: *Multiple Choice*

Circle the letter corresponding to the most appropriate terms/phrases. There may be more than one correct answer. Circle all that apply.

2.1 *Atomic Bonds and Molecular Interactions*

1. The six common atoms in living systems: hydrogen, carbon, nitrogen, phosphorus, oxygen, and sulfur, each form a characteristic number of covalent bonds. For carbon, this number is:

    a. 1

    b. 2

    c. 4

    d. 6

2. Covalent bonds between which of the following pairs of atoms are polar?

    a. C-C

    b. C-H

    c. O-H

    d. O=P

3. Which type of noncovalent interaction can occur between any two atoms?

    a. ionic interactions

    b. hydrogen bonding

    c. hydrophobic effect

    d. Van der Waals interactions

2.2 *Chemical Building Blocks of Cells*

4. Which amino acid can form disulfide bonds?

    a. arginine

    b. cysteine

    c. proline

    d. tyrosine

5. The synthesis of biological macromolecules by polymerization of monomeric subunits typically proceeds through what type of reaction?

    a. dehydration

    b. hydrolysis

    c. phosphorylation

    d. epimerization

6. Which of the following pairs are stereoisomers?

    a. UDP and UTP

    b. adenine and thymine

    c. glucose and galactose

    d. D-glucofuranose and D-glucopyranose

    e. D-glycine and L-glycine

7. In a transmembrane protein, which of the following amino acids is most likely to be found in contact with the phospholipid tails?

    a. alanine

    b. phenylalanine

    c. aspartate

    d. asparagine

    e. methionine

8. Naturally occurring phospholipids can form which of the following in solution?

    a. micelles

    b. liposomes

    c. bilayers

    d. none of the above

9. Which of the following molecules contains a high-energy phosphoanhydride bond?

    a. AMP

    b. dAMP

    c. ADP

    d. dADP

    e. none of the above

2.3 Chemical Equilibrium

10. Which of the following parameters affects the equilibrium constant for a given chemical reaction?

    a. temperature

    b. pressure

    c. initial concentration of reactants

    d. enzymatic activity

    e. none of the above

11. Which buffer maintains the physiologic pH of the cytosol?

    a. acetate

    b. carbonate

    c. lactate

    d. phosphate

2.4 Biochemical Energetics

12. The phosphoanhydride bonds in a molecule of ATP represent what form of energy?

    a. thermal

    b. radiant

    c. mechanical

    d. chemical potential

    e. electric potential

13. The addition of phosphate residues to proteins is a common and important biochemical reaction with a large positive ΔG value. In a cell, this reaction is driven forward by coupling to a reaction with a _____ _____ ΔG value.

    a. larger positive

    b. larger negative

    c. smaller positive

    d. smaller negative

14. In the above question, the coupled reaction that drives the energetically unfavorable phosphorylation of proteins is most likely to be:

    a. formation of a peptide bond

    b. formation of a disulfide bond

    c. epimerization of glucose

    d. hydrolysis of ATP

15. In the reaction, $FADH_2 \rightarrow FAD+2H^++2e^-$, $FADH_2$ becomes

    a. hydrolyzed

    b. oxidized

    c. dehydrated

    d. reduced

PART C: Reviewing Concepts

2.1 Atomic Bonds and Molecular Interactions

16. Proteins may be affinity purified from a complex cellular extract by absorption to an antibody that binds that protein specifically. Antibodies bind proteins based on molecular complementarity through a variety of noncovalent interactions. Name three noncovalent bonds that might mediate antibody-protein binding. Suggest three mechanisms to separate the protein from the antibody after the absorption step.

17. Explain why ethanol (CH_3CH_2OH) is readily soluble in water, whereas methane (CH_3CH_3) is insoluble in water.

2.2 Chemical Building Blocks of Cells

18. Humans can digest starch but not cellulose even though both of these carbohydrates are polymers of the monosaccharide glucose. What is the molecular distinction between starch and glucose that affects its digestibility?

19. Omega 3 and omega 6 fatty acids, also known as linoleic acid and linolenic acid, respectively, are essential fats in the human diet and can be found in vegetable and fish oils and in legumes. Are these fatty acids saturated or unsaturated? What is the molecular distinction between saturated and unsaturated fatty acids?

20. Cellular membranes are composed of a phospholipids bilayer and associated proteins.

Describe the amphipathic nature of the phospholipid molecule. What noncovalent interactions function to promote the association and stabilization of phospholipids into bilayers?

2.3 Chemical Equilibrium

21. In a cell, do chemical reactions typically reach equilibrium? Why or why not?

22. The pK_a for the dissociation of acetic acid is 4.74. Calculate the molar concentration of acetic acid [HAc] and of sodium acetate [Ac⁻] you would need to make a buffer that is 0.1 M in total acetate and at pH 5. What would be the final pH if 1 mL 1M HCl were added to 100 ml of this buffer?

2.4 Biochemical Energetics

23. A solution of 8 M urea is sometimes used in the isolation of protein molecules. When the solution is prepared by dissolving urea in water at room temperature, it becomes cold. What can you infer about the ΔH, ΔS and ΔG values for the dissolution of urea in water?

24. Hydrolysis of ATP to produce has a ΔG value of −7.3 kcal/mol ADP under standard conditions. The hydrolysis of ATP is often used as a source of energy to "pump" substances against a concentration gradient into or out of a cell. Assume that the hydrolysis of one molecule of ATP is coupled to the transport of one molecule of substance A from inside to the outside of the cell. At 25°C, if the concentration of substance A inside the cell is 100 μm, what is the maximum concentration of substance A outside the cell against which the pump can export it?

25. The $\Delta E°'$ for the reaction $NAD^+ + H^+ + 2e^- \rightarrow$ NADH is −0.32 volts. Is this reaction energetically favorable or unfavorable? Calculate the $\Delta G°'$ for this reaction.

PART D: Analyzing Experiments

26. Lipopolysaccharide (LPS) is a major component of the outer leaflet of the outer membrane of *Escherichia coli* and other gram-negative bacteria. LPS is anchored to the bacterial outer membrane through its lipid A domain. Lipid A contains six fatty acyl chains. As shown in Table 2-1, the composition of the lipid A domain of LPS varies when *E. coli* is grown at 37°C versus 12°C. For reference, the chemical composition of common fatty acids is summarized in *Molecular Cell Biology,* Fifth Edition, Table 2-3 and space filling models of example saturated and unsaturated fatty acids are illustrated in MCB, Figure 2-18.

   a. What is the expected effect of increased palmitoleate levels on the entropic state of the *E. coli* outer membrane?

   b. Biological membranes typically have a fluid-like state. What is the expected effect of increased palmitoleate levels on the viscosity of the *E. coli* outer membrane?

   c. How does varying the relative levels of laurate and palmitoleate over the temperature range of 12°C to 37°C contribute to maintaining a relatively constant membrane fluidity?

   d. Various amino acids, ions, and sugars are transported across lipid bilayers by transport proteins. How may maintaining membrane fluidity be important to the function of transport proteins?

LPS is medically important as endotoxin molecules that produce fever during infections.

	E. coli	
Fatty Acid	Grown at 12°C (μmol/mg LPS)	Grown at 37°C (μmol/mg LPS)
Laurate	0.05	0.16
Palmitoleate	0.10	<0.01

Table 2-1

27. You have purified a protein toxin from a spider. Preliminary studies indicate that the toxin associates with a neurotransmitter receptor expressed on the surface of neurons. To better characterize the toxin/receptor interactions, you perform the following experiments.

Radioactively labeled toxin is incubated with neurons expressing this receptor. The culture of neurons is washed and the amount of bound toxin is quantified by scintillation counting of the radioactivity that remains associated with the neurons.

a. In Figure 2-1, the binding curve in the presence of varying concentrations of the neurotransmitter is shown. What is the effect of neurotransmitter on toxin-receptor interactions?

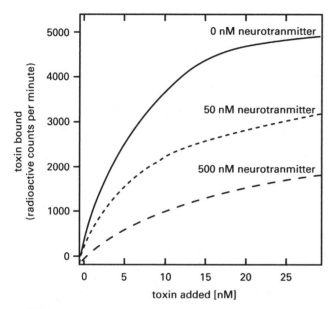

Figure 2-1

b. In Table 2-2, the K_d for neurotransmitter-receptor and for toxin-receptor binding at various salt concentrations is shown. Which compound, neurotransmitter or toxin, has a higher affinity for the receptor? What is the optimal salt concentration for toxin-receptor binding? What do these data tell you about the nature of the toxin-receptor interactions?

[NaCl]	K_d toxin	K_d neurotransmitter
10 mM	5.0×10^{-9} M	4.6×10^{-8} M
50 mM	7.3×10^{-8} M	9.2×10^{-8} M
100 mM	5.3×10^{-2} M	5.7×10^{-2} M
500 mM	3.4×10^{-2} M	3.9×10^{-2} M

Table 2-2

c. You have identified the putative binding site for the toxin on the receptor and have determined that it contains a serine residue that can be phosphorylated by a protein kinase. In cells that do not normally contain the receptor, you express receptors in which this serine residue has been mutated to another amino acid. In Table 2-3, the K_d values for toxin-receptor binding are shown for the different mutated receptors. What do these data tell you about the importance of the serine residue in toxin-receptor interactions?

Amino acid substitution in receptor	K_d toxin
None (wild-type) Ser-> Ser	5.0×10^{-9} M
Ser -> Ala	1.7×10^{-3} M
Ser -> Asp	3.6×10^{-9} M
Ser -> Lys	5.4×10^{-2} M

Table 2-3

28. Transport proteins are an interesting example of energy utilization in biological systems. Transport proteins are involved in the movement of amino acids, ions, and sugars across biological membranes. One such example is CzcA, a protein which mediates resistance to millimolar concentrations of Co^{2+}, Zn^{2+}, and Cd^{2+}. With varying kinetics, CzcA mediates the efflux of each of these divalent cations from cells and maintains the low intracellular concentration of each. Figure 2-2 shows the dependence of cation transport by CzcA on the concentration of metal ion.

a. What evidence does this provide for cation binding to CzcA?

b. Is the ΔG value for cation transport positive, negative, or zero and why?

c. Cation transport is not directly linked to ATP consumption. Rather, CzcA is also a transporter for protons as well as metal cations. The protons are transported in the opposite direction to metal cations. What must be the ΔG value for proton transport for the overall energetics of CzcA to be favorable?

d. Mutation of a negatively charged amino acid to a positively charged amino acid in position 402 of CzcA results in a greatly decreased rate on zinc transport. What is the likely basis of this?

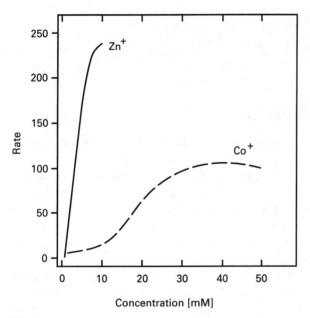

Figure 2-2

Answers

1. c

2. c, d

3. d

4. b

5. a

6. e

7. a, b, e

8. b, c

9. c, d

10. a, b

11. d

12. d

13. b

14. d

15. b

16. A variety of noncovalent interactions including ionic bonds, hydrogen bonds, hydrophobic interactions and Van der Waals interactions can mediate molecular complementarity between proteins. To separate the antibody from the protein, these interactions must be disrupted. This can be achieved in a variety of ways including changing the pH to disrupt ionic and hydrogen bonds, raising the salt concentration to disrupt ionic bonds, adding a detergent to disrupt hydrophobic interactions, or applying heat to increase the enthalpy of the system.

17. Because of its hydroxyl (-OH) group, methanol is a polar molecule that can form hydrogen bonds with water, and is therefore hydrophilic, that is, water soluble. Methane, on the other hand, is a nonpolar molecule that cannot form hydrogen bonds with water and is therefore insoluble in water.

18. Whereas starch is a polymer of the β anomer of glucose, cellulose is a polymer of the β anomer of glucose. Animals contain enzymes capable of digesting α but not β glycosidic bonds, and therefore can digest starch but not cellulose.

19. Linoleic and linolenic acids are both polyunsaturated fatty acids. Unsaturated fatty acids contain double bonds in the hydocarbon chains.

20. Phospholipids are called amphipathic because they contain polar, charged, phosphate-containing head groups and nonpolar fatty acyl tails. In a phospholipids bilayer, the polar head groups interact with each other by hydrogen and ionic bonds and the fatty acyl tails associate with each other by Van der Waals interactions.

21. Chemical reactions in a cell rarely reach equilibrium but rather, are said to be in steady state, e.i. the rate of formation of a product is equal to the rate of its consumption. Steady state dynamics occur because of linked biochemical reactions that generate reactants or consume products and because of the pumping of substances into and out of the cell.

22. The concentration of sodium acetate [Ac$^-$] = 0.0645M and of acetic acid [HAc] = 0.0355M. These values are calculated using the Henderson-Hasselbach equation.

$$pH = pK_a + \log \frac{[A^-]}{[HA]} \quad \text{or}$$

$$pH - pK_a = \log \frac{[Ac^-]}{[HAc]}$$

Since [Ac$^-$] + [HAc] = 0.1M, the following substitutions can be made:

$$5 - 4.74 = \log \frac{[Ac^-]}{0.1\,M - [Ac^-]}$$

Taking the antilog of both sides and solving for [Ac$^-$] gives [Ac$^-$] = 0.0645 M; thus [HAc] = 0.0355 M.

Addition of 1 ml 1 M HCl to 100 ml of this buffer would add 0.01 M H$^+$ ions. The H$^+$ would react with Ac- to form HAc, thus converting the [HAc] to 0.0455 M (0.0355 M + 0.01 M) and the [Ac$^-$]

to 0.0545 M (0.0645 M – 0.01 M). Substituting these values into the Henderson-Hasselbach equation:

$$pH - 4.74 = \log \frac{0.0545 \text{ M}}{0.0455 \text{ M}}$$

$$pH = 4.74 + 0.08$$

$$pH = 4.82$$

Addition of HCl results in a rather small change in pH, illustrating the buffering capacity of acetic acid.

23. The fact that the solution becomes cold means that heat is absorbed; that is, the reaction is endothermic, and ΔH for the reaction is positive. Since urea in fact dissolves under these conditions, ΔG must be negative. In order for ΔG to be negative when ΔH is positive, ΔS must be positive. Indeed, the increase in the degree of disorder when urea is dissolved in water is the driving force of the dissolution reaction.

24. The concentration outside could be as high as 22.6 M. The energy available to power the transport of substance A is the 7.3 kcal/mol available from the ATP hydrolysis. To answer the question, you need to calculate the value of C_2 (concentration of A outside the cell) for $\Delta G = +7.3$ kcal/mol and $C_1 = 100$ µM. If the outside concentration were any greater than the value calculated in this manner, the ΔG value for the transport process would be > +7.3 kcal/mol and, coupled with ATP hydrolysis at –7.3 kcal/mol, ΔG for the overall process would be positive; in this case, the transport of A from inside to outside could not occur.

The ΔG associated with a concentration gradient is given by the expression

$$\Delta G = RT \ln \frac{C_2}{C_1}$$

where a molecule is being transported from C_1 to C_2; in this case, C_2 = outside concentration and C_1 = inside concentration. Rearranging and substituting into this expression gives

$$\ln \frac{C_2}{C_1} = \frac{\Delta G}{RT}$$

$$= \frac{7300 \text{ cal/mol}}{(1.987 \text{ cal/degree mol})(298 \text{ K})} = 12.3$$

Taking the natural antilog of both sides and solving for C_2 when $C_1 = 100$ µM,

$$\frac{C_2}{C_1} = 2.26 \times 10^5$$

$$C_2 = (100 \times 10^{-6} \text{ M})(2.26 \times 10^5) = 22.6 \text{ M}$$

Thus under these conditions, the hydrolysis of one mole of ATP would provide energy for one mole of substance A (with an intracellular concentration of 100 µM) to be exported from the cell, as long as the concentration of A outside the cell remained below 22.6 M!

25. The reaction has a negative $\Delta E°$ ', which means that the reaction is energetically unfavorable. From the textbook, Equation 2-11, ΔG (in cal/mol) = -n (23,064) ΔE (in volts). For this reaction, $\Delta G°$ ' = -2 (23,064) -0.32 = 14,761 cal/mol.

26a. Because palmitoleate has a cis double double bond which introduces a kink into the structure, it can not pack in a lipid bilayer in as orderly a manner as a saturated fatty acid. Therefore, it introduces disorder and increases the entropic state of the E. coli outer membrane.

26b. Because of the disorder introduced, increased palmitoleate levels should lead to an increased viscosity at a constant temperature.

26c. As temperature is decreased, increased palmitoleate levels help to maintain normal membrane viscosity.

26d. Transport proteins bind to different amino acids, ions, and sugars. In doing so and effecting transport, they often have transition states that displace portions of the transporter protein molecule with respect to each other. Molecular displacements are facilitated by a viscous membrane.

27a. Neurotransmitter inhibits the binding of the toxin to the receptor.

27b. The toxin has a smaller K_d than the neurotransmitter, indicating that the toxin has a higher affinity for the receptor. Among the NaCl concentrations tested, 10 mM is optimal for toxin-receptor binding since it yields the smallest K_d. Increasing salt decreases noncovalent ionic interactions, suggesting that ionic bonds contribute to toxin-receptor binding.

27c. Since the K_d is higher when other amino acids are substituted, particularly alanine and lysine, the serine residue is important for toxin-receptor interactions. Since substitution of aspartate, a negatively charged amino acid, has a minimal effect on the K_d, it is likely that phosphorylated, and thereby negatively charged, serine normally contributes significantly to toxin-receptor interactions.

28a. The transport velocity for both Zn^{2+} and Co^{2+} saturates with increasing concentration of metal. This strongly indicates that the metal binds to the CzcA protein.

28b. CzcA keeps the intracellular metal concentration low relative to the extracellular concentration. This means moving the cation against an unfavorable concentration gradient. Everything else equal to the ΔG value is likely to be positive.

28c. If the energetics for moving the metal cation are unfavorable, then the ΔG for proton movement must offset this. The negative ΔG for proton movement must have a greater absolute value than the positive ΔG for cation movement.

28d. The divalent cations and protons must be able to bind to CzcA during their transport. The substitution of a positively charged amino acid for a negatively charged one might interfere with binding by charge repulsion.

3

Protein Structure and Function

PART A: *Chapter Summary*

Proteins are the working molecules of a cell. This chapter describes the structure of proteins and how structure gives rise to function. A protein is a linear polymer of amino acids linked together by peptide bonds. The three dimensional structure of the protein is dictated by its amino acid sequence and is stabilized mainly by non-covalent bonds. Proteins are folded with the assistance of specialized proteins called molecular chaperones and chaperonins. The degradation of many proteins is mediated by the addition of ubiquitin molecules to internal sequences followed by proteolysis in macromolecular assemblies called proteasomes.

Enzymes are catalytic proteins that accelerate the rate of cellular reactions by lowering the activation energy and stabilizing transition state intermediates. Some enzymes exhibit allostery, in which the binding of one ligand molecule affects the binding of subsequent ligand molecules. Molecular motors are proteins that convert energy released by the hydrolysis of ATP or from ion gradients into a mechanical force. Motor proteins also possess the ability to translocate along a cytoskeletal filament, nucleic acid strand, or protein complex and show net movement in a given direction.

Proteins can be separated by a number of methods based on mass, density or charge. Some of the common methods include centrifugation, electrophoresis, and liquid chromatography. Centrifugation separates proteins based on their mass and shape. Gel electrophoresis separates proteins based on their rate of movement in an applied electric field. Liquid chromatography separates proteins based on their size, charge, or binding affinity. The primary amino acid sequence of a protein can be determined by the chemical Edman degradation method, computer analysis of the nucleotide sequence of a cloned DNA, or by analysis of peptide fragments by mass spectrometry. In addition, protein conformation can be determined using x-ray crystallography, cryoelectron microscopy, and NMR spectroscopy.

PART B: *Multiple Choice*

Circle the letter corresponding to the most appropriate terms/phrases. There may be more than one correct answer. Circle all that apply.

3.1 *Hierarchical Structure of Proteins*

1. Which level of protein structure is held together by hydrophobic interactions between non-polar side chains and hydrogen bonds between polar side chains?

a. the primary structure

b. the secondary structure

c. the tertiary structure

d. the quaternary structure

2. A structural polypeptide domain

   a. is a discrete region in the tertiary structure of a protein

   b. consists of all positively-charged or negatively-charged amino acid residues

   c. consists of 100 to 150 amino acid residues

   d. usually consists of amino acids from multiple polypeptide chains

3. Which of the following statements about macromolecular assemblies is true?

   a. contain one very large polypeptide

   b. can exceed one mDa in mass

   c. can be 30-300 nm in size

   d. can contain nucleic acids in addition to amino acids

3.2   Folding, Modification and Degradation of Proteins

4. Molecular chaperones are a class of proteins that play a role

   a. in the degradation of unfolded proteins

   b. in the proper folding of proteins

   c. in the formation of peptide bonds

   d. in the transport of proteins out of the cell

5. Which of the following plays a role in the degradation of proteins?

   a. ribosome

   b. chaperonin

   c. proteasome

   d. ubiquitin

3.3   Enzymes and the Chemical Works of Cells

6. Which of the following properties is characteristic of enzymes?

   a. stabilize transition state intermediates

   b. increase the activation energy of a reaction

   c. increase the rate of a reaction

   d. react with many different substrates

7. The active site of an enzyme

   a. contains a substrate binding region

   b. contains a catalytic region

   c. is typically located at the amino terminus of the protein

   d. can be formed from amino acids located throughout the protein

3.4   Molecular Motors and the Chemical Work of Cells

8. Which of the following is a property of a molecular motor?

   a. generate linear or rotary motion

   b. translocate along a cytoskeletal filament

   c. generate net movement in a given direction

   d. function in an energy-independent manner

9. Which of the following is an example of action by molecular motors?

   a. phosphorylation of a substrate by protein kinase

   b. translation of mRNA by a ribosome

   c. movement of chromosomes during mitosis

   d. rotation of flagella

3.5   Common Mechanisms for Regulating Protein Function

10. Regulation of protein activity by phosphorylation/dephosphorylation requires

    a. kinase

    b. ubiquitin

    c. protease

    d. phosphatase

11. Positive cooperativity

    a. results in a linear plot of reaction velocity versus substrate concentration

    b. occurs in monomeric proteins, such as myoglobin

    c. results in a more efficient response to small changes in substrate concentration

    d. leads to inhibition of the sequential binding of ligands

3.6 Purifying, Detecting and Characterizing Proteins

12. Which of the following methods can separate particles based on density?

    a. centrifugation

    b. ion exchange chromatography

    c. SDS polyacrylamide gel electrophoresis

    d. affinity chromatography

13. During ion exchange chromatography with a column of positively charged beads

    a. positively-charged proteins will bind to the beads

    b. negatively-charged proteins will bind to the beads

    c. uncharged proteins will bind to the beads

    d. both positively and negatively charged proteins will bind to the beads

14. Which of the following methods can be used to determine the conformation of a protein?

    a. x-ray crystallography

    b. cryoelectron microscopy

    c. NMR spectroscopy

    d. mass spectroscopy

15. Starting with 1 mCi of a phosphorus-32 labeled compound, how many mCi are left after 28.6 days?

    a. 1 mCi

    b. 0.5 mCi

    c. 0.25 mCi

    d. 0.125 mCi

PART C: Reviewing Concepts

3.1 Hierarchical Structure of Proteins

16. Describe the four hierarchical levels of protein structure.

17. Describe the relationship between amino acid sequence, three-dimensional structure, and the function of a protein.

3.2 Folding, Modification and Degradation of Proteins

18. What role do molecular chaperones and chaperonins play in mediating protein folding?

19. Describe the ubiquitin-mediated proteolytic pathway.

3.3 Enzymes and the Chemical Works of Cells

20. Describe the structural and functional properties of the active site of an enzyme.

21. Describe how the K_d for antibody-antigen binding is similar to the K_m for an enzyme-catalyzed reaction.

3.4 Molecular Motors and the Chemical Work of Cells

22. Describe the cycle of events that are involved in the movement of myosin along an actin filament.

3.5 Common Mechanisms for Regulating Protein Function

23. Describe how phosphorylation/dephosphorylation can regulate the activity of a protein.

24. Describe how cyclic AMP can activate protein kinase A.

3.6   Purifying, Detecting and Characterizing Proteins

25. What methods can be used to separate proteins based on mass or charge?

26. What methods can be used to determine the primary amino acid sequence of a protein?

PART D: *Analyzing Experiments*

27. Two enzymes, enzyme X and enzyme Z, can both use substrate D to produce product E. The assay results are shown in Table 3-1.

a. Using the data shown in Table 3-1, plot the reaction velocity (μmol E produced/min) versus substrate concentration ([D] mM) for both enzyme X and enzyme Z using the graph shown in Figure 3.1. What type of enzyme kinetics do enzymes X and Z show?

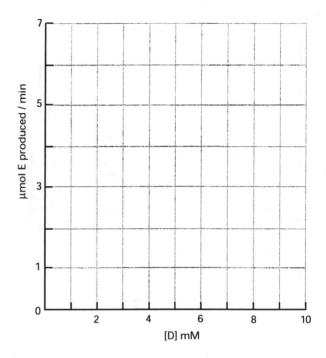

FIGURE 3-1

TABLE 3-1

| | Reaction velocity | |
| | μmol E produced/min | |
[D] mM	Enzyme X	Enzyme Z
1.0	1.4	0.2
2.0	2.7	0.4
3.0	3.6	1.3
4.0	4.3	3.7
5.0	4.9	5.5
7.5	5.7	5.9
10.0	6.0	6.0

b. Estimate the K_m for both enzyme X and enzyme Z for substrate D.

c. How do the activities of enzyme X and enzyme Z differ in the range of 3 to 5 mM D?

d. Based on this data, what can you conclude about the binding of substrate D to enzymes X and Z?

28. Protein A and protein B were analyzed using two-dimensional gel electrophoresis as shown in Figure 3-2.

a. What can you conclude about the size and net charge of protein A and protein B at neutral pH?

b. How can you separate protein A from protein B using an ion exchange column that consists of positively charged beads (an anion exchanger)?

c. Using gel filtration chromatography, protein B was found to elute from the column at an apparent molecular weight of 60 kDa. How can you explain the discrepancy between the molecular weight of protein B when determined by two-dimensional gel electrophoresis and that determined by gel filtration chromatography?

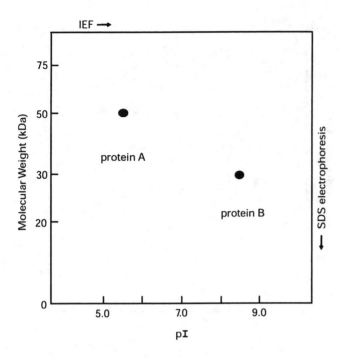

FIGURE 3-2

Answers

1. c

2. a, c

3. b, c, d

4. b

5. c, d

6. a, c

7. a, b, d

8. a, b, c

9. b, c, d

10. a, d

11. c

12. a

13. b

14. a, b, c

15. c

16. The four hierarchical levels of protein structure are the primary, secondary, tertiary, and quaternary structure. The primary structure is the sequence of amino acid residues. The secondary structure is the localized organization of parts of a polypeptide chain. The tertiary structure is the overall conformation of a polypeptide chain and the quaternary structure describes the number and relative positions of the subunits of a multimeric protein.

17. The primary amino acid sequence of a protein determines its three-dimensional structure. Proteins with the same three-dimensional structure have a similar function, even if they have different primary amino acid sequences.

18. Two general families of chaperones help fold proteins: molecular chaperones and chaperonins. Chaperones bind and stabilize unfolded or partially folded proteins and thereby prevent these proteins from being degraded. Molecular chaperones are thought to bind to all nascent peptides as they are being synthesized on ribosomes. Chaperonins directly facilitate the folding of proteins.

19. Proteins targeted for degradation by the ubiquitin-mediated pathway are degraded in a two step process. First, a chain of ubiquitin molecules is added to an internal lysine side chain of the target protein. Additional ubiquitin molecules are added forming a polyubiquitin chain. The ubiquitin-tagged protein is then degraded in a large cylindrical multisubunit complex called a proteosome.

20. The active site of an enzyme consists of two functionally important regions: the substrate binding region and the catalytic region. In some enzymes, the catalytic region is part of the substrate-binding region; in other enzymes, the regions are structurally separated. The active site of an enzyme is not necessarily formed by the folding of contiguous amino acids but can also be formed by folding amino acids from different parts of the protein.

21. The K_d for antibody-antigen binding is a measure of the binding affinity of the antigen for the antibody. In a similar fashion, the K_m for an enzyme-catalyzed reaction is also a measure of the affinity of the substrate for its enzyme. In both cases low K_d or K_m values indicate high affinity of the antigen for the antibody or substrate for the enzyme.

22. In the absence of ATP, a myosin head binds tightly to an actin filament. Upon binding ATP, the myosin head dissociates from the actin filament. The myosin head then hydrolyzes the bound ATP, causing a conformational change that moves the head towards the (+) end of the actin filament, where it rebinds to the actin filament. As phosphate (Pi) dissociates from the ATP-binding pocket, a second conformational change occurs that moves the actin filament (the power stroke). In the final step, ADP is released.

23. One of the most common mechanisms for regulating protein activity is the addition and removal of phosphate groups from serine,

threonine, or tyrosine residues. Protein kinases catalyze phosphorylation and phosphatases catalyze dephosphorylation. The activity of the target protein cycles between active and inactive depending upon the phosphorylation state of the protein.

24. Protein kinase A exists as an inactive tetramereric protein composed of two catalytic subunits and two regulatory subunits. In this conformation, the regulatory subunits block the active site in the catalytic subunits. Binding cyclic AMP (cAMP) to the regulatory subunits induces a conformational change that releases the catalytic subunits, resulting in active protein kinase A.

25. Proteins can be separated based on their mass (molecular weight) or charge using a variety of techniques. Centrifugation, gel electrophoresis, gel filtration, and time-of-flight mass spectrometry are four methods used for separating proteins based on their mass. Ion exchange chromatography separates proteins based on their charge.

26. The primary amino acid sequence of a protein can be determined by the chemical Edman degradation method. In this method, the amino acid sequence is determined by the stepwise cleavage and identification of amino acids from the amino terminal end of the protein. A second method involves using mass spectroscopy to determine the amino acid sequence of small peptides, which are derived from proteolytic cleavage of the protein. The most common method, however, for determining the amino acid sequence of a protein is computer analysis of the nucleotide sequence of the cloned gene.

27a. A plot of the reaction velocity (μmol E produced/min) versus substrate concentration ([D] mM) for both enzyme X and enzyme Z is shown in Figure 3-3. Enzyme X shows the standard hyperbolic curve characteristic of an enzyme that follows Michaelis-Menten kinetics. In contrast, enzyme Z shows sigmoidal kinetics, indicative of cooperative substrate binding.

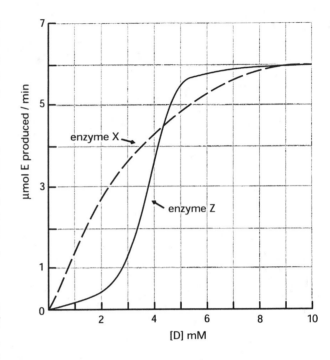

FIGURE 3-3

27b. K_m is a measure of the affinity of the substrate for the enzyme. The K_m for an enzyme is defined as the concentration of substrate that gives half maximal velocity (1/2 V_{max}). The V_{max} for both enzymes is 6 μmol E produced/min; however, the shapes of the two curves are different. From the curves, it can be estimated that the K_m for enzyme X is 2.3 mM D and for enzyme Z it is 3.7 mM D.

27c. The shapes of the curves for enzyme X and enzyme Z differ dramatically within the 3-5 mM D range. For enzyme X, the velocity of the reaction increases approximately 36% (from 3.6 to 4.9 μmol E produced/min; see Table 3-1). In contrast, for enzyme Z, the velocity of the reaction increases 4.2-fold or 420% (from 1.3 to 5.5 μmol E produced/min). For enzyme Z, a small change in substrate concentration results in a large change in enzyme activity.

27d. The simple Michaelis-Menten kinetics of enzyme X suggest that substrate D binds to only one site in enzyme X. In contrast, the sigmoidal kinetics of protein Z indicate that there are multiple binding sites for substrate D, and binding of substrate D shows cooperative allostery. During positive cooperative allostery, the binding of one substrate molecule facilitates the binding of subsequent substrate molecules.

28a. From the two-dimensional gel, you can determine that the molecular weight of protein A is 50 kilodaltons (kDa) and protein B is 30 kDa. The isolectric point (pI) is the pH at which the protein has a net charge of zero. The pI for protein A is 5.5. Therefore at pH 5.5, protein A has a net charge of zero, and at pH 7.0 (neutral pH), protein A would have a net negative charge. Similarly, at pH 7.0, protein B would have a net positive charge.

28b. As shown in 28a, protein A has a net negative charge and protein B has a net positive charge at pH 7.0. Therefore, at neutral pH, protein A will bind to the positively charged beads used in ion exchange chromatography. Protein B will not bind and will flow through the ion exchange column. Protein A can then be eluted from the column using increasing concentrations of salt.

28c. Whereas SDS gel electrophoresis separates a protein into its individual subunits, gel filtration chromatography does not disrupt the subunit structure of a protein. Therefore, we can conclude that protein B normally exists as a dimer of two identical 30 kDa subunits (a homodimer). This would explain why protein B appears to be 60 kDa by gel filtration chromatography but only 30 kDa by SDS gel electrophoresis.

4

Basic Molecular Genetic Mechanisms

PART A: *Chapter Summary*

Geneticists from Gregor Mendel through Thomas Hunt Morgan have studied the hereditary units of information that control the identifiable traits of an organism. The exact duplication of this information in any species from generation to generation assures the genetic continuity of that species. Chapter 4 focuses on the way plans stored in deoxyribonucleic acid (DNA) are used to create functioning proteins. Discovery of the structure of DNA in 1953 and the subsequent elucidation of the steps in which DNA, RNA, and proteins are synthesized are the monumental achievements of the early days of molecular biology.

DNA directs synthesis of RNA, which then directs assembly of proteins in a chain of events known as the *central dogma* of molecular biology. The structure of nucleic acids, linear sequences of nucleotides, makes these processes possible. There are a variety of types of RNA with different conformations that allow them to serve specific functions in the cell. Messenger RNA, transfer RNA and ribosomal RNA serve important functions during protein synthesis.

The "decision" to initiate transcription of a gene is the major mechanism for controlling production of encoded proteins in the cell. Prokaryotes provide a good example of the regulatory process because they constantly govern gene expression in order to accommodate changes in their environment. Once the decision to create a protein is made, mRNA molecules carry information from DNA in a three-letter code, tRNA helps decipher the code, and rRNA catalyzes the assembly of amino acids into polypeptide chains. These components come together to carry out the stepwise synthesis of proteins. After becoming familiar with how genetic information encoded by DNA is translated into the structure of proteins that perform most cell functions, take a look at how DNA sequences are copied precisely in DNA replication.

Lastly, because viruses have been important models for studying DNA replication, RNA transcription and protein translation, the chapter closes with a basic description of viral growth cycles. Since the processes described in this chapter are so central to all biological functions—growth control, differentiation, and the specialized chemical and physical properties of the cell—they will arise again and again in later chapters. A firm grasp of the fundamentals of DNA, RNA and protein synthesis is necessary to follow the subsequent discussions.

PART B: *Multiple Choice*

Circle the letter corresponding to the most appropriate terms/phrases. There may be more than one correct answer. Circle all that apply.

4.1   *Structure of Nucleic Acids*

1. The percentage of G-C base pairs in a DNA molecule is related to the T_m of that molecule because:

   a. the stability of G-C and A-T base pairs is intrinsically different.

   b. A-T base pairs require a higher temperature for denaturation.

   c. the triple bonds of G-C base pairs are less stable than the double bonds of A-T base pairs.

   d. the G-C content equals the A-T content.

2. An investigator would be able to distinguish a solution containing RNA from one containing DNA by

   a. heating the solutions to 82.5 °C and measuring the absorption of light at 260 nm.

   b. comparing the T_m of each solution.

   c. monitoring the change in absorption of light at 260 nm while elevating the temperature.

   d. measuring the absorption of light at 260 nm.

3. What happens to a supercoiled, circular DNA molecule if it becomes *nicked*?

   a. torsional stress increases

   b. torsional stress decreases

   c. a relaxed circle forms

   d. a linear molecule forms

4.2   *Transcription of Protein-Coding Genes and Formation of Functional mRNA*

4. Which of the following are substrates used by RNA polymerase in the process of transcription?

   a. DNA template

   b. ribonucleotide triphosphates

   c. deoxyribonucleotide triphosphates

   d. inorganic phosphate

5. What protein factors are required for transcription?

   a. general transcription factors

   b. RNA polymerase

   c. DNA polymerase

   d. ribosome

6. Alternative splicing can result in which of the following?

   a. production of different mRNA molecules from the same gene

   b. production of different proteins from the same gene

   c. production of 2 or more proteins which differ in their function

   d. production of novel peptide domains

4.3   *Control of Gene Expression in Prokaryotes*

7. Which of the following mutations can result in a reduction of β-galactosidase?

   a. a mutation in adenylate cyclase

   b. a mutation in catabolite activator protein (CAP)

   c. a mutation in the CAP site in the *lac* control region

   d. a mutation in the repressor binding site in the operator

8. The *trp* operon in *E. coli*

   a. allows for a decrease in production of genes required for tryptophan synthesis.

   b. allows for an increase in production of genes required for tryptophan utilization.

   c. is controlled by tryptophan binding to the *trp* repressor.

   d. is controlled by tryptophan binding to the *trp* operator.

   e. involves conformational changes in the *trp* repressor.

9. The NtrC protein from *E. coli*

   a. is a σ^{54} RNA polymerase.

   b. stimulates transcription of the *glnA* gene.

   c. is activated by a protein kinase called NtrB.

   d. binds both to an enhancer site upstream of the *glnA* gene and to RNA polymerase.

   e. has ATPase activity.

4.4 The Three Roles of RNA in Translation

10. Using the Genetic Code found in Table 4-1 in the textbook, determine which of the following sequences has the potential to encode this peptide sequence: Val-Leu-Leu-Gln-Asp.

    a. GTGCTCTTGCAAGACA

    b. AGTATTGTTGCAGGATT

    c. TAGTTCTTTTACAGGAC

    d. ATAGTTCTTTTACAGGA

11. In which of the following sets of organisms would translation of sequence a (found above) yield similar peptides?

    a. plants, mammals

    b. Acetabularia, mammals, yeast

    c. Acetabularia, yeast

    d. mammals, yeast

12. Each of the 20 different aminoacyl-tRNA synthetases

    a. links an amino acid to the 3' terminus of a tRNA molecule.

    b. recognizes multiple amino acids.

    c. requires ATP to catalyze reactions.

    d. sometimes make mistakes.

13. The ribosome

    a. is an enzyme complex made entirely of protein molecules.

    b. directs elongation of polypeptides.

    c. is organized into 2 subunits whose sizes are designated in Svedburg (S) units.

    d. is not used by cells that secrete large amounts of protein.

4.5 Stepwise Synthesis of Proteins on Ribosomes

14. Which of the following occurs when the eukaryotic translational machinery encounters the TAG codon:

    a. The bound preinitiation complex stops scanning and positions the Met-tRNA$_i^{Met}$ at this site.

    b. The termination factors recognize this codon and translation ends.

    c. This codon is recognized by the corresponding anticodon of an empty tRNA molecule that is not linked to an amino acid.

    d. This codon is not recognized by any factors which ultimately causes the translational machinery to stop.

4.6 DNA Replication

15. DNA replication

    a. requires a DNA template, deoxynucleotides, primers and DNA polymerase.

    b. can be initiated de novo.

    c. requires the addition of de-oxynucleotides to the 5' free hydroxl group.

    d. occurs on only 1 strand of DNA.

    e. utilizes Okazaki fragments on the leading strand.

PART C: *Reviewing Concepts*

4.1   *Structure of Nucleic Acids*

16. If the adenine content of DNA from an organism is 36 percent, what is the guanine content?

17. Which of the common Watson-Crick base pairs in DNA is most stable? Why? How does this property affect the melting temperature of DNA?

4.2   *Transcription of Protein-Coding Genes and Formation of Functional mRNA*

18. What is the major difference in the genome organization of prokaryotic and eukaryotic genes?

19. Describe three features of eukaryotic transcripts that generally are not shared with prokaryotic transcripts.

4.3   *Control of Gene Expression in Prokaryotes*

20. Regulation of the *lac* operon is under both negative and positive control. Describe how these two transcriptional control mechanisms regulate expression of the *lac* operon in the presence of glucose and/or lactose.

4.4   *The Three Roles of RNA in Translation*

21. What is one conclusion that can be drawn from the observation that the genetic code is nearly identical in all cells on earth?

22. What is one possible reason why nonstandard base pairing (wobble) is allowed during protein synthesis?

23. What purpose is served by having mRNA, aminoacyl-tRNAs, and various enzymes associated with a large, complicated structure (the ribosome) during protein synthesis?

4.6   *DNA Replication*

24. How does the enzyme pyrophosphatase participate in DNA replication, in transcription, and in protein synthesis?

25. Why is DNA synthesis discontinuous; that is, why is DNA ligase needed to join fragments of one strand of DNA?

4.7   *Viruses: Parasites of the Cellular Genetic System*

26. Why did the discovery of reverse transcriptase go against the so-called *central dogma*? Which organism was utilized in the discovery of reverse transcriptase and why is this enzyme crucial to this organism's life cycle?

PART D: *Analyzing Experiments*

27. In the experiments that led to the deciphering of the genetic code, synthetic mRNAs such as polyuridylate were incubated with a cell-free *E. coli* translation system. Although these synthetic polynucleotides were translated slowly (relative to the rates observed for biological mRNAs), the corresponding peptides were produced in sufficient quantity to be analyzed. You are probably wondering why these synthetic mRNAs were ever translated, since they do not contain start codons. The answer lies in the relatively high concentrations of Mg^{2+} (0.02 M) used by Nirenberg and his coworkers in these experiments.

The effects of Mg^{2+} were demonstrated by incubating bacterial ribosomes, a synthetic polyribonucleotide, initiation and elongation factors, tRNAs, and nucleotide triphosphates with 0.005 M Mg^{2+} or 0.02 M Mg^{2+} for 2 min. A portion of each mixture was then centrifuged for 2 h at 100,000g on a 15–40 percent sucrose density gradient. This procedure separates macromolecules on the basis of mass, or S value. After centrifugation, the centrifuge tubes were punctured and the contents allowed to drip slowly into a series of collection tubes. These fractions were assayed for RNA content by measuring the absorbance at 260 nm, a wavelength at which RNA absorbs quite strongly (the bulk of the RNA in these preparations is rRNA). Results of such an analysis, called a *shift assay,* are shown in Figure 4-1. Translation assays also were conducted by adding amino acids to another portion of each incubation mixture and measuring the amount of protein formed. Protein synthesis was observed at the higher magnesium concentration; no protein synthesis could be detected at the lower magnesium concentration.

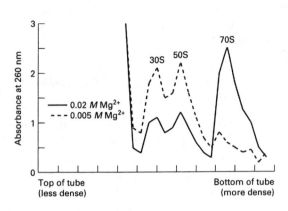

Figure 4-1

a. What difference in the interactions among the components in the incubation mixtures with 0.005 M Mg^{2+} and 0.02 M Mg^{2+} is likely to provide the basis for the observed differences in the RNA profiles of the two mixtures?

b. What would be the predicted profile of a similar fractionation performed on a mixture of bacterial ribosomes, initiation and elongation factors, tRNAs, nucleotide triphosphates, and a biologically synthesized mRNA in 0.005 M Mg^{2+}? Why?

28. One of the classic studies contributing to our understanding of the *lac* operon was the PaJaMo experiment. In this experiment, diploid *E. coli* cells (merozygotes) were formed by conjugation of $I^+ Z^+$ (donor) cells with $I^- Z^-$ (recipient) cells in the absence of inducer. The levels of β-galactosidase activity in the merozygotes were monitored as a function of time and of inducer addition. The basic experimental protocol and the results observed are summarized in Figure 4-2.

   a. Explain why an increased rate of β-galactosidase synthesis was observed initially in the diploid bacteria and why, at later times, inducer was required for rapid β-galactosidase synthesis.

   b. What can you conclude about the nature of the inducer from this experiment?

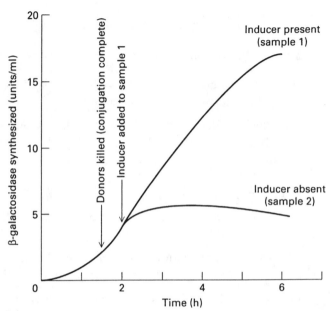

Figure 4-2

Answers

1. a

2. c

3. b, c

4. a, b

5. a, b, d

6. a, b, c, d

7. a

8. a, c, e

9. b, c, d, e

10. a, b, c

11. a, b, c, d

12. a, c, d

13. b, c

14. b

15. a

16. Since A = T, the A + T content = 72 percent and G + C content = 28 percent. Since G = C, the content of G is 14 percent.

17. Guanine-cytosine (G-C) base pairs are more stable than adenine-thymine (A-T) pairs, because G and C form three hydrogen bonds, whereas A and T form only two. The greater stability conferred by the additional hydrogen bonds in G-C pairs means that DNA rich in G-C pairs requires more energy for denaturation than does DNA rich in A-T pairs. Thus, the melting (denaturation) temperature of G-C-rich DNA is higher than that of A-T-rich DNA.

18. In prokaryotes, genes devoted to a single metabolic goal are most often found in a contiguous array called an operon. These operons are regulated as a unit from a single promoter. In eukaryotes, this clustering of genes does not occur, and such genes are most often physically separated in the DNA. Each gene is transcribed from its own promoter producing single RNA molecules. In addition, eukaryotes have a significant amount of non-coding DNA present in their genomes. Prokaryotes genes are closely packed with very few non-coding gaps. Lastly, eukaryotic genes can contain non-coding sequences called introns within their coding regions. Prokaryotic genes do not contain introns.

19. Unprocessed eukaryotic transcripts contain introns which must be spliced out before translation occurs. Eukaryotic transcripts also are modified at the 5' and 3' ends. The 5' end contains a 7-methylguanylate cap that is connected to the terminal nucleotide of the RNA molecule. In addition, processing at the 3' end produces a free hydroxyl group to which a string of adenylic acid residues is added. This is called the poly(A) tail. While some prokaryotic transcripts do contain a poly(A) tail, this is still considered a hallmark of eukaryotic mRNA transcripts.

20. In the presence of glucose and the absence of lactose, cAMP levels are low. The *lac* repressor is bound to the *lac* operator which blocks the synthesis of *lac* mRNA. In the presence of glucose and lactose, the *lac* repressor is bound to lactose and is unable to bind to the *lac* operator. However, because cAMP is low in the presence of glucose, very little *lac* mRNA is synthesized. In the absence of glucose and the presence of lactose, the *lac* repressor is again unable to bind to the *lac* operator because it is bound to lactose. However, cAMP levels are high in the absence of glucose. A cAMP-CAP complex forms which binds to the CAP site and stimulates transcription of *lac* mRNA.

21. A strong conclusion from this observation is that life on earth evolved only once.

22. Wobble may speed up protein synthesis by allowing the use of alternative tRNAs. If only one codon-anticodon pair was permitted for each amino acid insertion, protein synthesis might be temporarily halted until a reasonable level of that particular activated tRNA was regenerated. In fact, a slowdown of protein synthesis, due to the lack of a particular aminoacyl tRNA, is used to regulate the levels of enzymes involved in synthesis of some amino acids. This process is called *attenuation*.

23. The highly specific chemical reactions of translation take place at a much higher rate if the individual components (mRNA, aminoacyl-tRNAs, and the appropriate enzymes) are confined by mutual binding to a common structure, the ribosome. This interaction limits diffusion of one component away from the rest and enables protein synthesis to proceed at the rate of nearly 1 million peptide bonds per second in the average mammalian cell. (Similarly, electron donors and acceptors, in a highly organized array, such as that found in the inner membrane of the mitochondrion or the plasma membrane of a bacterium, can operate much more efficiently than they would if diffusion in three dimensions occurred.) Ribosomes not only provide a site at which the necessary components for protein synthesis are assembled, but at least one ribosomal component, the (prokaryotic) 23S rRNA, is involved in catalyzing the formation of peptide bonds.

24. Pyrophosphatase catalyzes the breakdown of pyrophosphate (PPi) to two molecules of inorganic phosphate. In DNA replication and in transcription, the α-phosphate of a nucleotide triphosphate is attached to the 3′-hydroxyl of the pentose on the preceding residue, releasing a pyrophosphate. The breakdown of the pyrophosphate by pyrophosphatase is an energetically favorable reaction that helps to drive nucleic acid synthesis. (Synthesis of large polymers from monomers is generally energetically unfavorable due to a large decrease in entropy upon polymer formation.) In protein synthesis, the breakdown of ATP is used to drive the activation of the tRNAs by tRNA-aminoacyl synthetase. ATP is first broken down to AMP as the tRNA is aminoacylated. The reaction is driven further in the direction of tRNA acylation as the released pyrophosphate is broken down by pyrophosphatase. Thus, in all these cases, pyrophosphatase acts to make a synthetic process more energetically favorable by allowing the energy present in both phosphoanhydride bonds of a nucleotide triphosphate to be used to drive an unfavorable reaction.

25. DNA synthesis is discontinuous because the double helix consists of two antiparallel strands and DNA polymerase can synthesize DNA only in the 5′ to 3′ direction. Thus one strand is synthesized continuously at the growing fork, but the other strand is synthesized in fragments that are joined by DNA ligase.

26. The *central dogma* states that information flows from DNA to RNA. The discovery of reverse transcriptase goes against the *central dogma* because this enzyme catalyzes the synthesis of DNA from a viral RNA template. Reverse transcriptase was first identified from retroviruses, viruses which contain an RNA genome. Retroviruses need reverse transcriptase to copy the viral genome into double-stranded DNA complementary to the virion RNA. This DNA is integrated into the chromosomal DNA of the infected cell. This DNA is then transcribed by the cell's own machinery into RNA, which is either translated into viral proteins or is packaged within the virion coat proteins to form progeny.

27a. The analytical technique described separates the 30S (first peak) and 50S (second peak) ribosomal subunits from each other and from fully associated 70S (third peak) prokaryotic ribosomes. At the higher Mg^{2+} concentration, the 30S and 50S subunits form a ternary 70S complex with the synthetic polyribonucleotide, whereas at the lower concentration, they do not (see Figure 4-1). Because this association must occur before protein synthesis is initiated, translation occurs only at the higher Mg^{2+} concentration.

27b. The fractionation profile for a mixture containing a biological mRNA should resemble the profile depicted with the solid line in Figure 4-1; that is, the ribosomal subunits and the mRNA are associated, and protein synthesis can proceed on the 70S ribosomes. This productive association occurs only after the 30S subunit forms a complex with GTP, N-formylmethionyl tRNA, and initiation factors and finds an AUG codon on the biological mRNA. The complex of the 30S subunit, GTP, initiation factors, N-formylmethionyl tRNA, and mRNA can form at low Mg^{2+} concentrations if an AUG triplet is present on the ribonucleotide to be translated.

28a. Since the *I* gene codes for repressor and the *Z* gene codes for β-galactosidase, the recipient I^-Z^- cells are incapable of synthesizing either repressor or enzyme; these cells do, however, have an intact operator to which repressor can bind. As the donor I^+Z^+ genome enters the I^-Z^- cells, some of the repressor bound to the donor operators is released and binds to free operators in recipient cells; transcription of the Z^+ gene on the donor genome can then proceed. Gradually, however, new repressor is synthesized in the diploid, leading to re-repression of the Z^+ gene unless inducer is added.

28b. If the explanation above is correct, then the *lac* repressor is diffusible and limiting in amount with respect to regulation of the *lac* operon. Moreover, the repressor can dissociate from the donor operator at a sufficient rate to support the initial increase in β-galactosidase synthesis observed.

5

Biomembranes and Cell Architecture

PART A: *Chapter Summary*

At both the molecular and cellular level, structure and function are intimately related. This chapter focuses on the basic structures of eukaryotic cells and the methods used to identify and characterize cellular structures. Much of our knowledge of cell architecture is derived from direct observations of cells and subcellular structures by light and electron microscopy. Another basic approach to the study of cell structure and function is to break open cells and separate them into subcellular components for further characterization. Together, microscopy and cell fractionation techniques have provided clear evidence for the universality of biomembrane structure in all cells, and have helped define the specific structure and function of the plasma membrane, the cytoskeleton, and each of the organelles present in eukaryotic cells.

PART B: *Multiple Choice*

Circle the letter corresponding to the most appropriate terms/phrases. There may be more than one correct answer. Circle all that apply.

5.1 Biomembranes: Lipid Composition and Structural Organization

1. Which of the following statements describe the typical biomembrane?

   a. It may include different types of lipid molecules.

   b. The lipid molecules in the membrane are attached to each other though covalent bonds.

   c. All lipids and proteins in the biomembrane are equally mobile.

   d. It contains a hydrophilic core.

   e. It has a cytosolic face and an exoplasmic face.

2. Which of the following increase membrane fluidity?

   a. phospholipids with long, saturated fatty acyl chains

   b. phospholipids with short, unsaturated fatty acyl chains

   c. lower temperatures

   d. cholesterol at the usual concentrations found in biomembranes

   e. lipid rafts

3. Membrane lipids may

   a. spontaneously flip-flop from one membrane leaflet to the other

   b. move laterally within one membrane leaflet

   c. aggregate with membrane proteins to form "lipid rafts"

   d. be unequally distributed in the two membrane leaflets

   e. be present in varying amounts in different cellular membranes

5.2 Biomembranes: Protein Components and Basic Functions

4. A GPI anchor

   a. serves to attach some proteins to the cytosolic face of the plasma membrane

   b. contains phosphotidylserine

   c. may be cleaved to release the protein from the membrane

   d. contains sugar residues in addition to lipid components

   e. contains phosphoethanolamine

5. Membrane proteins may interact with biomembranes through

   a. covalently-attached fatty acid molecules

   b. covalently-attached lipid molecules

   c. non-covalent protein-protein interactions

   d. hydrophobic α-helical domains

   e. non-covalent protein-lipid interactions

6. Which of the following is true of transmembrane proteins?

   a. all contain a hydrophobic α helix

   b. some serve as membrane attachment sites for peripheral membrane proteins

   c. some contain more than one membrane-spanning domain

   d. all are asymmetrically oriented in the lipid bilayer

5.3 Organelles of the Eukaryotic Cell

7. Cells that produce large amounts of secretory proteins are abundant in

   a. mitochondria

   b. lysosomes

   c. smooth endoplasmic reticulum

   d. nuclei

   e. rough endoplasmic reticulum

8. Which of the following contain enzymes that degrade biopolymers into their subunit form?

   a. microtubules

   b. rough endoplasmic reticulum

   c. endosomes

   d. lysosomes

   e. peroxisomes

5.4 The Cytoskeleton: Components and Structural Functions

9. The cytoskeleton of eukaryotic cells is

   a. often arranged as bundles or networks

   b. highly conserved in evolution

   c. found only in the nucleus

   d. typically composed of three filament systems

   e. not related to any bacterial structures or proteins

10. Specialized cell junctions such as desmosomes and hemidesmosomes require

    a. microfilaments

    b. microtubules

    c. intermediate filaments

    d. MTOCs

    e. spectrin

5.5 Purification of Cells and Their Parts

11. If a cellular homogenate was subjected to differential centrifugation, which of the following would be expected to pellet first?

    a. nuclei

    b. cytosol

    c. ribosomes

    d. microsomes

    e. mitochondria

12. Prior to cell fractionation, cells may be ruptured by placing them in

    a. an isotonic solution

    b. a hypotonic solution

    c. a hypertonic solution

    d. a high speed blender

    e. a sonicator

13. Although many types of vesicles are similar in size and density, it is possible to isolate specific types of vesicles through the use of

    a. antibodies attached to bacterial cells.

    b. electron microscopy.

    c. differential centrifugation.

    d. detergents.

    e. a fluorescence-activated cell sorter.

5.6 Visualizing Cell Architecture

14. Fluorescent imaging of thick, living specimens is best accomplished by

    a. phase contrast microscopy

    b. differential interference microscopy

    c. immunofluorescence microscopy

    d. confocal scanning microscopy

    e. deconvolution microscopy

15. Which of the following techniques may be used to determine the location of a specific protein inside a cell?

    a. bright-field microscopy

    b. GFP-tagging / fluorescent microscopy

    c. phase contrast microscopy

    d. immunofluorescence microscopy

    e. scanning electron microscopy

PART C: Reviewing Concepts

5.1 Biomembranes: Lipid Composition and Structural Organization

16. The lipid molecules that form biomembranes are amphipathic molecules. What is an amphipathic molecule, and how do such molecules generate the fundamental structure of a biomembrane?

17. The lipid bilayer serves as the foundation of every biomembrane. What two important properties does the bilayer structure provide for biomembrane function?

5.2 Biomembranes: Protein Components and Basic Functions

18. Describe some of the general functions of protein domains on the extracellular membrane surface, within the membrane, and on the cytosolic face of the membrane.

19. What is the difference between integral and peripheral membrane proteins?

5.3 Organelles of the Eukaryotic Cell

20. Just as the evolution of organelles allows separation of diverse biochemical processes into specific compartments, the evolution of multicellular organisms allows particular cell types to specialize in different activities. This cellular specialization is often accompanied by organelle-specific specialization; that is, cells optimized for a particular role often have an abundance of the particular organelle or organelles involved in that role. Based on what you know about organelle functions, which organelle(s) or membrane(s) would you predict would be over-represented in each of the following cell types?

    a. Osteoclast (involved in degradation of bone tissue)

    b. Anterior pituitary cell (involved in secretion of peptide hormones)

    c. Palisade cell of leaf (involved in photosynthesis)

    d. Brown adipocyte (involved in lipid storage and metabolism, as well as thermogenesis)

    e. Ceruminous gland cell (involved in secreting earwax, which is mostly lipid)

    f. Schwann cell (involved in making myelin, a membranous structure that envelops nerve axons)

g. Intestinal brush border cell (involved in absorption of food materials from gut)

h. Leydig cell of testis (involved in production of male sex steroids, which are oxygenated derivatives of cholesterol)

21. Certain organelles contain proteins that are not encoded by nuclear DNA. What organelles contain such proteins, and where are these proteins synthesized?

5.4 The Cytoskeleton: Components and Structural Functions

22. Actin, tubulin, lamin and keratin are examples of cytoskeletal proteins. What filaments do each of these proteins form and what is the function of each resulting filament?

5.5 Purification of Cells and Their Parts

23. Techniques used in the fractionation of cells to obtain preparations of subcellular structures (organelles and plasma membranes) include differential-velocity centrifugation, equilibrium density-gradient centrifugation, binding to antibody-coated bacterial cells, and sonication. In what order would these techniques normally be used during cell fractionation and why?

5.6 Visualizing Cell Architecture

24. The ability of a microscope to discriminate between two objects separated by a distance D is dependent upon the wavelength of the radiation (λ), the numerical aperture of the optical apparatus (a), and the refractive index of the medium between the specimen and the objective lens (N), according to the following equation:

$$D = (0.61\,\lambda) \div (N \sin a)$$

a. If you are viewing an object with visible light at a wavelength of 600 nm, in an instrument with a numerical aperture of 70°, what would be the resolution if air (refractive index = 1) was the medium between the specimen and the objective?

b. What would be the resolution if immersion oil with a refractive index of 1.5 were placed between the specimen and the objective?

c. What would be the resolution if you used the same immersion oil and viewed the specimen in blue light (wavelength of 450 nm)?

d. Under any of the conditions specified above, could you see a mitochondrion (average size $1 \times 2\ \mu$m)?

25. Why does electron microscopy require the specimen to be placed in a vacuum? What consequence does this requirement have for the specimen?

PART D: *Analyzing Experiments*

26. A fluorescence-activated cell sorter (FACS) can be used to identify and purify cells that have a specific fluorescent antibody bound to them. In addition, this instrument can also identify and purify cells with varying amounts of DNA. For example, cells that have just divided and have X amount of DNA can be separated from cells that have duplicated their DNA (2X) and are preparing to divide again. This is done by incubating the cells in a fluorescent dye that binds strongly to DNA. Because the fluorescence of this dye is directly proportional to the DNA content, cells containing 2X DNA are twice as fluorescent as those containing only 1X DNA. Data from two such analyses are shown in Figure 5-1. The x-axis (channel number) indicates the level of fluorescence of a given cell; a higher channel number means that more fluorescence was measured. The y-axis indicates the number of cells with a given fluorescence level.

a. In panels A and B of Figure 5-1, which portions of the graph correspond to the following cells: G_1 cells, which have not replicated their DNA (1X DNA); G_2 and M phase cells, which have replicated but not segregated their DNA (2X DNA); and S phase cells that are in the process of replicating? See MCB Figure 21-1 for reference.

b. Were the cells used in analysis A (panel A) dividing more or less rapidly than those used in analysis B (panel B)? Explain your answer.

c. Certain drugs can stop or arrest cells in specific phases of the cell cycle. How would the pattern shown in panel A change if the cells had been treated with a drug that arrested cells prior to DNA replication? How would the pattern change if the cells had been treated with a drug that arrested cells in the middle of mitosis? Explain your answers.

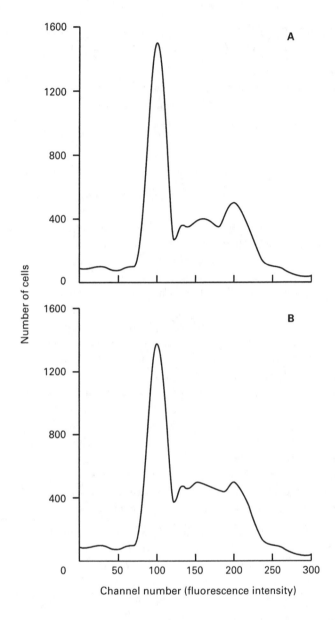

Figure 5-1

for many days in the cells but apparently had no ill effects on cellular metabolism.

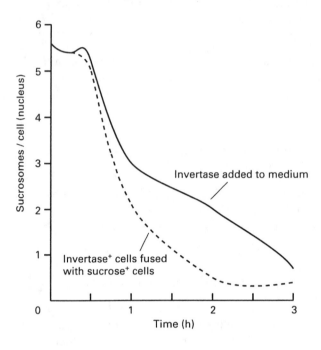

Figure 5-2

27. Mammalian cells grown in culture usually contain a representative population of organelles such as mitochondria, Golgi vesicles, lysosomes, etc. These cells can be used as a model system in which to study the formation and dynamics of organelles. In one such study, it was found that cultured hamster cells grown in the presence of 0.03 M sucrose accumulated numerous refractile (very bright in the phase-contrast microscope), sucrose-containing vacuolar structures called *sucrosomes*. Cytochemical staining techniques demonstrated the presence of the enzyme acid phosphatase in these structures, indicating that they were derived from lysosomes. The structures persisted

a. In one experiment, hamster cells were grown in the presence of sucrose, and then the enzyme yeast invertase, which catalyzes the cleavage of sucrose into its monosaccharide components, was added to the medium. The number of sucrosomes per cell was monitored following addition of invertase by phase-contrast microscopy, as shown by the solid line in Figure 5-2. Explain why these data indicate that invertase is internalized by these cells. What can you conclude about its subcellular localization? What additional experiments could you perform in order to test this hypothesis?

b. In a second type of experiment, hamster cells grown in the presence of sucrose (sucrose+ cells) were mixed with hamster cells grown in the presence of invertase (invertase+). The two cell types were fused together by the addition of a fusogenic agent, which produces hybrid cells containing two or more nuclei. During the course of cell fusion, the number of sucrosomes per cell (actually sucrosomes per nucleus) was monitored by phase-contrast microscopy, as shown by the dashed line in Figure 5-2. From these data, what can you conclude about the subcellular localization of invertase in the fused cells? Are these data consistent with your hypothesis regarding the fate of invertase in unfused cells? Do the results of this experiment suggest that there is a dynamic exchange of molecules between subcellular organelles?

Answers

1. a, e

2. b

3. b, c, d, e

4. c, d, e

5. a, b, c, d, e

6. b, c, d

7. e

8. d

9. a, b, d

10. c

11. a

12. b, d, e

13. a

14. d, e

15. b, d

16. Amphipathic molecules contain both hydrophobic and hydrophilic regions. In the case of the lipid molecules found in biomembranes, the hydrophobic tail groups self-associate via the hydrophobic effect and van der Waals forces to produce a bilayer structure with the hydrophilic head groups exposed toward the aqueous environment on either side of the bilayer.

17. The two important properties provided by the bilayer structure are the impermeable barrier created by the hydrophobic core of the bilayer and the stability the bilayer. The barrier property restricts the diffusion of hydrophilic molecules across the membrane and allows specific transport proteins to control the movement of these molecules, thus allowing cells to generate different chemical environments on opposite sides of the bilayer. The stability property means that the structure of the bilayer is resistant to a range of chemical and physical perturbations.

18. Protein domains on the cell surface are generally involved in cell-cell signaling or adhesion. Domains within the membrane, particularly those that form channels or pores, move molecules in and out of cells. Domains on the cytosolic membrane surface anchor cytoskeletal filaments to the membrane or may be involved in cell signaling.

19. Integral membrane proteins, also called transmembrane proteins, have one or more segments buried in the phospholipid bilayer. Most integral membrane proteins contain hydrophobic amino acids that pass through, and interact with, the fatty acyl groups of the lipid bilayer. Peripheral membrane proteins do not interact with the hydrophobic core of the lipid bilayer directly, but instead are bound to the membrane either indirectly by interactions with integral membrane proteins or directly by interactions with lipid head groups.

20a. Osteoclasts contain an above-average amount of lysosomal enzymes. In a strict sense, these cells do not have more lysosomes, as the degradative activity occurs outside the cell; osteoclasts could be considered to have a large external lysosome.

20b. Anterior pituitary cells contain an above-average amount of rough endoplasmic reticulum (ER) and Golgi complex; these organelles are involved in the synthesis and processing of secretory proteins.

20c. Palisade cells of the leaf contain an above-average amount of chloroplasts, which are solely responsible for photosynthesis in eukaryotes.

20d. Brown adipocytes contain an above-average amount of peroxisomes, which are involved in fatty acid degradation and heat production. In addition, these cells contain above-average amounts of the other oxygen-utilizing, fatty acid-catabolizing organelles, the mitochondria.

20e. Ceruminous gland cells contain above-average amounts of smooth ER, which is the site of lipid biosynthesis.

20f. Schwann cells contain above-average amounts of both smooth and rough ER; these organelles are the site of biosynthesis of the proteins and lipids that compose the myelin sheath.

20g. Intestinal brush border cells contain an above-average amount of plasma membrane (the microvilli), which is the site of nutrient uptake.

20h. Leydig cells of the testis contain above-average amounts of smooth ER, which is the site of cholesterol biosynthesis.

21. Mitochondria and chloroplasts contain proteins that are not coded by nuclear DNA. Each of these organelles contains DNA that codes for a subset of the proteins present in that organelle and the transcription and translation machinery that produces proteins from the organelle DNA.

22. Actin assembles to form microfilaments (or actin filaments), while tubulin, specifically $\alpha\beta$ tubulin dimers, assembles to form microtubules. Lamin and keratin each form a type of intermediate filament: lamins form the nuclear lamina and keratins form cytosolic keratin filaments. Microfilaments provide structural support, particularly in the plasma membrane, and are involved in cell motility. Microtubules provide structural support and cytoplasmic organization and are also involved in cell motility and cell polarity. The nuclear lamina supports the nuclear envelope and keratin filaments provide structural support and are important for cell adhesion.

23. The order of use for these techniques would be: sonication, differential-velocity centrifugation, equilibrium density-gradient centrifugation, and binding of specific structures or molecules to antibody-coated bacterial cells. Sonication is used to break open the plasma membrane to allow access to organelles and cytosolic components. Differential-velocity centrifugation of the disrupted cells provides partial fractionation of cell components and subsequent equilibrium density-gradient centrifugation of selected fractions (enriched for the component of interest) would further improve the purity of the component. Finally, use of antibody-coated bacterial cells where the antibody is specific to a certain protein (e.g., one found on the surface of a specific organelle) would produce a highly-purified component.

24a. $D = \dfrac{0.61 \times 600 \text{ nm}}{1 \times \sin 70°}$

$= \dfrac{366 \text{ nm}}{0.94}$

$= 390 \text{ nm or } 0.39 \text{ } \mu\text{m}$

24b. 260 nm or 0.26 µm

24c. 190 nm or 0.19 µm

24d. A mitochondrion should be visible under all of the conditions in 24a–c, since its dimensions are considerably larger than the value of D.

25. Electron microscopy requires that the specimen be placed in a vacuum because air molecules absorb electrons. Observation in a vacuum, along with procedures necessary to withstand the vacuum and the need to section specimens to facilitate penetration of electrons, means that the specimen is typically fixed or frozen at very low temperatures. This means that no living material can be visualized by electron microscopy.

26a. G_1 cells with 1X DNA are represented by the large peak centered on channel number 100, while G_2 and M phase cells with 2X DNA are represented by the smaller peak centered on channel number 200. S phase cells in the process of replicating are represented in the intermediate channel numbers (120-180).

26b. Comparison of A and B indicates that the cells used in A were dividing more slowly; that is, a smaller proportion were synthesizing DNA (channel numbers 120–180) and a larger number were in the G_1 and G_2/M phases of the cell cycle.

26c. Cells arrested prior to DNA synthesis would be in the G_1 phase of the cell cycle and would have 1X DNA. Analysis with a FACS machine would reveal essentially all cells in a single peak centered around channel number 100. Cells arrested in M phase, which follows DNA replication, would have 2X DNA, and would produce a single peak centered around channel number 200.

27a. The disappearance of the sucrosomes suggests that the added invertase is taken up by the cells and becomes localized in lysosomes, where it catalyzes breakdown of the sucrose. Additional experiments to test this hypothesis might include labeling invertase with a fluorescent or radioactive marker, adding it to cells, isolating the lysosomes, and determining whether the invertase activity co-purified with acid phosphatase or some other lysosomal marker molecule. In fact, if you prepared a fluorescent invertase preparation, you might even be able to detect lysosomal fluorescence using a fluorescence microscope.

27b. Invertase in the fused cells is also located in a lysosomal compartment; this conclusion is consistent with the hypothesis discussed above. The observations may or may not be consistent with other hypotheses that you might have formulated. Since the sucrosomes in the fused cells disappeared over time, the invertase in the lysosomes derived from invertase[+] cells must come into contact with and break down the sucrose in the lysosomes derived from sucrose[+] cells, indicating that lysosomes inside cells mix rapidly, probably by fusion of their membranes.

6 Integrating Cells into Tissues

PART A: *Chapter Summary*

In the development of multicellular organisms, progenitor cells differentiate into distinct cell "types" that have characteristic compositions, structures and functions. Cells of any given cell type often aggregate or develop into a *tissue* that cooperatively performs a similar function. Different tissues can be organized into an *organ*, again to perform one or more specialized functions. In this chapter, we examine the ability of cells to adhere tightly and interact specifically with each other, and the assembly of distinct tissues. This assembly could not occur without molecular interactions at the cellular level which are mediated by the temporally and spatially regulated cell-adhesion molecules (CAMs). These molecules enable many animal cells to adhere tightly and specifically with cells of the same, or similar type. The chapter begins with a brief overview of the various types of CAMs, their major functions in organisms, and their evolutionary origins. This overview also includes the elaborate specialized cell *junctions* which promote local communication between adjacent cells, and the animal extracellular matrix (ECM).

The organization of epithelial tissue in animal cells is the next topic. In general, the surfaces of animal organs are covered with a sheetlike layer of cells called epithelium. Cells that form epithelial tissues are polarized because their plasma membranes are organized into discrete regions called the apical, basal and lateral surfaces. To achieve this polarization, epithelial cells require three classes of specialized junctions: anchoring junctions (adherens junctions, desmosomes and hemidesmosomes), tight junctions and gap junctions. These junctions can be facilitated by either cadherin proteins (calcium binding molecules) or integrins, heterodimeric integral membrane proteins. To become organized into tissues, epithelial sheets require several ECM components. The next sections detail the ECM of both epithelial and nonepithelial tissues. Molecules of the ECM discussed include: highly viscous proteoglycans, which cushion cells; insoluble collagen fibers, which provide strength and resilence; and soluble multi-adhesive matrix proteins such as laminin, which binds these components to receptors on the cell surface. Different combinations of these components tailor the strength of the extracellular matrix for different purposes. There are also adhesive interactions in nonepithelial cells, such as various cell-surface structures in nonepithelial cells that mediate long-lasting adhesion and transient adhesive interactions especially adapted for the movement of cells.

Plant cells are surrounded by a cell wall that is thicker and more rigid than the animal ECM. Although the plant cell wall and the animal ECM serve many of the same functions, they are structurally very different. Because of these differences, the plant cell wall and its interactions with plant cells is presented in a separate section. Lastly, because our understanding of animal cells depends greatly on the study of isolated cells, we study ways of working with populations of cells removed from tissues and organisms.

PART B: *Multiple Choice*

Circle the letter corresponding to the most appropriate terms/phrases. There may be more than one correct answer. Circle all that apply.

6.1 *Cell-Cell and Cell-Matrix Adhesion: An Overview*

1. The major classes of CAMs include

    a. cadherins.

    b. fibronectin.

    c. Ig-superfamily CAMs.

    d. P-selectins.

6.2 *Sheetlike Epithelial Tissues: Junctions and Adhesion Molecules*

2. Desmosomes

    a. associate with actin filaments on the cytoplasmic side.

    b. contain the transmembrane cadherin proteins desmoglein and desmocollin.

    c. interconnect fibroblasts.

    d. contain integrins.

3. An example of an integrin containing cell junction is:

    a. a hemidesmosome.

    b. a gap junction.

    c. an adherens junction.

    d. a tight junction.

4. Integrins

    a. are heterodimeric proteins.

    b. bind to the tripeptide sequence Arg-Gly-Asp (RGD).

    c. are expressed in a cell-specific manner.

    d. have high-affinity binding sites ($K_D > 10^{-9}$ mol/L) for their ligands.

5. Multiple integrin forms arise from

    a. differential splicing of mRNA precursors.

    b. differential splicing of proteins.

    c. variations in proteolytic processing.

    d. multiple genes.

6.3 *The Extracellular Matrix of Epithelial Sheets*

6. Which of the following is a major extracellular molecule in mammals containing no protein portion?

    a. fibronectin

    b. hyaluronan

    c. laminin

    d. proteoglycans

7. Common sugars found in proteoglycans include all of the following except:

    a. N-acetyl-D-glucosamine

    b. D-glucose

    c. D-glucuronic acids

    d. L-iduronic acid

8. The collagen triple-helix domain

    a. is rich in glycine.

    b. is rich in proline.

    c. is rich in hydroxyproline.

    d. is an alpha helix.

9. Collagen is exported from the endoplasmic reticulum as

    a. a folded monomer.

    b. a folded dimer.

    c. a folded triple helix.

    d. a folded quadramer.

10. Which of the following constituents are *not* found in most basal laminae?

    a. type I collagen

    b. type IV collagen

    c. glycosaminoglycans

    d. laminin

11. Proteoglycans are a group of cell-surface and extracellular-matrix substances that

    a. are highly positively charged.

    b. have a molecular weight less than 1000.

    c. may contain heparan sulfate as a constituent.

    d. bind to collagens and fibronectin.

6.4   *The Extracellular Matrix of Nonepithelial Tissues*

12. Which of the following are included as functions of the extracellular matrix?

    a. supporting differentiation

    b. synthesis of procollagen

    c. binding growth hormones

    d. filtering

6.5   *Adhesive Interactions and Nonepithelial Cells*

13. Cells involved in extravasation include all of the following except:

    a. B lymphocytes

    b. epithelial cells

    c. monocytes

    d. T lymphocytes

6.6   *Plant Tissues*

14. Which of the following can be found in plant cell walls?

    a. collagen

    b. pectin

    c. cellulose

    d. hemicellulose

15. Cellulose is synthesized:

    a. by free ribosomes in the cytosol.

    b. at the cell surface by a synthase.

    c. by membrane bound ribosomes associated with the rough endoplasmic reticulum.

    d. within the Golgi complex by a glucosyltransferase.

6.7   *Growth and Uses of Cultured Cells*

16. Cultured cells from a population of primary cells that have undergone oncogenic transformation are:

    a. able to grow for only seven generations.

    b. able to grow indefinitely.

    c. immortal.

    d. a cell line.

PART C: *Reviewing Concepts*

6.2   *Sheetlike Epithelial Tissues: Junctions and Adhesion Molecules*

17. How do cadherins in adherens junctions and spot desmosomes interact with cytoskeletal proteins?

18. What distinguishes a hemidesmosome from a focal adhesion?

6.3   *The Extracellular Matrix of Epithelial Sheets*

19. Heating of calf type I collagen fibers to 45°C denatures the triple helices and separates the three chains from each other. Collagen that has been treated in this way does not renature to form a normal collagen triple helix. Why?

20. Even low levels of expression of mutant α1(I) collagen cause severe growth abnormalities in transgenic mice. Why does low-level expression suffice to cause abnormalities?

21. Type IX collagen does not form fibrils, but does associate with collagen fibrils composed of type II collagen. What structural characteristics of type IX

collagen are responsible for this functional distinction?

22. What are the two major classes of soluble extracellular matrix proteins in vertebrates?

6.4 The Extracellular Matrix of Nonepithelial Tissues

23. Extracellular matrix components in animals are often mostly carbohydrate. How do the saccharide portions of these molecules contribute to their biological role?

6.5 Adhesive Interactions and Nonepithelial Cells

24. What is the role of selectins in leukocyte extravasation?

25. Cell-matrix interactions through integrins hold cells in place. How can such associations be modulated to promote cell motility?

26. How does the structure of fibronectins support their role as a multi-adhesive matrix proteins?

6.6 Plant Tissues

27. What properties of plant cells are thought to be responsible for the fact that all plant hormones are small and water soluble?

28. What is the role of expansin and pH in auxin-induced plant cell growth?

29. Plant cell-wall molecules have many functional, but not necessarily structural, homologies with molecules found in the extracellular matrices around animal cells. Three important plant cell-wall constituents are (a) cellulose, (b) pectin, and (c) lignin. Describe the functions performed by each of these plant cell wall compounds.

6.7 Growth and Use of Cultured Cells

30. What are the advantages and disadvantages of using cultured cells rather than a whole organ, such as the liver, in molecular cell biology research?

31. What are the differences between monoclonal and polyclonal antibodies?

PART D: Analyzing Experiments

32. Gap junctions, which form between adjacent cells in an epithelium, mediate the exchange of small molecules and ions between cells. This exchange may be important in many cellular regulatory processes, including those that regulate cell growth. Some researchers have conducted experiments to determine whether cancerous cells exhibit junctional communication.

In one experiment, depicted in Figure 6-1a, four normal cells were surrounded by cancerous cells. A small amount of fluorescent dye (MW = 310) was injected into one of the normal cells. After a few minutes, the cells were examined. A fluorescence microscope was used to determine whether the dye had spread to other cells. Fluorescent cells are indicated by crosshatching.

In a second experiment, depicted in Figure 6-1b, a small amount of the fluorescent dye was injected into one of the cancerous cells surrounding the four normal cells. Again, the cells were examined under a fluorescence microscope after a few minutes.

a. Are these data consistent with the hypothesis that normal cells are coupled to each other via gap junctions? Why or why not?

b. Are these data consistent with the hypothesis that cancerous cells are coupled to each other via gap junctions? Why or why not?

c. Are these data consistent with the hypothesis that cancerous cells and normal cells are coupled to each other via gap junctions? Why or why not?

d. Describe an experimental approach that could be used to determine whether the cellular communication demonstrated with this fluorescence assay occurs via gap junctions.

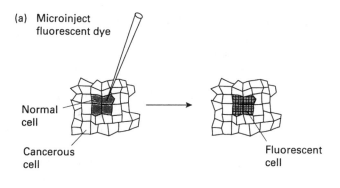

(a) Microinject
fluorescent dye

Normal
cell

Cancerous
cell

Fluorescent
cell

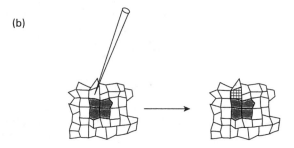

(b)

Figure 6-1

33. In order to prepare single-cell suspensions of most
epithelial cells grown in culture, scientists add
calcium-chelating reagents such as EDTA or EGTA
to confluent monolayers of these cells. In many cases
trypsin must be added as well.

   a. How do these agents act to dissociate epithelial
   cells?

   b. What enzymes would you choose if you wanted
   to dissociate cultured plant cells?

34. Many connective tissue diseases result from synthesis
of aberrant, but secreted, collagens. Some of these
collagens can be distinguished from native collagens
by their different migratory pattern on a denaturing
(SDS) gel. SDS gel electrophoresis of collagen from a
child who exhibits a possible collagen-deficiency
disease produces a profile indicative of a type I
collagen deficiency. When the separated chains are
extracted from the gel and recombined in an
appropriate buffer, they fail to reassociate into a
native triple helix. Do these data indicate that the
molecular defect in the child involves the inability of
collagen to form its triple helix once secreted into
the extracellular matrix?

Answers

1. b, c, d

2. b

3. a

4. d

5. d

6. b

7. b

8. a, b, c

9. c

10. a

11. c

12. a, c, d

13. b

14. b, c, d

15. b

16. b, c, d

17. Cadherins bridge between neighboring cells in adherens junctions and desmosomes. The cadherins link to adapter proteins, which then link to cytoskeletal proteins. In adherens junctions, the link is to F-actin and myosin in the circumferential belt located below the tight junction of the epithelial cell layer and in desmosomes, to the intermediate filament protein keratin. In the case of desmosomes the adapter complex appears distinguishable as a cytoplasmic plaque. The plaque proteins are plakoglobin and desmoplakins. In adherens junctions, adapter proteins include α- and β-catenin.

18. Focal adhesions and hemidesmosomes are cell-matrix junctions that consist of clustered integrin molecules. The extracellular domain of integrins in a focal adhesion binds to fibronectin, and in hemidesmosomes, to laminin in the basal lamina. The intracellular domain binds to adapter proteins. In focal adhesions the adapter proteins link to actin-rich stress fibers. In hemidesmosomes, the adapter proteins link to keratin containing intermediate filaments.

19. The N-terminal and C-terminal propeptides present in newly synthesized collagen monomers assist in alignment of the peptides to form the triple helix. These propeptides are removed after the trimers are transported to the extracellular matrix, and thus are not available to perform the same function in denatured calf type I collagen. In addition, inappropriate disulfide bridges can be generated during renaturation; these will also inhibit the generation of a normal triple helix.

20. Each type I collagen molecule contains two α1(I) and one α2(I) chains. Hence expression at low frequency for the α1(I) chain will have more effect than the same frequency of expression of an α2(I) chain mutation. The polymerization of the triple helix and its formation into fibrils is affected.

21. Type IX collagen consists of two triple-helical domains separated by a flexible kink. A proteoglycan is attached at and protrudes from this kink region. The interruption in the triple helix, as well as the presence of the proteoglycan, prevents this molecule from self-associating to form collagen fibrils. Type IX collagen associates with other type II collagen fibrils but cannot form fibrils itself.

22. The two major classes of soluble extracellular matrix proteins are multi-adhesive matrix proteins, which bind cell-surface adhesion receptors, and proteoglycans, a diverse group of macromolecules containing a core protein with multiple attached polysaccharide chains.

23. Because of their high content of charged polysaccharides, proteo-glycans and free hyaluronan, which is entirely carbohydrate, are highly hydrated. The swelled, hydrated structure of proteoglycans is largely responsible for the volume of the extracellular matrix and also acts to permit diffusion of small molecules between cells and tissues. The swelling forces of water associated with hyaluronan, for example, give connective tissues their ability to resist compression forces.

The loose, hydrated nature of hyaluronan is particularly important in permitting cell migration.

24. Extravasation is the movement into tissues of four types of leukocytes (white blood cells): monocytes, the precursors of macrophages, which ingest foreign particulates; neutrophils, which release several antibacterial proteins; and T and B lymphocytes, the antigen-specific cells of the immune system. These cells are present in the blood stream and carry selectin-specific ligands (the sialyl Lewis-x antigen) on their surfaces. Endothelial cells facing onto the blood vessel become activated in response to various inflammatory signals released by surrounding cells in areas of infection or inflammation. Activated endothelial cells exocytose P-selectin to their cell surface. As a consequence, passing leukocytes adhere weakly to the endothelium; because of the force of the blood flow, these trapped leukocytes are slowed but not stopped and seem to roll along the surface of the endothelium.

25. Various de-adhesion factors promote cell migration. One class of de-adhesion factors comprises small peptides called disintegrins. They contain the inegrin-binding RGD sequence present in many ECM proteins. By binding to integrins on the surface of cells, disintegrins competitively inhibit binding of cells to matrix components. A second class of de-adhesion factors includes two types of proteases, fibrinogens and matrix-specific metalloproteinases. Both proteases degrade matrix components, thereby permitting cell migration.

26. Fibronectins are an important class of soluble multi-adhesive matrix proteins. The multi-adhesive properties of fibronectins arise from the presence on different domains of high-affinity binding sites for collagen and other ECM compenents and for certain integrins on the surface of cells. Hence, the structure leads to the ability to bind to multiple matrix components and to cell surface proteins.

27. The porosity of the plant-cell wall permits soluble factors to diffuse and interact with receptors on the plant plasma membrane. Water and ions diffuse freely in cell walls but diffusion of even relatively small proteins is restricted. This then is one of the reasons that plant hormones are small, water-soluble molecules.

28. Auxin stimulates proton secretion at the growing end of the cell. The low pH activates a class of cell wall proteins, termed expansins. Expansins disrupt the hydrogen bonding between cellulose fibrils causing the laminate structure of the cell wall to loosen. With the rigidity of the wall reduced, the cell can elongate.

29. (a) Cellulose is analogous to collagen. Both are long, insoluble, fibrous polymers, and in both cases the fibers are generated extracellularly. Both confer tensile strength to their respective tissues, even though one is a carbohydrate and one is a protein. (b) Pectin is analogous to hyaluroran. Both are multiple negatively charged polysaccharides, with a high binding capacity for water. Both bind to other matrix molecules and allow tissues to resist compressive forces. (c) Lignins are analogous to proteoglycans. Both bind to other components of the matrix, enabling tissues to resist compressive forces.

30. Cultured cells are preferable to whole organs because they consist of a single cell type and can be derived from organisms that are not routinely used as experimental animals (e.g., humans). In addition, environmental and genetic variables can be more closely monitored (if not controlled) with cultured cells than with organs. A disadvantage of cultured cells is that cell-cell interactions, which are present in an organ and which may be important determinants of the process under study, are abnormal in cell cultures. Also, biosynthesis, especially of tissue-specific macromolecules (e.g., glutamine synthetase in the liver), may be low or nonexistent in cultured cells and very high in the intact organ.

31. Monoclonal antibodies react with only one antigenic determinant (epitope) on a protein molecules, even though multiple epitopes are present. Polyclonal antibody preparations are a mixture of monoclonal antibodies.

32a. Yes. The fluorescent dye spreads to adjacent cells; this is consistent with the known properties of gap junctions.

32b. No. The dye is confined to the injected cell. If the cells contained functional gap junctions, the dye should spread to adjacent cells.

32c. No. Dye injected into a normal cell does not spread to adjacent cancerous cells. Dye injected into a cancerous cell does not spread to adjacent normal cells.

32d. The simplest approach involves injecting a dye with a molecular weight greater than 2000. A dye this large should not transfer to adjacent normal cells via gap junctions, which only permit ready passage of quite small molecules. Molecules with a molecular weight of about 1200 pass easily, whereas those with molecular weights above 2000 are excluded. Passage of intermediate-sized molecules across gap junctions is limited and variable. Other approaches such as inhibitory antibodies against connexin could also be used.

33a. Calcium chelators remove calcium from E-cadherin, thus facilitating dissociation. Trypsin degrades cadherin and other cell-surface adhesion molecules, thus preventing reassociation.

33b. Adhesion of plant cells is affected by interactions between components of the cell wall around the cells. Because cellulose and pectin are major components of the cell wall, pectinase and cellulase would be the most appropriate enzymes for dissociating plant cells.

34. No. Although some collagen diseases are thought to result from an inability to assemble triple-helical collagen in the extracellular matrix, all collagens need N- and C-terminal propeptides to aid in the formation of the triple helix. These propeptides are cleaved and not present in the mature, extracellular matrix collagen. Thus even denatured type I collagen from normal individuals would not renature to form the native triple helix under the experimental conditions described.

7

Transport of Ions and Small Molecules Across Cell Membranes

PART A: *Chapter Summary*

The plasma membrane is the essential, selectively permeable barrier between the cell and its extracellular environment. The selective permeability of the plasma membrane allows the cell to maintain a constant internal environment. Metabolites such as glucose, amino acids, and lipids readily enter the cell, metabolic intermediates remain in the cell, and waste components leave the cells. Transport of glucose, amino acids, and ions into or out of cells requires different mixtures of transport proteins. Similarly, organelles within the cell contain specific transport proteins in their membranes that are essential to creating the internal environment of the organelle. The movement of virtually all molecules and ions across cellular membranes requires selective membrane transport proteins embedded in the phospholipids bilayer.

We consider first the general properties of transport across membranes and an overview of the three major types of transport proteins. Second, after characterization of specific examples of the structure and function of these proteins, we describe how membranes of homologous protein families have different properties that enable different cell types to function. We review how combinations of transport proteins in different subcellular membranes enable cells to carry out important physiological processes, including the maintenance of cytosolic pH, accumulation of sucrose and salts in plant cell vacuoles, and the directed flow of water in both plants and animals.

In animals, sheets of cells, termed epithelial cells, line all the body cavities. Epithelial cells frequently transport ions or small molecules from one side to the other. Those lining the small intestine transport the products of digestion into the blood and those lining the stomach secrete hydrochloric acid into the stomach lumen. For epithelial cells to carry out these transport functions, their plasma membranes are organized into two discrete regions, each with specialized transport proteins. In addition, epithelial cells interconnect by junctional complexes that prevent material from moving between cells and from one side to the other. Nerve cells generate and conduct along their entire length a type of electrical signal called an action potential. This process and the transmission of these signals to other cells, inducing a change in the electrical properties of the receiving cell, requires a large number of transport proteins.

PART B: *Multiple Choice*

Circle the letter corresponding to the most appropriate terms/phrases. There may be more than one correct answer. Circle all that apply.

1. Uniport transport

   a. occurs at a higher rate for D-glucose than for L-glucose in erythrocytes

   b. is proportional at all glucose concentrations to the concentration of extracellular glucose

   c. is the mechanism by which K^+ is taken up by mammalian cells

   d. is the mechanism by which the acidic pH of lysosomes is maintained

2. Which of the following factors increase the rate of simple (passive) diffusion of a solute across a phospholipid bilayer?

   a. increased ability of the solute to dissolve in the membrane

   b. increased hydrophilicity of the solute

   c. increased membrane thickness

   d. increased level of transport protein in the membrane

3. A channel protein

   a. moves ions at a rate similar to that of a transporter

   b. can be open at all times

   c. can move ions up an electrochemical gradient

   d. generally moves both anions and cations through the same channel

4. An artificial membrane composed only of phospholipids has significant permeability to

   a. gases

   b. Na^+

   c. lysine

   d. sucrose

5. Facilitated diffusion can be distinguished from passive diffusion by

   a. the equilibrium concentration of solute reached

   b. the linear dependence of the rate of transport on the concentration gradient across the membrane

   c. the specificity of the transport process for particular molecular species

   d. the ΔG for the transport process

6. Which of the following properties are typical of uniport proteins?

   a. the ability to "flip" within the membrane, releasing the transported molecule at the opposite surface

   b. a permanent open pore through the membrane

   c. multiple transmembrane α-helical segments

   d. an ATPase activity for the direct coupling of ATP hydrolysis to transport

7. The generation and maintenance of a membrane electric potential requires which of the following:

   a. active pumping of ions

   b. ion-specific membrane channel proteins

   c. ATP

   d. a generally permeable membrane

8. ATP hydrolysis is directly coupled to the movement of

   a. glucose across the animal-cell plasma membrane

   b. sucrose into the plant vacuole

   c. Cl^- and HCO_3. across the erythrocyte membrane

   d. Ca^{2+} to the extracellular medium or sarcoplasmic reticulum lumen

9. Which of the following properties are characteristic of P-class ion-motive ATPases?

   a. localization in vacuoles

   b. localization in chloroplasts and mitochondria

   c. binding of ATP on the exoplasmic side of the membrane

   d. a conserved amino acid sequence about the covalent phosphorylation site of the enzyme

10. Differences between cotransport and active ion transport include

    a. the directed nature of molecule movement

    b. the restriction of typical active transporters to movement of ions such as H^+, Na^+, K^+, and Ca^{2+}

    c. the inability of cotransport to move a molecule against a concentration gradient

    d. the location of cotransporters peripherally associated with the membrane lipid bilayer

11. Polarized cells in animals

    a. are a general feature of epithelial tissue, which lines body cavities

    b. include erythrocytes and most other blood cells

    c. have the same protein molecules exposed on their apical and basolateral surfaces

    d. may be sealed together by plasmodesmata to form a barrier to the passage of most small molecules

12. Osmosis

    a. can cause cells to burst.

    b. is not regulated in protozoa or plant cells.

    c. is the process by which water flows from a solution with a high solute concentration to a solution with a lower solute concentration.

    d. may involve large-scale movement of lipid molecules.

13. Typical ion/ligand concentrations are

    a. 1×10^{-8} M H^+ in the plant vacuole

    b. $>1 \times 10^{-6}$ M Ca^{2+} in the cytosol of an animal cell

    c. 1×10^{-1} M H^+ in the lysosomes of cultured animal cells

    d. 145 mM K^+ in the cytosol of a mammalian cell

14. Action potentials are sudden membrane depolarization during which

    a. generalized membrane swelling will occur

    b. voltage-gated Na^+ channels open

    c. the cell will burst

    d. channel opening and closing occurs on a time scale of 1 h

15. Voltage gated Na^+ and K^+ channels are

    a. very dissimilar in molecular structure

    b. very rigid proteins in which the ball portion of a K^+ channel, for example, is held in place by a rigid coiled-coil domain

    c. used also to mediate an influx of Ca^{++} at the synapse of a motor neuron and a striated muscle cell

    d. consist of a familiarly similar domain organization which can be divided between four polypeptides or one large protein

16. Postsynaptic neurons generate action potentials only when

    a. innervated by a minimum of 50 neurons

    b. the summation of a series of multiple small depolarizations has dissipated

    c. depolarized to the threshold potential

    d. localized K^+ channel openings exceeds the threshold potential

PART C: *Reviewing Concepts*

7.1 *Overview of Membrane Transport*

17. Which aspects of the selective permeable biomembrane may be attributed to its phospholipid components, and which to its protein components?

18. Both uniporter proteins and ion channel proteins facilitate the movement of a single substance down a concentration gradient. Ion channels typically move 10^7–10^8 ions/s while uniporters more typically move 10^2–10^4 molecules. Why are uni-porters so much slower than ion channels?

19. Why is "cotransporter" a better term to describe a symport or antiport than "active transporter"?

20. How do artificial liposome systems contribute to analysis of the functional properties of membrane-transport proteins?

21. Glucose levels in the human bloodstream vary between 3 to 7 mM. GLUT1 is a glucose uniport found in many cell types. It has a K_m of 1.5 mM and a V_{max} of 500 µmol glucose/ml packed cells/h. How does the rate of glucose transport by GLUT1 change, going from a glucose concentration of 3 mM, a starvation condition, to 5 mM, the normal condition, to 7 mM, a condition found after a feast?

22. What effect does a uniporter have on the chemistry of the transported substance?

7.2 *ATP-Powered Pumps and Intracellular Ion Environment*

23. Compare and contrast the substrate specificity of ABC pumps with that of P, F, and V class pumps.

24. Chemotherapy with one drug, (e.g. colchicine), frequently selects for cells resistant to several chemotherapeutic agents (e.g., colchicine, vinblastine, and adriamycin). What is the molecular basis of this phenomenon?

7.3 *Nongated Ion Channels and the Resting Membrane Potential*

25. In mammalian cells, the intracellular K^+ concentration is about 140 mM and the extracellular K^+ concentration is about 5 mM. Assuming that the plasma membrane is permeable only to K^+, calculate the potassium equilibrium potential E_K across the membrane.

26. Describe how the structural properties of resting K^+ channels result in selectivity for K^+ versus the smaller Na^+ ion.

7.4 *Cotransport by Symporters and Antiporters*

27. Cholera toxin, produced by *Vibrio cholerae*, a pathogenic intestinal bacterium, causes an indirect reduction in the activity of the Na^+/K^+ ATPase in intestinal epithelial cells. This results in reduced uptake of small sugars and amino acids from the intestine. How is this reduction in uptake coupled to impaired Na^+/K^+ ATPase function?

28. Explain the relationship between HCO_3^- and anti-porters in regulating cytosolic pH in animal cells.

7.5 *Movement of Water*

29. Frog oocytes and eggs, which have an internal salt concentration of 150 mM, do not swell when placed in water with a very low solute concentration. Erythrocytes that have a similar internal solute concentration swell rapidly and burst in solutions of very low osmolarity. What could account for this difference?

30. The concentrations of solutes in the cytosol and vacuole of plant cells are generally much higher than those typical in the cytosol and lysosome of mammalian cells. Why don't plant cell plasma membranes undergo osmotic lysis?

7.6 *Transepithelial Transport*

31. How is glucose transported from the lumen of the intestine across the intestinal epithelium into the blood?

7.7 *Voltage-Gated Ion Channels and the Propagation of Action Potentials in Nerve Cells*

32. What is the role of a dendrite versus an axon in neuron function?

33. Describe the 3 phases of an action potential and how each phase corresponds to the flow of ions.

7.8 *Neurotransmitters and Receptor and Transport Proteins in Signal Transmissions at Synapses*

34. Neurotransmitters are small molecules that could be pumped across membranes in response to a signal. In actuality, how are neurotransmitters released upon arrival of an action potential at axon termini?

35. The acetylcholine receptor on skeletal muscle cells has a known three-dimensional structure. Describe how the opening of this receptor creates an ion channel.

PART D: *Analyzing Experiments*

36. When present in the culture medium, ouabain inhibits the Na⁺/K⁺ ATPase in cultured cells, but when the drug is microinjected into cells, it exerts no inhibitory effect. Although ouabain itself possesses no negative charge, ouabain treatment causes the α subunits of the Na⁺/K⁺ ATPase to become more negatively charged but does not alter the charge properties of the β subunits.

   a. On which side of the plasma membrane are the ouabain- and ATP-binding sites of the enzyme located?

   b. During the transport process, the Na⁺/K⁺ ATPase becomes transiently phosphorylated. What do these observations suggest about a likely mechanism for ouabain inhibition of this ion pump?

   c. Which of the two subunits, α or β, is most likely to be phosphorylated?

37. You have conducted a series of experiments measuring the rate of glucose uptake by erythrocytes from two different patients (DZ and KD) and a normal control group. The data are shown in Figure 7-1, in which the uptake rate (in μmol glucose/ml packed cells/h) is plotted as a function of the external glucose concentration.

   a. Estimate the V_{max} and K_m for glucose uptake by erythrocytes from the control population and each of the patients.

   b. What is the nature of the defect(s) in patient DZ?

   c. Propose two explanations for the defect in patient KD.

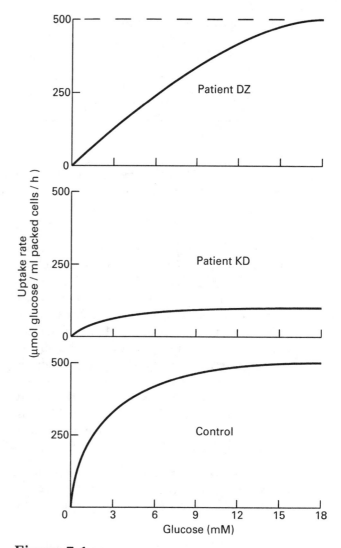

Figure 7-1

38. The intestinal two-Na⁺/one-glucose symporter was sequenced by an expression cloning approach. Messenger RNA was expressed by microinjection into a frog oocyte. Three days following injection the level of Na⁺–dependent glucose transport was assessed. mRNA coding for the transporter was found to be about 2.2 kilobases. A cDNA library corresponding to mRNAs of this size was created. Progressively smaller pools of the 2000 cDNA clones positive for Na⁺-dependent glucose transport were assayed. In the end, a single clone was isolated, and the cDNA was sequenced. The deduced protein sequence was then analyzed to predict the membrane topology of the protein. The initial approach was based on the outcome of hydropathy plots. In hydropathy plots, the hydrophobicity/hydrophilicity of amino acids are plotted relative to their position within a protein

sequence. The hydropathy plot for the two-Na$^+$/one-glucose symporter is presented in Figure 7-2.

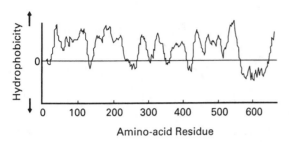

Figure 7-2

a. Expression cloning is one approach to screening cDNA libraries. Other approaches include antibodies or synthetic oligonucleotide probes derived from partial amino acid sequences. Based on the hydrophobicity plot and the fact that the intestinal two-Na$^+$/one-glucose symporter accounts for less than 0.2% of plasma membrane protein, why might one predict that antibody or partial amino acid sequence derived oligonucleotide probe approaches would prove to be impractical? Note: The plasma membrane is typically about 0.5% of total cellular protein and there is about 2×10^{-4} µg of protein per cell.

b. From analysis of the hydrophobicity plot, how many transmembrane domains do you predict the intestinal two-Na$^+$/one-glucose symporter to have?

c. From your prediction of transmembrane domains, do you expect the amino and carboxyl terminus of the protein to be on the same or different sides of the membrane?

d. In the actual publication in *Nature*, amino acids 314 through 334 were predicted to be a transmembrane domain. If modeled as an α helix, this portion of amino acids would be an amphipathic helix with a lysine residue (K) located approximately half-way along the helix. What thermodynamic problem is posed by K in this position? What is the likely solution to this problem?

39. The introduction of claudin as an expressed protein in mouse L cells is sufficient to cause the formation of morphologically identifiable tight junctions. The tight junctions form over extensive but relatively limited regions of cell-cell contact. Mouse L cells do not normally express tight junctions.

a. To what extent do these experiments provide evidence that a single protein is sufficient to produce a recognizable tight junction?

b. Should these junctions be effective in blocking nutrient movement across a cell layer or in separation of apical and basolateral cell surfaces? Explain why or why not.

c. An epitope tagging approach can be taken to ask the question of whether or not claudin from both contacting cells contributes to the formation of the junctional complex. In an epitope tagging approach the epitope tagged expressed protein is engineered to include one of another epitope sequences coding for the antibody recognition regions for well-characterized monoclonal antibodies. Describe how the approach could be applied to the experimental question raised here.

Answers

1. a

2. a

3. b

4. a

5. c

6. c

7. a, b, c

8. d

9. d

10. b

11. a

12. a

13. b

14. b

15. d

16. c

17. The phospholipids establish the permeability properties of biomembranes, e.g., lack of permeability to charged versus hydrophobic substances. The phospholipids determine the hydrophobic core of the membrane. The membrane proteins serve as various pumps and transporters and allow for the selective transport of various ions, amino acids and sugars. Proteins can create local hydrophilic environments within a hydrophobic membrane.

18. Uniporter proteins are slower than channels because they mediate a more complicated process. The transported substrate both binds to the uniporter and elicits a conformational change in the transporter. A uniporter transports one substrate molecule at a time. In contrast, channel proteins form a protein-lined passageway through which multiple water molecules or ions move simultaneously, single file, at a rapid rate.

19. Symporters and antiporters, like ATP pumps, mediate coupled reactions in which an energetically unfavorable reaction is coupled to an energetically favorable reaction. However, unlike pumps, neither symporters nor antiporters hydrolyze ATP or any other molecule during transport. Hence, these transporters are better referred to as cotransporters rather than active transporters. The term active transporter should be restricted to the ATP pumps where ATP is hydrolyzed in the transport process.

20. Artificial liposomes provide an experimental situation designed to test for the effect of a single gene product, i.e., protein. A specific transport protein may be extracted and purified. It may then be reincorporated into pure phospholipid bilayers, i.e., liposomes, to have its transport properties assessed. The functional properties of the various proteins can be examined in isolation with little, if any, ambiguity.

21. This problem can be solved using the Michaelis equation:

$$v = \frac{V_{max}}{1 + K_m/C} \quad \text{(see } MCB, \text{ eq. 7-1)}$$

where [C] is the concentration of the transported molecule, in this case, glucose. By substituting the values for V_{max}, K_m, and the three glucose concentrations [C], the initial velocities in mmol glucose/ml packed cells/h can be calculated: at 3 mM glucose, $v = 333$; at 5 mM glucose, $v = 385$; and at 7 mM glucose, $v = 412$.

Hence, although the glucose concentration increases over a range of more than 2-fold, v increases by only 24%.

22. A uniporter, like all other transporters, catalyzes the rate of movement of ions or molecules across a membrane. It does not catalyze the making or breaking of covalent bonds, and hence the transported substance is unchanged in chemistry.

23. Members of the ABC (ATP binding cassette) class pumps may either transport ions or small molecules such as amino acids, sugars, vitamins, peptides, or hydrophobic molecules. F and V class

pumps transport only protons (H^+) and P class pumps transport only one of four cations (H^+, Na^+, K^+, or Ca^{2+}).

24. The first ABC class pump described in eukaryotes was the multidrug-resistance (MDR) transport protein known as MDR1. MDR1 exports a large variety of relatively hydrophobic drugs from the cytosol to the extracellular medium. Hence cells that produce (overproduce) MDR1 and survive one drug because of the presence of MDR1 are apt to be resistant to several drugs. The initial drug serves as a selective agent for cells having MDR1 in their plasma membranes.

25. $E_K = -0.089$ V or -89 mV. The electric potential across a membrane due to differences in the K^+ concentration can be calculated from the Nernst equation

$$E_K = \frac{RT}{ZF} \ln \frac{[K^+_{out}]}{[K^+_{in}]} \qquad \text{(see MCB, eq. 7-2)}$$

by substituting the following quantities: $R = 1.987$ cal/degree = mol; $T = 310$ K; $Z = 1$; $F = 23,062$ cal/mol = V; $[K^+_{out}] = 5$ mM; and $[K^+_{in}] = 140$ mM. Note that the electric potential across the membrane is described in reference to the exteriorpotential that is arbitrarily defined as zero. Thus the extracellular K^+ concentration is placed in the numerator.

26. As K^+ ion enters the narrow selectivity filter of a resting K^+ channel, it loses its water of hydration but becomes bound to eight backbone carbonyl oxens on glycine residues found in an analogous position in the P segment of every known K^+ channel. Dehydrated Na^+ is smaller and hence too small to bind to all eight carbonyl oxygens that line the selectivity filter. Because of this, the energy of activation for the passage of Na^+ is relatively large and K^+ channels are K^+ selective.

27. The Na^+/K^+ ATPase uses energy from ATP to establish Na^+ and K^+ ion gradients across the intestinal epithelial cell plasma membrane. Cotransporters couple the energetically unfavorable movement of glucose and amino acids into epithelial cells to the energetically favorable

movement of Na^+ into these cells. Reduction of Na^+/K^+ ATPase activity results in a less energetically favorable distribution of Na^+ for powering the cotransport of sugars and amino acids from the intestinal lumen into the intestinal epithelial cell.

28. Three different cotransporters work together in a coordinated manner in most animal cells to regulate intracellular pH. A Na^+/H^+ antiport removes excess protons from the cell and hence raises pH. A $Na^+HCO_3^-/Cl^-$ cotransporter serves to raise pH. At the pH at which the cotransporter is active, the imported bicarbonate anion combine with protons to produce CO_2 that diffuses out of the cell. A Cl^-/HCO_3^- antiporter is active at slighly elevated pH and serves to lower pH by exporting OH^- as HCO_3^- (= OH^- + CO_2).

29. The permeability of the erythrocyte membrane to water is roughly tenfold greater than that of a phospholipid bilayer. Erythrocytes contain an integral membrane protein, called aquaporin, that acts as a water channel, allowing water to enter the erythrocyte cytoplasm. Frog eggs and oocytes apparently lack such a protein. Frog oocytes micro-injected with mRNA encoding aquaporin swell in hypotonic solution, demonstrating that this protein is a water channel.

30. The high osmotic pressure inside plant cells results in an inward flow of water and outward turgor pressure against the plant-cell plasma membrane. If the same osmotic pressure could be experimentally induced inside mammalian cells, they would lyse. However, plant cells are surrounded by a rigid cell wall that is able to withstand the high turgor pressure from within the cell. Consequently, the plasma membrane does not lyse. If the cell wall were removed, the cell would lyse.

31. As glucose is moved across the intestinal epithelium, it is transported up a concentration gradient to enter epithelial cells and then down a concentration gradient to the blood. A two-Na^+/glucose symport protein mediates transport of glucose into epithelial cells from the intestinal lumen. The energetically favorable cotransport of Na^+ powers glucose import. The GLUT2 uniport

protein facilitates the movement of glucose into the blood stream at the basolateral surface of epithelial cells. The Na⁺/K⁺ ATPase establishes the Na⁺ concentration gradient across the epithelial cell membrane.

32. In simple terms, the dendrite is the neuron extension that receives signals at synapses and the axon is the neuron extension that transmits signals to other neurons or muscle cells. Typically each neuron has multiple dendrites that radiate out from the cell body and only one axon that extends out from the cell body.

33. An action potential has three phases: depolarization, in which the local negative membrane potential goes to a positive membrane potential; repolarization, in which the membrane potential goes from positive to negative; and hyperpolarization, in which the resting negative membrane potential is exceeded. Depolarization corresponds to the opening of voltage-gated Na⁺ channels and a resulting influx of Na⁺. Repolarization corresponds to the opening of voltage-gated K⁺ channels. Hyperpolarization corresponds to a period of closure and inactivation of voltage-gated Na⁺ channels. During action potential propagation there is a sequential opening and closing of voltage-gated Na⁺ and K⁺ channels.

34. Rather than being pumped across a pre-synaptic membrane by a transporter, neurotransmitters such as acetylcholine are accumulated in synaptic vesicles in a process mediated by several transport proteins and then released by exocytotic fusion of synaptic vesicles with the presynaptic plasma membrane of the nerve ending.

35. The acetylcholine receptor from skeletal muscle is a pentameric protein. In muscle it is a ligand-gated channel for both K⁺ and Na⁺. The actual cation channel is a tapered central pore line by homologous segments from each of the five subunits. Each of these M2 α helices has negative charges that ring either side of the gate. These negative charges attract hydrated K⁺ and Na⁺. Binding of two molecules of acetylcholine to two separate sites on the extracellular side of the receptor that are 4-5 nm distal from the gate

produce conformational changes in the receptor in a presently unknown manner. This opens the gate to hydrated K⁺ and Na⁺ ions.

36a. The ouabain-binding site of the Na⁺/K⁺ ATPase must be located on the exoplasmic face of the plasma membrane, as microinjected ouabain does not inhibit the ATPase. The ATP-binding site is located on the cytoplasmic face of the plasma membrane, where it has access to ATP in the cytosol.

36b. The addition of a phosphate group to the enzyme would cause it to become more negatively charged. In the presence of extracellular ouabain, this negative charge, which normally is temporary, becomes permanent. Most likely, ouabain treatment inhibits the Na⁺/K⁺ ATPase by blocking the reaction at the point when phosphate has been added to the enzyme. Thus the phosphorylated form of the enzyme accumulates, blocking further ion transport. See *Molecular Cell Biology*, Figure 7-7.

36c. The observation that ouabain treatment results in the α subunit becoming more negatively charged but has no effect on the charge of the β subunit suggests that the α subunit is phosphorylated. This hypothesis could be tested by looking at the effect of a phosphatase on the charge properties of the α subunit following ouabain treatment. Phosphatase treatment should result in removal of the negatively charged phosphate group.

37a. The V_{max} and K_m values for glucose uptake can be estimated by inspection of the kinetic curves in Figure 7-1. The values are as follows:

Subject	K_m	V_{max}
DZ	6.0 mM	500 µmol glucose/ml-cells/h
KD	1.5 mM	100 µmol glucose/ml-cells/h
Control	1.5 mM	500 µmol glucose/ml-cells/h

37b. In the erythrocytes from patient DZ, the K_m for glucose uptake is about fourfold higher than in the controls. The higher K_m indicates a decreased affinity of the glucose uniporter for glucose, most likely due to an alteration in the glucose-binding site.

37c. In the erythrocytes from patient KD, the V_{max} for glucose uptake is about fivefold lower than in the controls. The lower V_{max} could be due to a fivefold decrease in the number of normal glucose uniporters present in the erythrocyte membrane. Alternatively, a normal number of defective glucose uniporters may be present. An alteration in uniporter amino acid sequence or in secondary modifications could lead to a decrease in V_{max}.

38a. The two-Na^+/one-glucose symporter is predominantly hydrophobic as shown by the hydrophobicity plot. As a very hydrophobic protein with little in the way of exposed non-membrane buried sequence, the protein is likely to be poorly antigenic. Consequently, an antibody approach is not likely. Purifying the protein is difficult because it is not very abundant. Assuming a molecular weight of 65,000 daltons ($\approx$650 amino acids, each amino $\approx$100 daltons), there are fewer than 20,000 copies of the protein per cell. Expression cloning is a good approach here.

38b. This is real world data. It is somewhat noisy. The exact prediction of transmembrane segment is therefore difficult and somewhat arbitrary. Transmembrane segments are typically amino acid stretches of about 20 hydrophobic amino acids in length. A reasonable estimate is 13 transmembrane domains.

38c. 13 transmembrane domains is an odd number of domains, and the N-terminal sequence itself is not hydrophobic. With an odd number of transmembrane domains, the N-terminus and C-terminus of the protein should be on opposite sides of the membrane. If the number of trans-membrane domains predicted were even, then the two ends of the protein would be on the same side of the membrane.

38d. The lysine side chain is positively charged and highly hydrophilic. Thermodyamically, the placement of a positive charge in a hydrophobic lipid environment is unfavorable. The positive charge is likely neutralized by association with a negatively

charged amino acid residue in another transmembrane domain of the protein.

39a. Assuming that no endogenous mouse L cell protein contribution is made to the tight junctions formed in the transfected cells, then the expression of the single protein claudin must be sufficient to generate tight junctions between cells.

39b. The tight junctions form over regions of cell-cell contact. The junctions do not go completely around the cell. Therefore, they cannot form an effective block to nutrient movement between cells within the cell layer, or maintain separation between apical and basolateral cell surfaces.

39c. The tight junctions should consist of claudin-claudin protein complexes between neighboring cells. If the cell population were a mixed population of cells expressing either claudin tagged with epitope A (e.g., myc) or claudin tagged with epitope B (e.g., HA), then the claudin contribution of each cell could be distinguished by immunofluorescence using appropriate antibodies to the respective epitope tags. Note that antibodies to the core claudin protein sequence would not allow for distinctions to be made between claudins contributed by neighboring cells.

8 Cellular Energetics

PART A: *Chapter Summary*

Cells use the energy released by hydrolysis of the terminal high-energy phosphoanhydride bond of ATP to power many otherwise energetically unfavorable processes. ATP is the most important molecule for capturing and transferring free energy in biological systems and can be thought of as the universal "currency" of chemical energy. ATP is generated from ADP and P_I in an endergonic reaction (requiring an input of 7.3 kcal/mol) during two main processes—aerobic oxidation and photosynthesis. In aerobic oxidation, oxygen and carbohydrates (principally glucose) are metabolized to CO_2 and H_2O and the released energy is used to synthesize ATP. The initial steps in the oxidation of glucose, called glycolysis, occur in the cytosol in both eukaryotes and prokaryotes and do not require O_2. The final steps, which require O_2, occur in the mitochondria of eukaryotes and at the plasma membrane of prokaryotes. In photosynthesis, light energy is used to synthesize ATP and is also stored in the chemical bonds of carbohydrates. In plants and eukaryotic single-celled algae, photosynthesis occurs in chloroplasts, and several prokaryotes also carry out photosynthesis by a mechanism similar to that in chloroplasts. The O_2 generated during photosynthesis is the source of virtually all O_2 in the air, and the carbohydrates photosynthesis produces are the ultimate source of energy for virtually all nonphotosynthetic organisms.

Although aerobic oxidation and photosynthesis appear to have little in common, there are many striking similarities between these two processes. Bacteria, mitochondria, and chloroplasts all use the process of chemiosmosis to generate ATP from ADP and P_i. A transmembrane proton concentration gradient and electric potential (voltage gradient), termed the proton-motive force, is generated by the stepwise movement of electrons from higher to lower energy states via membrane-bound electron carriers. In mitochondria and nonphotosynthetic bacterial cells, electrons are transferred to O_2, the ultimate electron acceptor. In the thylakoid membrane of chloroplasts, energy absorbed from light removes electrons from water (forming O_2) and eventually these electrons are donated to CO_2 to synthesize carbohydrates. All these systems, however, couple electron transport to the pumping of protons to generate the proton-motive force. Moreover, all cells utilize essentially the same kind of membrane protein, an ATP Synthase termed the F_0F_1 complex, to use the energy stored in the proton-motive force to synthesize ATP. Protons flow through the F_0F_1 complex from the exoplasmic to the cytosolic face of the membrane, driven by a combination of the proton concentration gradient and the membrane electric potential, leading to ATP formation on the cytosolic face of the membrane. Chemiosmotic coupling thus illustrates an important principle: the membrane potential, the concentration gradients of protons (and other ions) across a membrane, and the phosphoanhydride bonds in ATP are equivalent and exchangeable forms of chemical potential energy.

PART B: *Multiple Choice*

Circle the letter corresponding to the most appropriate terms/phrases. There may be more than one correct answer. Circle all that apply.

8.1 Oxidation of Glucose and Fatty Acids to CO_2

1. Which of the following acts as an electron acceptor during glycolysis?

    a. FAD^+

    b. NAD^+

    c. glucose

    d. ADP

    e. O_2

2. Which of the following statements describe glycolysis?

    a. Glycolysis occurs in the cytosol.

    b. Glycolysis produces more molecules of ATP than can be generated by pyruvate oxidation in mitochondria.

    c. Some reactions in the glycolytic pathway require an energy input.

    d. Glycolysis converts glucose into glycogen.

    e. The NADH generated during glycolysis can be used to produce ATP in the mitochondria.

3. Which of the following are mechanisms by which glycolysis is regulated to control intracellular ATP levels?

    a. phosphofructose kinase-1 inhibition by ATP

    b. pyruvate kinase inhibition by ATP

    c. phosphofructose kinase-1 activation by AMP

    d. hexokinase inhibition by glucose 6-phosphate

    e. pyruvate kinase activation by AMP

8.2 Electron Transport and Generation of the Proton-Motive Force

4. The ability of cytochromes to accept and donate electrons depends on covalently attached

    a. heme molecules

    b. ubiquinone molecules

    c. flavin molecules

    d. copper ions

    e. Coenzyme Q

5. The complete aerobic respiration of glucose produces

    a. water

    b. oxygen

    c. carbon dioxide

    d. glycerol

    e. fatty acids

6. Electron transport in mitochondria

    a. is used to produce NADH

    b. generates a voltage gradient across the inner membrane

    c. occurs in the mitochondrial matrix

    d. provides energy to create a proton gradient across the inner membrane

8.3 Harnessing the Proton-Motive Force for Energy-Requiring Processes

7. ADP entry into the mitochondrial matrix depends on

    a. simple diffusion through porins

    b. channel proteins

    c. an ATP/ADP antiporter

    d. an OH^- antiporter

    e. ATP hydrolysis

8. ATP synthesis in animal cells takes place in the

   a. mitochondrial inner membrane

   b. mitochondrial outer membrane

   c. mitochondrial intermembrane space

   d. cytosol

   e. mitochondrial matrix

8.4 Photosynthetic Stages and Light-Absorbing Pigments

9. In chloroplasts, light absorption and electron transport occur in the

   a. thylakoid lumen

   b. thylakoid membrane

   c. stroma

   d. outer membrane

   e. inner membrane

10. When chlorophyll *a* is in its oxidized state, it

    a. serves to drive H_2O synthesis.

    b. directly reduces $NADP^+$.

    c. can accept electrons from quinone Q.

    d. drives removal of protons from H_2O.

8.5 Molecular Analysis of Photosystems

11. During noncyclic electron transport, purple bacteria remove electrons from _____ and therefore produce no O_2.

    a. ATP

    b. H_2

    c. NAD^+

    c. H_2O

    e. H_2S

12. Both P_{680} and P_{700} are

    a. present in PSII

    b. located in the "reaction centers"

    c. specialized chlorophyll *a* molecules

    d. required for the splitting of H_2O

    e. contribute to the Emerson effect

13. In linear electron flow, electrons are transferred from PSII to PSI by

    a. ferredoxin

    b. the malate shuttle

    c. cytochrome c

    d. NADPH

    e. plastocyanin

8.6 CO₂ Metabolism During Photosynthesis

14. In the cytosol, glyceraldehyde 3-phosphate is used as a substrate for the production of

    a. CO_2

    b. sucrose

    c. glucose

    d. ATP

    d. NADPH

15. Which of the following are substrates of ribulose 1,5-bisphosphate carboxylase?

    a. fructose 1,6-bisphosphate

    b. O_2

    c. CO_2

    d. phosphoenolpyruvate

    e. ribulose 1,5-bisphosphate

PART C: *Reviewing Concepts*

8.1 *Oxidation of Glucose and Fatty Acids to CO₂*

16. In the left-hand column of Table 8-1, the reactions involved in the breakdown of glucose to carbon dioxide are listed. Fill in the numbers showing the net production of the indicated molecules by each pathway listed on the left.

17. Write the chemical equation for the overall glycolytic pathway.

Overall reaction	Net production of			
	CO_2	ATP or GTP	NADH	$FADH_2$
1 glucose → 2 pyruvate				
2 pyruvate → 2 acetyl CoA				
2 acetyl CoA → 4 CO_2				
glycolysis and citric acid cycle total				

Table 8-1

8.2 *Electron Transport and Generation of the Proton-Motive Force*

18. The electron-transport protein complexes present in the mitochondrial inner membrane do not directly interact with one another. How then are electrons transferred from one complex to another so that the proton motive force may be generated?

19. The four mitochondrial electron-carrier complexes are listed in the left-hand column of Table 8-2. Indicate the order in which a pair of electrons from NADH or $FADH_2$ would move through these multiprotein complexes to O_2 by writing 1 (first), 2, or 3 (last) on the appropriate lines in the middle two columns. In the right-hand column, indicate how many protons are pumped by each carrier complex per electron pair.

Carrier Complex	Order of electron transfer		No. protons pumped
	NADH → O_2	$FADH_2$ → O_2	
CoQH₂-cytochrome c reductase			
NADH-CoQ reductase			
Cytochrome c oxidase			
Succinate-CoQ reductase			

Table 8-2

8.3 Harnessing the Proton-Motive Force for Energy-Requiring Processes

20. Describe the experiment with chloroplast thylakoid membranes that demonstrated the role of a proton-gradient in ATP synthesis.

8.4 Photosynthetic Stages and Light-Absorbing Pigments

21. Sometimes the phrase "dark reactions" is used to refer to carbon fixation. Why is this phrase somewhat inappropriate?

22. Describe how the absorption of light by chlorophyll leads to the production of O_2.

8.5 Molecular Analysis of Photosystems

23. Listed in the left-hand column of Table 8-3 are various components of the photosynthetic machinery in higher plants. During photosynthesis, these components are involved in linear and/or cyclic electron (e^-) flow.

   a. During linear electron flow, electrons move from P_{680} in PSII to $NADP^+$, indicated by a 1 and 8 in the middle column of Table 8-3. Indicate the order in which electrons move through the remaining components of this pathway by writing in the appropriate numbers (2–7).

   b. During cyclic electron flow, electrons move from P_{700} in PSI (1) through four other components and back to P_{700} (6). In the right-hand column of Table 8-3, indicate the order in which electrons move through the other components of this pathway by writing in the appropriate numbers (2–5).

PSI/PSII component	Linear e⁻ flow	Cyclic e⁻ flow
$NADP^+$	8	
P700		1,6
Cytochrome b/f		
P680	1	
FAD		
Ferredoxin		
Quinone		
Plastocyanin		

Table 8-3

8.6 CO_2 Metabolism during Photosynthesis

24. In hot, dry environments, plants must keep their stomata closed much of the time to conserve water; as a result, the CO_2 level in the cells exposed to air falls below the K_m of ribulose 1,5-bisphosphate carboxylase for CO_2. What is the implication of this phenomenon? What special adaptation do some plants have that allow them to fix CO_2 under such conditions?

25. How is the sucrose produced as a result of photosynthesis delivered to the entire plant?

PART D: *Analyzing Experiments*

26. The F_1 portion of the ATP synthases found in thylakoids are composed of five different polypeptides. These polypeptides, which can be distinguished from each other using SDS gel electrophoresis, are named α, β, γ, δ, and ε in order of their decreasing molecular weights. The subunit stoichiometry in the thylakoid F_1 particle is 3:3:1:1:1 with a particle molecular weight of approximately 400,000. It is known that the β subunits bind ADP and ATP and contain the catalytic sites, while the γ polypeptide acts as an inhibitor of ATP hydrolysis.

One experiment designed to improve our understanding of the thylakoid F_0F_1 particle subjected thylakoid membranes to three extraction methods (A, B, and C), with the goal of removing part(s) of the F_0F_1 particle in order to define the function of the extracted and unextracted portions. After extraction, the chloroplasts were exposed to a flash of light, which generated a pH gradient across the thylakoid membranes. In order to test the function of the F_0F_1 complex, the ability of the membrane preparations to dissipate the proton gradient was monitored. Figure 8-1a depicts the decrease in the pH gradient across unextracted thylakoid membranes and across those prepared with the three extraction techniques. These data show that method A produced membranes that dissipated the pH gradient at nearly the same rate as the no-extraction control. Method C produced membranes that were very leaky to protons, and method B produced membranes that dissipated the pH gradient at an intermediate rate.

a. Suggest a method to determine which portion of the F_0F_1 complex was removed by each extraction method.

b. Immunoelectrophoresis of thylakoid membranes showed that methods A and B each extracted 13 percent of the F_1 from the membranes, while method C extracted 55 percent of the F_1. Because methods A and B had similar extraction efficiencies, but produced membranes that differed in proton permeability, the F_1 particles extracted by each method were partially purified and subjected to SDS gel electrophoresis. The results are shown in Figure 8-1b. What do these data suggest about the function of the δ subunit in the thylakoid F_0F_1 particle?

c. When the F_0 inhibitor N,N'-dicyclohexylcarbodiimide (DCCD) is added to membranes extracted by method C, these membranes exhibit the same profile for dissipation of the pH gradient as the unextracted control membranes. DCCD does not affect the extraction of the F_1 particle. Why is measurement of the proton flux in the presence of DCCD an important control in the investigation of the role of δ subunit in regulating proton flux through the thylakoid F_0F_1 particle?

d. Given the data suggesting that removal of the δ subunit results in membranes that are leaky to protons, propose more definitive experiments to test the hypothesis that this subunit acts as a "stopcock" of the F_0 portion.

27. Ribulose 1,5-bisphosphate carboxylase (rubisco) catalyzes the initial reaction in carbon fixation, the addition of CO_2 to ribulose 1,5-bisphosphate (RuBP), as well as the competing reaction of O_2 with RuBP. In order for rubisco to be active, it must be carbamylated on a lysine residue by CO_2. In light, the enzyme rubisco activase activates rubisco by carbamylation; this reaction requires ATP hydrolysis. Activation of rubisco is strongly inhibited by tight binding of RuBP to rubisco. RuBP binding to inactive rubisco induces an isomerization of the enzyme to a form in which RuBP is so tightly bound that its spontaneous release occurs with the extremely slow half-time of 35 min.

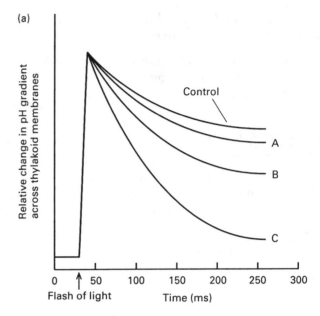

(a)

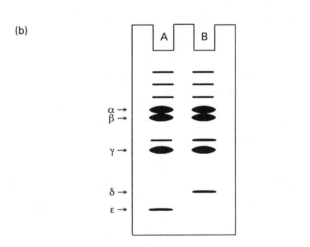

(b)

Figure 8-1

In one set of experiments designed to examine the regulation of rubisco activity, RuBP was added to rubisco so that tight binding occurred. Then at time 0, an incubation mixture with Mg^{2+}, ATP, and rubisco activase was added. Another mixture was prepared without rubisco activase. The carbon-fixing activity of rubisco was then measured at various time points, as shown in Figure 8-2a. In another experiment, the tightly bound RuBP-rubisco complex was prepared; at time 0, an incubation mixture with Mg^{2+}, ATP or its nonhydrolyzable analog, ATP-γ-S, and rubisco activase was added to the complex. Again, another mixture was prepared without rubisco activase. The release of RuBP from the RuBP-rubisco complex was monitored, as shown in Figure 8-2b.

a. Was rubisco activase necessary for activation of rubisco under the conditions used in these experiments?

b. Was rubisco activase, ATP hydrolysis, or both necessary for the dissociation of RuBP from rubisco?

c. Comparison of the kinetics of RuBP dissociation from rubisco and those of rubisco activation showed that 30 sec after exposure of the tightly bound RuBP-rubisco complex to rubisco activase, Mg^{2+}, and ATP, the specific activity of rubisco was 0.08 μmol CO_2 incorporated/min/mg; at this time,

65 percent of the RuBP had been released. At 150 sec, the specific activity was 0.79 μmol CO_2/min/mg; at this time, 98 percent of the RuBP had dissociated from the enzyme. What do these data suggest about the order in which dissociation of RuBP from rubisco and activation of rubisco took place?

d. Based on the data presented in this problem, summarize the events involved in the activation of inactive rubisco with tightly-bound RuBP, in the presence of Mg^{2+}, ATP, and rubisco activase.

(a)

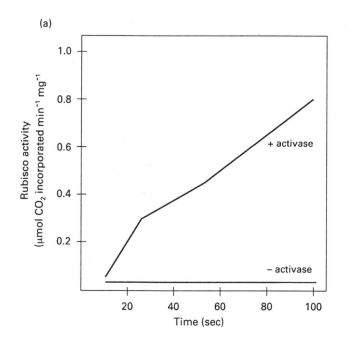

(b)

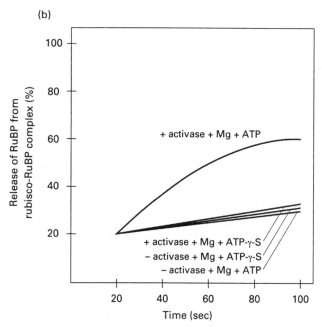

Figure 8-2

Answers

1. b

2. a, c, e

3. a, b, c, d

4. a

5. a, c

6. b, d

7. c

8.   e

9.   b

10. d

11. e

12. b, c, e

13. e

14. b

15. b, c, e

	Net production of			
Overall reaction	CO_2	ATP or GTP	NADH	$FADH_2$
1 glucose → 2 pyruvate	0	2 ATP	2	0
2 pyruvate → 2 acetyl CoA	2	0	2	0
2 acetyl CoA → 4 CO_2	4	2 GTP	6	2
glycolysis and citric acid cycle total	6	4	10	2

Table 8-4

16. See Table 8-4.

17. The overall reaction for glycolysis is as follows:

$$C_6H_{12}O_6 + 2NAD^+ + 2ADP^{3-} + 2P_i^{2-} \rightarrow$$

$$2C_3H_4O_3 + 2NADH + 2ATP^{4-}$$

18. Electrons are transferred from one electron-transport protein complex to the next by mobile electron shuttles. CoQ, a lipid soluble molecule, shuttles electrons from NADH-CoQ reductase (complex I) or succinate-CoQ reductase (complex II) to $CoQH_2$-cytochrome c reductase (complex III), while cytochrome c, an intermembrane space protein, shuttles electrons from $CoQH_2$- cytochrome c reductase (complex III) to cytochrome c oxidase (complex IV).

19. See Table 8-5.

Carrier Complex	Order of electron transfer		
	$NADH \rightarrow O_2$	$FADH_2 \rightarrow O_2$	No. protons pumped
$CoQH_2$-cytochrome c reductase	2	2	4
NADH-CoQ reductase	1		4
Cytochrome c oxidase	3	3	2
Succinate-CoQ reductase		1	0

Table 8-5

20. Isolated chloroplast thylakoid vesicles were equilibrated in the dark with a solution at pH 4.0 to allow the pH of the thylakoid lumen to reach 4.0. The vesicles were then rapidly exchanged into a solution at pH 8.0 that contained ADP and P_i. The resulting proton gradient (high in the lumen, low in the surrounding solution) caused protons to move through the F_0F_1 complexes in the membrane and provided the energy to synthesize ATP from the ADP and P_i in the surrounding solution. This experiment thus provided direct evidence that a chemiosmotic mechanism could power ATP synthesis.

21. The Calvin cycle reactions that result in carbon fixation can occur both in dark and light conditions. However, these reactions, which depend on the light-dependent reactions to provide ATP and NADPH, are inhibited by dark conditions. The activity of certain of these "dark reaction" enzymes is lower in the dark than in the light for several reasons: a decrease in the pH and Mg^{2+} concentration of the stroma in the dark; oxidation of the stromal enzymes in the dark; and a lack of activation of ribulose 1,5-bisphosphate carboxylase by rubisco activase under dark conditions.

22. A chlorophyll molecule that absorbs light will move to an excited, higher energy state. In the reaction center, the excited chlorophyll donates an electron to an intermediate acceptor, which passes the electron on to quinone Q, located on the stromal surface of the thylakoid membrane. The result of this photoelectron transport is a positively charged chlorophyll molecule close to the luminal surface and a negatively charged molecule near the stromal surface. The positively charged chlorophyll molecule is a powerful oxidizing agent, which ultimately removes electrons from water, leaving protons and O_2.

23. See Table 8-6.

PSI/PSII component	Linear e^- flow	Cyclic e^- flow
$NADP^+$	8	
P700	5	1, 6
Cytochrome b/f	3	4
P680	1	
FAD	7	
Ferredoxin	6	2
Quinone		3
Plastocyanin	4	5

Table 8-6

24. If the level of CO_2 falls below the K_m of ribulose 1,5-bisphosphate carboxylase, photorespiration, in which ribulose 1,5-bisphosphate carboxylase utilizes O_2 rather than CO_2 as a substrate, is favored over carbon fixation. Photorespiration is a wasteful process that consumes ATP. To avoid this problem, plants such as corn, sugar cane, and crabgrass, termed C_4 plants, have developed a

two-step mechanism of CO_2 fixation in which the initial assimilation of CO_2 can occur in a relatively low CO_2, high O_2 environment. The leaves of C_4 plants possess two types of chloroplast-containing cells: mesophyll cells, which are directly exposed to air, and bundle sheath cells, which underlie the mesophyll cells. (The leaves of other plants, termed C_3 plants, possess chloroplast-containing mesophyll cells but lack bundle sheath cells.) In C_4 plants, the bundle sheath cells generally contain more chloroplasts (and hence Calvin cycle enzymes) than the mesophyll cells. In the mesophyll cells, CO_2 reacts with phosphoenolpyruvate to produce the four-carbon compound oxaloacetate, which is reduced to malate. The enzyme catalyzing this reaction is active even at low CO_2 levels. Malate is shuttled to the bundle sheath cells, where it is decarboxylated, releasing CO_2 and thereby producing a relatively high CO_2, low O_2 environment. Under these conditions, ribulose 1,5-bisphosphate carboxylase operates to fix CO_2 and the Calvin cycle can operate as usual.

25. Sucrose, which is produced in photosynthetic cells, is delivered to other cells via the cellular phloem. The phloem, which is composed of sieve-tube cells and companion cells, forms a continuous tube of cytosol throughout the entire plant. Sucrose moves through the phloem due to force generated by osmotic pressure differences.

26a. The membranes could be separated from the extracted material by centrifugation. SDS gel electrophoresis of the extracted material in the supernatant and of the pelleted membranes would reveal which subunits of the complex were present in the membranes and which were extracted. Immunoblotting of the gel could be used to verify the identification of the bands, corresponding to the ATP synthase subunits.

26b. Whereas the material extracted by method B contains the δ subunit, that extracted by method A does not. Thus the membranes left after extraction by method B are deficient in the δ subunit. The finding that thylakoid membranes extracted by method B, and thus deficient in δ, are leaky suggests that the δ subunit blocks proton flow through the membrane-bound F_0 particle.

26c. Addition of DCCD, which is known to interact with F_0, allows one to distinguish proton movement through F_0 from proton movement through other integral membrane proteins or parts of the (leaky) membrane. A full block (i.e., DCCD-dependent return to control levels) indicates that the observed loss of the pH gradient upon extraction by method C is probably all mediated through the F_0 particle.

26d. This hypothesis could be tested by isolating the δ subunit by chromatography and then adding it to the membranes resulting from extraction method B. If δ acts as a stopcock, then the pH gradient profile of membranes extracted by method B with added δ subunit should be the same as that of unextracted membranes. Another approach would be to examine the function of the F_0F_1 complex in thylakoid membranes isolated from plants with nonlethal mutations in the gene encoding the δ subunit. The finding that changes in the amino acid sequence of the δ subunit alters the permeability of the thylakoid membrane to protons would support the hypothesis.

27a. The data in Figure 8-2a show that CO_2 fixation by rubisco requires activation of the enzyme by rubisco activase.

27b. As shown in Figure 8-2b, both were probably necessary. Dissociation appears to have been greater in the presence of ATP than in the presence of its nonhydrolyzable analog, suggesting that hydrolysis is necessary. An alternative explanation of the data is that dissociation requires binding of ATP to rubisco activase but not its hydrolysis and that ATP-γ-S did not bind to rubisco activase. However, other experimental evidence indicates that the analog does bind, so ATP hydrolysis most likely is required.

27c. These data suggest that dissociation of RuBP from rubisco occurred before activation of the enzyme. At 30 sec, dissociation was already 65 percent complete, whereas the rubisco CO_2 fixation activity was only about 10 percent of the activity at 150 sec.

27d. Rubisco activase catalyzes the dissociation of tightly bound RuBP from rubisco. This dissociation requires ATP hydrolysis. Dissociation is followed by activation of rubisco in a reaction also catalyzed by rubisco activase. Data not presented in this problem have demonstrated that this latter reaction requires ATP hydrolysis too.

9

Molecular Genetic Techniques and Genomics

PART A: *Chapter Summary*

Classical genetics, recombinant DNA technology and genomics are powerful tools that can be used together to analyze the function of genes and proteins in cells and organisms. The function of a gene product can be deduced by assessing phenotypic changes resulting from natural or induced mutations of the gene. Mutation analysis can be used to determine the number of functionally related genes involved in a process or the order in which proteins function in a biochemical or signaling pathway. In a method known as reverse genetics, a specific mutation can be introduced into a cloned DNA sequence and then the mutant gene can be introduced into the genome of a cell or organism to observe its phenotypic effects.

Restriction enzymes and DNA ligase are two key enzymes used in DNA cloning. Restriction enzymes cut double stranded DNA at specific 4-8 base pair, palindromic sequences, generating blunt or "sticky" ends. DNA ligase covalently links these DNA fragments into cloning vectors such as plasmids or lambda phage. A collection of DNA fragments cloned into a vector constitutes a library. A genomic DNA library consists of random fragments of an organism's entire genome, whereas a cDNA library consists of copies of expressed messenger RNAs. Specific clones can be identified from the library using hybridization screening. DNA sequencing, polymerase chain reaction (PCR) and Southern/Northern blotting are powerful tools used for the characterization of cloned DNA fragments.

Genomics is the genome-wide analysis of gene structure and expression. Potential genes can be identified by comparing them with DNA or amino acid sequences that are stored in a gene bank (e.g., GenBank). Comparison of related sequences from different species can reveal evolutionary relationships among proteins. The function of a gene can be determined by inhibiting its expression in cells or animals using gene knockout technology, dominant negative mutations, or RNA interference. The expression of genes can be examined individually or on a more global scale using DNA microarrays. DNA microarrays permit an analysis of the expression of thousands of genes simultaneously and can potentially identify coregulated genes. Genes for human diseases can be mapped to specific chromosomal regions by linkage analysis of the disease gene with DNA polymorphisms.

PART B: *Multiple Choice*

Circle the letter corresponding to the most appropriate terms/phrases. There may be more than one correct answer. Circle all that apply.

9.1 Genetic Analysis of Mutations to Identify and Study Genes

1. A recessive allele usually results from a mutation that

a. disrupts expression of the gene

b. alters the structure of the encoded protein leading to gain of function

c. alters the structure of the encoded protein leading to loss of function

d. increases the activity of the encoded protein

2. Temperature-sensitive mutations would be particularly useful for studying which of the following processes?

   a. DNA synthesis

   b. hormone secretion

   c. protein synthesis

   d. mitosis

9.2   DNA Cloning by Recombinant DNA Methods

3. Which of the following is an essential element of plasmids used as cloning vectors?

   a. a centromere

   b. a telomere

   c. an origin of replication

   d. a drug resistance gene

4. The restriction enzyme *Bgl*II has the recognition sequence (A*GATCT) and cuts between the A and the G as indicated by the asterisk. DNA cut by which of the following restriction enzymes (given with their recognition sequences and cut sites) could be cloned into a plasmid cut with *Bgl*II?

   a. *Bam*HI (G*GATCC)

   b. *Hin*dIII (A*AGCTT)

   c. *Pvu*I (CGAT*CG)

   d. *Xba*I (T*CTAGA)

5. Which of the following DNA sequences can be used to screen a cDNA library for the gene that encodes a protein with the amino acid sequence Phe-Gly-Cys-Ser? Use the genetic code; see Table 4-1 of the text.

   a. 5' TTCCAATGGTCG 3'

   b. 5' TTTGGATGTTCA 3'

   c. 5' TTTGAGTGCAGA 3'

   d. 5' TTCGGGTGCAGC 3'

9.3   Characterizing and Using Cloned DNA Fragments

6. Which of the following is used in the polymerase chain reaction (PCR) technique?

   a. single-stranded DNA primers

   b. deoxynucleotide triphosphates (dNTPs)

   c. a thermostable DNA polymerase

   d. DNA ligase

   e. a DNA template

7. Northern blotting is used to detect

   a. DNA

   b. RNA

   c. protein

   d. polysaccharides

9.4   Genomics: Genome-wide Analysis of Gene Structure and Expression

8. An open reading frame (ORF) is defined as a DNA sequence that

   a. begins with a start codon

   b. ends with a stop codon

   c. contains 50 codons

   d. contains approximately an equal frequency of A, T, G, and C

9. Two related genes that are derived from a gene duplication event are considered to be

   a. homologous

   b. paralogous

   c. orthologous

   d. members of a gene family

9.5 Inactivating the Function of Specific Genes in Eukaryotes

10. Which of the following steps is required to generate a knockout mouse?

    a. microinjection of DNA into the pronucleus of a fertilized mouse egg

    b. introduction of DNA into mouse embryonic stem cells

    c. selection for cells containing the gene targeted insertion

    d. expression of Cre protein during embryonic development

11. RNA interference is a method for inhibiting the function of specific genes using

    a. single-stranded DNA

    b. double-stranded DNA

    c. single-stranded RNA

    d. double-stranded RNA

9.6 Identifying and Locating Human Disease Genes

12. Huntington's disease is an example of

    a. an autosomal recessive disorder

    b. an autosomal dominant disorder

    c. a sex linked recessive disorder

    d. a sex linked dominant disorder

13. A microsatellite is a type of polymorphism that

    a. results from the change of a single nucleotide

    b. changes the recognition sequence for a restriction enzyme

    c. results from a variable number of repetitive sequences

    d. results from a chromosomal inversion

14. In the human genome, one centimorgan or a 1% recombination frequency represents, on average, a distance of about

    a. 5×10^3 base pairs

    b. 2.5×10^4 base pairs

    c. 7.5×10^5 base pairs

    d. 1×10^7 base pairs

PART C: Reviewing Concepts

9.1 Genetic Analysis of Mutations to Identify and Study Genes

15. Describe how you can differentiate between recessive and dominant mutations by analyzing segregation patterns in true breeding organisms.

16. Describe how suppressor mutations function.

9.2 DNA Cloning by Recombinant DNA Methods

17. Describe the basic features of plasmids and lambda phage as cloning vectors.

18. Describe the differences between cDNA and genomic DNA libraries.

9.3 Characterizing and Using Cloned DNA Fragments

19. Describe the steps in one cycle of the polymerase chain reaction (PCR) process.

20. Describe how Southern blotting differs from Northern blotting.

9.4 Genomics: Genome-wide Analysis of Gene Structure and Expression

21. What is a BLAST search and how can it be used to help determine the function of an unknown cloned gene?

22. What are the applications of DNA microarrays?

9.5   *Inactivating the Function of Specific Genes in Eukaryotes*

23. Compare the effectiveness of inhibiting gene function using knockout technology, dominant negative alleles or RNA interference.

24. Describe how a gene can be specifically knocked out in the liver of a mouse.

9.6   *Identifying and Locating Human Disease Genes*

25. What changes in the nucleotide sequence result in restriction fragment length polymorphisms (RFLPs), single nucleotide polymorphisms (SNPs), and simple sequence repeats (SSRs).

	A	B	C	D	E	F	G
A							
B	+						
C	-	+					
D	+	+	+				
E	+	-	+	+			
F	?	+	-	+	+		
G	+	+	+	+	+	+	

FIGURE 9-1

PART D: *Analyzing Experiments*

26. Yeast were screened for temperature-sensitive mutations in a regulatory pathway. The screening was designed to allow the mutant protein to be active at 23°C but completely inactive at 36°C. The wild type protein is active at both temperatures. Complementation analysis of the seven isolated temperature-sensitive mutants (A-G) was performed and is shown in Figure 9-1. Pairs of haploid temperature-sensitive mutants were mated to form diploids, and the diploids were tested for growth at 23°C and 36°C. All diploids grew at 23°C. In Figure 9-1, a "+" indicates that the diploid resulting from the mating of the two mutants grew at 36°C, whereas a "-" indicates that the diploid did not grow at 36°C.

a. Based on the results of the complementation analysis, how many genes are involved in this regulatory pathway?

b.

At 36°C, what phenotype would you predict for the diploid yeast generated by the mating of haploid mutant A and mutant F (indicated by the ?)? Why?

c. How could you use functional complementation to clone the wild type gene that corresponds to one of the temperature-sensitive mutations?

d. Assume that you have cloned the wild type gene affected in one of the temperature-sensitive mutants as described in part c. You perform a Southern blot comparing wild type and mutant genomic DNA digested with the restriction enzyme *Eco*RI and probed with a labeled wild type gene (Figure 9-2). Hypothesize what nucleotide change could have caused the temperature-sensitive mutation.

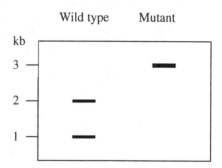

FIGURE 9-2

27. Pedigrees of families affected by Cystic fibrosis (CF), Huntington's disease (HD), and Duchenne muscular dystrophy (DMD) are shown in Figure 9-3.

a. Based on these case studies, determine whether each disorder is inherited autosomally or is sex linked and whether it is recessive or dominant. Explain your answers.

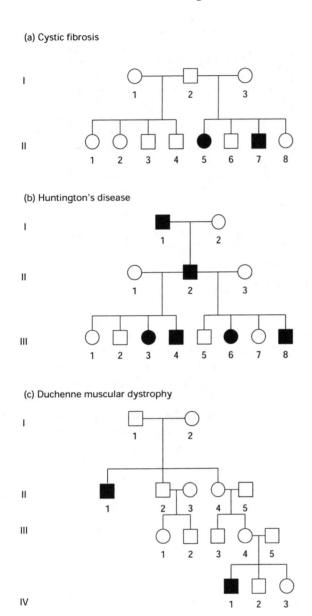

(a) Cystic fibrosis

(b) Huntington's disease

(c) Duchenne muscular dystrophy

☐ male, normal phenotype

○ female, normal phenotype

■ male, disease phenotype

● female, disease phenotype

Figure 9-3

b. Phenotypically unaffected members of the human population may be carriers for a genetic disease. What are the odds that the children of parents, both of whom are carriers for an autosomal recessive defect, will be affected? What are the odds that the children of parents, one of whom is a heterozygous carrier of an autosomal dominant defect, will be affected? What are the odds that children of parents, of which the mother is a carrier for an X-linked trait, will be affected?

c. How can a dominant lethal trait be maintained in the human gene pool?

Answers

1. a, c

2. a, c, d

3. c, d

4. a

5. b, d

6. a, b, c, e

7. b

8. a, b

9. a, b, d

10. b, c

11. d

12. b

13. c

14. c

15. When a homozygous mutant is crossed with a homozygous wild type, all F$_1$ progeny will be heterozygotes, i.e., contain one mutant and one wild type allele. If the heterozygous F$_1$ progeny exhibit the mutant trait, then the mutant allele is dominant; if the F$_1$ progeny exhibit the wild type trait, then the mutant allele is recessive.

16. Genetic suppression is a phenomenon whereby the effects of one mutation are counteracted by a second compensatory mutation. For example, assume that two wild type proteins can interact. A mutation in protein 1 disrupts the ability of mutant protein 1 to interact with wild type protein 2. A compensatory mutation in protein 2 can suppress the effects of the mutation in protein 1, thus allowing the two mutant proteins to interact. The compensatory mutation in protein 2 would be considered a suppressor mutation since it suppressed the mutation in protein 1.

17. Plasmids are circular, double-stranded, extrachromosomal DNA molecules. Plasmid cloning vectors contain an origin of replication, a drug resistance gene for selection, and a polylinker region that contains multiple restriction enzyme sites for cloning DNA fragments. Bacteriophage λ is a virus that infects certain bacteria. Used as a cloning vector, bacteriophage λ contains the genes for head and tail proteins as well as various proteins involved in phage DNA replication and cell lysis. Non-essential genes have been deleted from λ vectors to allow space for the insertion of foreign DNA.

18. Both cDNA and genomic DNA libraries are collections of random DNA fragments inserted into a cloning vector (e.g., plasmid or bacteriophage λ). A cDNA library consists of copies of all expressed messenger RNAs and reflects the levels of the expressed mRNAs. A genomic DNA library contains fragments of the total genome, including expressed regions (exons), introns, promoter sequences and intergenic DNA.

19. The polymerase chain reaction (PCR) is a technique used to amplify DNA regions of known sequences. In the first step of a PCR cycle, the reaction is heated to 95°C to denature the double-stranded DNA template and then lowered to 50-60°C to allow primers to hybridize (anneal) to their complementary sequences in the single-stranded template molecules. The hybridized primers serve as DNA synthesis primers in the next step. The reaction is then raised to 72°C where a heat-stable DNA polymerase (e.g., *Taq* DNA polymerase) extends the primers generating a double-stranded DNA molecule. This cycle of heating, annealing, and extension is then repeated many times.

20. Southern and Northern blotting combine separation of nucleic acid molecules by gel electrophoresis with hybridization to labeled probes. In the case of Southern blotting, DNA fragments are separated by gel electrophoresis and then transferred to a nylon membrane. A labeled probe is then used to detect by hybridization the presence of a specific DNA fragment. In the case of Northern blotting, RNA molecules are separated by gel electrophoresis prior to hybridization analysis.

21. The BLAST (basic local alignment and search tool) program searches the DNA/protein

databases for significant matches between a query sequence and stored sequences. The search program then assigns a score based on the extent of the match. For an unknown cloned gene, the DNA sequence or predicted amino acid sequence can be used as the query sequence for the BLAST search. If a known gene or protein in the database shows significant similarity to the query sequence, then it is likely that the unknown cloned gene is functionally similar to the known gene.

22. A DNA microarray consists of thousands of individual, closely packed, gene-specific sequences attached to a surface such as glass or a nylon membrane. The power of the microarray is its ability to examine the expression of thousands of genes simultaneously. It is the functional equivalent of performing thousands of Northern blots at one time.

23. The expression of a gene can be inhibited using gene knockouts, dominant-negative alleles, and RNA interference technology. However, only gene knockout technology can examine the effect of complete loss of the targeted gene product, because true null alleles are generated. In contrast, both dominant-negative alleles and RNA interference technology, although effective at inhibiting gene expression, cannot do so with 100% efficiency. Therefore, there would always be some residual wild type activity, which could complicate interpretation of the results.

24. Tissue-specific gene knockouts in mice can be generated using the Cre-loxP system coupled with standard embryonic stem (ES) cell gene knockout technology. An essential exon of the target gene is flanked with loxP sites. This loxP construct is then introduced into mouse ES cells and ES cells containing the gene-targeted insertion are selected. Mice homozygous for the loxP-modified gene can be generated and mated to mice that are heterozygous for the loxP-modified gene and also express Cre protein under the control of a liver-specific promoter. The resulting mice will have the target gene knocked out in the liver, because only in the liver will the Cre protein be expressed and cause recombination between the loxP sites and deletion of the essential exon of the target gene.

25. Restriction fragment length polymorphisms (RFLPs) arise due to a change in the nucleotide sequence that results in the formation or destruction of a restriction enzyme recognition sequence. This changes the pattern of restriction fragments generated. Single nucleotide polymorphisms (SNPs) are the most abundant type of polymorphism and are exactly what the name implies, i.e., a single nucleotide change. Simple sequence repeats (SSRs) or microsatellites are defined as a variable number of a short repeat unit.

26a. The first column of the complementation analysis shows that mutation A can complement mutations B, D, E, and G but cannot complement mutation C, indicating that mutation A and C are in the same gene. Using a similar approach, one can conclude that mutations B and E must be in the same gene. Similarly, mutations C and F must be in the same gene. Thus, in combination with the first conclusion, mutations A, C, and F must all be in the same gene. This leaves mutations D and G to be the mutations in two other genes. Thus there are four genes in total (D, G, B/E, and A/C/F).

26b. In part a, mutations in A, C, and F are shown to be in the same gene by complementation analysis. A diploid yeast generated between mutant A and mutant F would not be expected to show complementation, and thus would not grow at 36°C.

26c. To clone the wild type gene corresponding to one of the temperature-sensitive mutations by functional complementation, one first needs to generate a yeast genomic library from wild type genomic DNA. Recombinant plasmids from this genomic library are then introduced into temperature-sensitive yeast mutant cells. The transformed yeast are first incubated at 23°C, replica plated, and then checked for growth at 36°C. Only transformed yeast that grow at 36°C carry a plasmid that contains the wild type copy of the temperature-sensitive gene.

26d. The Southern blot shows that the wild type gene contains two hybridizing fragments of 2 and 1 kb after digestion with the restriction enzyme *Eco*RI and probing with a labeled wild type gene. In contrast, the mutant gene contains only a single hybridizing band at 3 kb. This suggests that the mutation causes a restriction fragment length polymorphism (RFLP) that results in loss of an *Eco*RI site. You can hypothesize that any one of the nucleotides in the *Eco*RI recognition sequence (5' GAATTC 3') of the wild type gene has been mutated (e.g., 5' GGATTC 3') such that the mutant no longer contains a recognition sequence for *Eco*RI.

27a. Cystic fibrosis must be an autosomal recessive disorder, since none of the parents in generation I show a phenotypic defect, but two out of the four children (both male and female) of parents I-2 and I-3 show a phenotypic defect. Parents I-2 and I-3 must be heterozygotes, whereas parent I-1 must be a homozygous normal since none of the four children of parents I-1 and I-2 show a phenotypic defect. Huntington's disease must be an autosomal dominant disorder, since at least one member of each generation shows a phenotypic defect, and in generation III this defect is shown irrespective of sex. Duchenne muscular dystrophy must be an X-linked trait, which is not expressed on the Y chromosome of males, since all individuals showing the phenotypic defect are male, yet no parent in generations I-IV is phenotypically defective. The defective allele must be carried on the X chromosome of heterozygous females in which it is recessive. For each generation, male children of a carrier female have a 50 percent probability of the defect. Females I-2, II-4, and III-4 must all be carriers. The number of male descendants of female II-3 is insufficient to determine whether or not she is a carrier.

27b. For the case of both parents being carriers of an autosomal recessive allele, the progeny have a 25% chance of being homozygous for the mutant allele and being affected. For the case of one parent being heterozygous for an autosomal dominant allele, the progeny have a 50% chance of inheriting the mutant allele and being affected. For the final case of one parent being a carrier for an X-linked trait, the probability that the progeny will be affected depends upon the sex of the progeny. Male progeny will have a 50% chance of inheriting the recessive allele from the mother and being affected. Female progeny will either be homozygous wild type or heterozygous but neither will express the mutant phenotype.

27c. An autosomal dominant trait can be maintained in the population only if the defect is minor with respect to reproductive performance or if the defect, although severe, is of late onset. The latter is true of Huntington's disease.

10 Molecular Structure of Genes and Chromosomes

PART A: *Chapter Summary*

The genomes of vertebrates are complex and contain both coding and noncoding regions. This chapter describes the organization of both the nuclear and organellar genomes of higher eukaryotes and the packaging of DNA into chromosomes. A gene is defined in molecular terms as the entire DNA sequence that is required for the synthesis of a functional protein or RNA molecule. This includes coding regions, control regions and sometimes introns. Eukaryotic protein-coding genes can exist in the genome as solitary genes or duplicated genes of a gene family and can be expressed as a simple or complex transcription unit. Only about 1.5% of the human genome corresponds to protein-coding sequences. Much of the genomes of higher eukaryotes contain nonfunctional DNA, some of which exists as repetitious DNA. Differences in the structural organization of this repetitious DNA allows us to identify individuals using a technique known as DNA fingerprinting. The size and coding capacity of organellar DNA varies among different organisms. Furthermore, the mitochondrial genes of mammals, invertebrates, and fungi use a different genetic code for translation of mitochondrial proteins.

The genome is not static. There are mobile elements such as insertion sequences, transposons, and retrotransposons, which can move around the genome. The two major classes of mobile elements are DNA transposons and retrotransposons, which move as DNA or RNA intermediates, respectively. Most mobile elements encode the enzymes, e.g., transposase or reverse transcriptase, necessary for mediating the transposition event. Mobile DNA elements most likely influenced evolution by inducing mutations due to insertion of the DNA element into or near a transcription unit or by serving as sites of recombination resulting in gene duplication and exon shuffling.

The DNA in the nucleus is packaged into a highly ordered structure. DNA is first wrapped around a histone core to form chromatin. Chromatin is then coiled into a solenoid structure, which is attached to a chromosome scaffold composed of non histone proteins. The DNA-protein scaffold is further compacted to form a metaphase chromosome. Differences in the localized organization of the condensed chromatin produce unique banding patterns that are useful for chromosome identification. A newly developed technique, called chromosome painting, identifies chromosomes using in situ hybridization with fluorescently tagged DNA probes. Chromosomes require three functional elements for replication and stable inheritance: a replication origin, a centromere, and telomeres. Yeast artificial chromosomes containing these elements have been constructed for cloning large DNA fragments.

PART B: *Multiple Choice*

Circle the letter corresponding to the most appropriate terms/phrases. There may be more than one correct answer. Circle all that apply.

10.1 Molecular Definition of a Gene

1. A typical eukaryotic gene consists of which of the following elements?

   a. internal ribosome binding sites

   b. introns

   c. exons

   d. enhancers

2. Multiple mRNAs can arise from a primary transcript by use of alternative

   a. splicing

   b. poly A sites

   c. promoters

   d. ribosome binding sites

10.2 Chromosomal Organization of Genes and Noncoding DNA

3. Which of the following is true of the human genome?

   a. 1.5% of the genome corresponds to protein coding sequences.

   b. The median length of an intron is 10 kilobases.

   c. 10% of the genome is transcribed into pre mRNA precursors.

   d. Most human exons contain 500-1000 base pairs.

4. Which of the following is encoded by multiple genes present in the human genome?

   a. ribosomal RNA

   b. transfer RNA

   c. histone

   d. lysozyme

5. DNA fingerprinting is a technique based on differences in the

   a. length of introns

   b. number of tandem copies of a simple sequence repeat

   c. number of tandem ribosomal RNA genes

   d. size of protein coding genes

10.3 Mobile DNA

6. Transposition by a bacterial insertion element

   a. occurs at a frequency of approximately 1 in 10^3 cells per generation

   b. can inactivate an essential gene

   c. is mediated through a RNA intermediate

   d. requires the enzyme transposase

7. Transposition by a retrotransposon requires which of the following enzyme activities?

   a. RNA polymerase

   b. reverse transcriptase

   c. DNA methylase

   d. DNA polymerase

8. LINES (long interspersed elements)

   a. are retrotransposons that lack LTRs

   b. are approximately 300 base pairs long

   c. are a rare class of mobile elements in mammals

   d. use the enzyme transposase for transposition

10.4 Structural Organization of Eukaryotic Chromosomes

9. A transcriptionally active gene compared with a transcriptionally inactive gene would be expected to

   a. contain acetylated histones

   b. contain unacetylated histones

   c. be sensitive to DNase I

   d. be resistant to DNase I

10. Scaffold-associated regions

   a. are the chromosome attachment points for the mitotic spindle

   b. can insulate transcription units from each other

   c. are the points at which DNA interacts with histone proteins

   d. are found between transcription units

10.5 Morphology and Functional Elements of Eukaryotic Chromosomes

11. Metaphase chromosomes can be identified

    a. by shape

    b. by the size and number of introns

    c. by banding patterns with Giemsa reagent

    d. by chromosome painting

12. Telomerase

    a. extends DNA strands during DNA synthesis

    b. has reverse transcriptase activity

    c. is a protein-RNA complex

    d. replicates repetitious DNA located at the centromere

10.6 Organelle DNAs

13. Human mitochondrial DNA

    a. uses the standard genetic code

    b. encodes its own ribosomal RNAs

    c. contains introns like nuclear genes

    d. is larger than yeast mitochondrial DNA

14. Plant mitochondrial DNA

    a. is the same size as human mitochondrial DNA

    b. encodes a 5S mitochondrial rRNA

    c. contains multiple copies that recombine with each other

    d. uses the standard genetic code

PART C: Reviewing Concepts

10.1 Molecular Definition of a Gene

15. Compare and contrast simple transcription units and complex transcription units.

16. Describe the three different ways that multiple mRNAs can arise from a complex transcription unit in eukaryotes.

10.2 Chromosomal Organization of Genes and Noncoding DNA

17. Describe the general organization of the human genome in terms of protein-coding and functional RNA genes, repetitious DNA and spacer DNA.

18. Describe the molecular basis for the DNA fingerprinting technique. How can this be used to differentiate between two individuals?

10.3 Mobile DNA

19. Describe the structural features of bacterial insertion sequences and transposons.

20. Describe the possible role that mobile DNA elements played in evolution.

10.4 Structural Organization of Eukaryotic Chromosomes

21. Describe the role that nonhistone proteins play in organizing chromosome structure.

22. How does chromatin differ between transcriptionally inactive and transcriptionally active regions of DNA?

10.5 Morphology and Functional Elements of Eukaryotic Chromosomes

23. Compare the effectiveness of inhibiting gene function using knockout technology, dominant negative alleles or RNA interference.

24. Describe how a gene can be specifically knocked out in the liver of a mouse.

10.6 Organelle DNAs

25. How does plant mitochondrial DNA differ from mammalian mitochondrial DNA?

PART D: Analyzing Experiments

26. The α-tropomyosin gene contains multiple exons and polyadenylation signals (indicated by an A). As a result of alternative splicing and polyadenylation, the primary transcript can be processed into multiple mRNAs, of which three are shown in Figure 10-1. These alternatively processed mRNAs

are expressed specifically in striated muscle, smooth muscle, or brain.

a. Is the α-tropomyosin gene classified as a simple or complex transcription unit? What is the advantage of having a gene that can be processed in multiple ways?

b. A mutation in the α-tropomyosin gene results in the loss of functional α-tropomyosin in striated muscle, smooth muscle, and brain. Where is a likely site for this mutation?

c. Another mutation in the α-tropomyosin gene results in the loss of functional α-tropomyosin mRNA in smooth muscle only. Where is a likely site for this mutation?

27. In maize, the C locus makes a factor required for synthesis of a purple aleurone pigment that colors the kernels (a) as shown in Figure 10-2. In this figure the black regions represent the purple colored regions. Insertion of the transposable element, Ds, into the C locus inactivates the C locus and thus makes the kernels colorless (b). The Ds element is unable to transpose by itself; it requires the presence of the Ac element for transposition. After introduction of the Ac element, kernels containing both colorless and purple spots are seen (c and d).

a. How do kernels containing both colorless and purple spots arise?

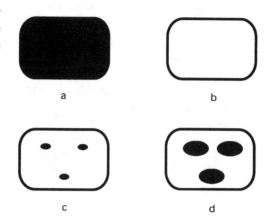

FIGURE 10-2

b. How can you explain the difference in size between the purple spots seen on kernels c and d?

c. Suppose that you have a different maize plant that contains kernels with 20 purple spots. What can you conclude about the frequency of transposition for the maize plant with kernels containing 3 spots versus the maize plant with kernels containing 20 spots?

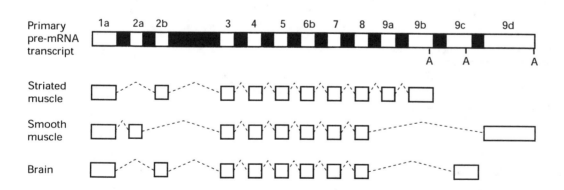

FIGURE 10-1

28. Histone genes in *Drosophila* are arranged as a tandem repeat of clustered H1, H2A, H2B, H3, and H4 genes. About 100 copies of this repeat are found at the cytological 39DE locus of the polytene chromosome. Laemmli and colleagues investigated how this repeat is bound to the nuclear scaffold by determining whether histone gene fragments were retained in a scaffold preparation following digestion of the preparation with a mixture of restriction enzymes. In brief, a scaffold preparation was obtained from nuclei containing polytene chromosomes, digested with restriction enzymes, and pelleted; the DNA from the pellet (P) and supernatant (S) was then analyzed by Southern blotting using a radiolabeled histone-gene repeat probe. Typical results are shown in Figure 10-3. In these experiments, the nuclei were incubated with Cu^{2+} or at 37°C to stabilize the scaffolds. The actual nuclear scaffolds were then prepared by extracting the pellet fraction with lithium diiodosalicylate (LIS).

a. In this experimental protocol, what result indicates that a histone-repeat fragment is bound to the scaffold?

b. What can you conclude from the data in Figure 10-3 about the location of the scaffold-binding site in the histone-gene repeat?

c. How could you more precisely locate the scaffold-binding site in the histone-gene repeat?

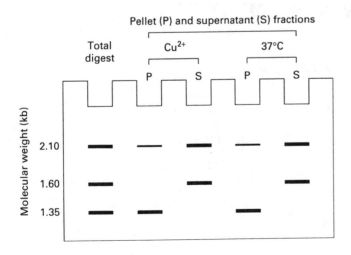

FIGURE 10-3

Answers

1. b, c, d

2. a, b, c

3. a

4. a, b, c

5. b

6. b, d

7. a, b

8. a

9. a, c

10. b, d

11. a, c, d

12. b, c

13. b

14. b, c, d

15. A simple transcription unit produces a single monocistronic mRNA, which is translated into a single protein. A complex transcription unit produces primary transcripts that can be processed in alternative ways and translated into multiple proteins.

16. A complex transcription unit is transcribed into multiple mRNAs using alternative promoters, poly A sites and splicing. The use of alternative promoters produces mRNAs with different 5' exons but common 3' exons. The use of alternative poly A sites produces mRNAs with common 5' exons but different 3' exons. The use of alternative splice sites produces mRNAs with common 5' and 3' exons but different combinations of internal exons.

17. The human genome consists of protein-coding and functional RNA genes, repetitious DNA and spacer DNA. Protein-coding genes and functional RNA genes make up approximately 30% of the genome. Protein-coding genes can exist as solitary genes or duplicated genes. Functional RNAs such as tRNA, rRNA, and snRNA are present as tandemly repeated genes. Repetitious DNA makes up almost half of the human genome and is usually concentrated at specific chromosomal locations, at the centromere for example. The remainder of the genome consists of spacer or intergenic DNA.

18. The DNA fingerprinting technique is based on differences in the length of simple-sequence DNAs. Simple sequence DNA usually occurs in tandem arrays. The number of simple sequence repeat units at a given genetic locus varies between individuals, and thus the total length of the tandem array differs. These differences in the length of tandem arrays throughout the genome are unique to individuals and is the basis for an individual's unique DNA fingerprint.

19. Bacterial insertion sequences (IS) are about 1-2 kilobases long. At each end of the IS element is an approximately 50 base pair inverted repeat. Between the inverted repeats is a region that encodes the enzyme transposase—which is required for transposition or the "cut and paste" operation of the element—to a new site in the genome.

20. Mobile DNA elements are hypothesized to have had a profound effect on the evolution of organisms. Insertion of a mobile DNA element into or near a gene can cause a mutation in the gene. Some evidence suggests that recombination between repeat sequences (mobile elements, for example), of two separate genes can generate a novel combination of exons, a process known as exon shuffling. Unequal crossing over between repeat sequences can result in gene duplication. Furthermore, transposition of DNA adjacent to mobile DNA elements can move transcriptional control elements to new regions of the genome.

21. Nonhistone proteins provide a structural scaffold for organizing chromatin loops. The chromosome scaffold, which has the shape of a metaphase chromosome, consists of nonhistone proteins that serve as binding sites for chromatin loops.

22. Experimental evidence indicates that transcriptionally inactive DNA has a condensed

chromatin structure. This structure makes the DNA inaccessible to RNA polymerase. In contrast, actively transcribed DNA is present in a more extended, "beads-on-a-string" form of chromatin. DNA with this more open conformation is more accessible to RNA polymerase and other proteins required for transcription.

23. The expression of a gene can be inhibited using gene knockouts, dominant-negative alleles, and RNA interference technology. However, only gene knockout technology can examine the effect of complete loss of the targeted gene product, because true null alleles are generated. In contrast, both dominant-negative alleles and RNA interference technology, although effective at inhibiting gene expression, cannot do so with 100% efficiency. Therefore, there would always be some residual wild type activity, which could complicate interpretation of the results.

24. Tissue-specific gene knockouts in mice can be generated using the Cre-loxP system coupled with standard embryonic stem (ES) cell gene knockout technology. An essential exon of the target gene is flanked with loxP sites. This loxP construct is then introduced into mouse ES cells and ES cells containing the gene-targeted insertion are selected. Mice homozygous for the loxP-modified gene can be generated and mated to mice that are heterozygous for the loxP-modified gene and also express Cre protein under the control of a liver-specific promoter. The resulting mice will have the target gene knocked out in the liver, because only in the liver will the Cre protein be expressed and cause recombination between the loxP sites and deletion of the essential exon of the target gene.

25. Plant mitochondrial and animal mitochondrial DNA are circular molecules that encode for tRNA, rRNA and essential mitochondrial proteins. In contrast to mammals, plants contain multiple mitochondrial DNAs that appear to recombine with one another. Plant mitochondrial DNAs are much larger and more variable in size than those of other organisms. Plant mitochondrial DNA also encode a 5S mitochondrial rRNA, which is present only in

plant mitochondrial ribosomes, and the α subunit of the F_1 ATPase. Furthermore, plant mitochondrial DNA uses the standard genetic code, whereas mammalian mitochondria use a modified code.

26a. The α-tropomyosin gene would be considered a complex transcription unit because multiple mRNAs are generated as a result of alternative splicing and polyadenylation usage. One advantage of a complex transcription unit is the generation of diversity. From a single gene, multiple mRNAs and thus multiple proteins can be synthesized.

26b. A mutation in the promoter/regulatory region of this gene, (i.e., upstream of exon 1a), would result in loss of α-tropomyosin mRNA transcription in all cells. A second possibility is a mutation that changes the amino acid sequence in a shared exon, i.e., exons 1a, 3, 4, 5, 6b, 7 or 8. A third possibility is a mutation that alters a splice site for a shared exon, resulting in an incorrectly spliced mRNA. These mutations could result in the synthesis of a nonfunctional α-tropomyosin in all cells.

26c. A mutation in an exon that is present only in smooth muscle α-tropomyosin mRNA could lead to loss of functional α-tropomyosin specifically in smooth muscle. In this case, exons 2a and 9b are unique to smooth muscle α-tropomyosin mRNA. A mutation that changes the amino acid sequence or a splice site in exons 2a or 9b could result in loss of functional α-tropomyosin only in smooth muscle cells.

27a. Insertion of the Ds transposable element into the C locus leads to loss of a factor required for synthesis of a purple aleurone pigment. In the presence of the Ac element, the Ds element is excised from the C locus, which allows resynthesis of the purple aleurone pigment. Therefore, a maize kernel that contains both colorless and purple spots is due to a combination of cells that still contain the Ds transposable element (colorless) and cells that lack Ds (purple) at the C locus.

27b. The size of the purple spots is a reflection of the time in development during which the Ds element was excised. In the case of kernel c, small purple spots indicate that excision of Ds occurred late in development. The clone of cells without the Ds element is small, and thus the size of the pigmented area is small. In contrast, the excision event occurred earlier in development in kernel d which generates larger purple spots.

27c. The number of purple spots is a reflection of the frequency of the excision (transposition) event. The maize plant that contains kernels with 20 spots shows a much higher frequency of excision of Ds from the C locus than the maize plant with kernels containing three spots.

28a. In this experiment, the nuclear scaffold and any associated DNA is located in the pellet fraction; however, free DNA and DNA fragments are located in the supernatant fraction. Thus detection of a histone-repeat restriction fragment in the pellet indicates that the fragment is bound to the scaffold.

28b. The Southern-blot pattern suggests that the 1.35-kb fragment is specifically bound to the scaffold preparation (the pellet fraction), as this fragment is found exclusively in the pellet fraction. The 1.60-kb fragment is completely released into the supernatant fraction by the restriction endonuclease digestion, as is almost all of the 2.10-kb fragment. Hence the 1.35-kb fragment, the only fragment preferentially retained with the scaffold preparation, appears to contain the scaffold-binding site in the histone gene repeat.

28c. The use of additional or alternative restriction endonucleases to produce smaller fragments would increase the precision of this determination.

11

Transcriptional Control of Gene Expression

PART A: *Chapter Summary*

The actions and properties of each cell type are determined by the proteins it contains. Theoretically, regulation at any one of the various steps in gene expression could lead to differential production of proteins in different cell types or developmental stages or in response to different conditions. In reality, control of the first step of gene expression, transcriptional initiation, is the most important mechanism for determining whether genes are expressed and how much of the encoded mRNAs and, consequently, proteins are produced. This chapter focuses on the molecular events that determine when the transcription of eukaryotic genes is initiated.

Eukaryotes contain three types of nuclear RNA polymerase which catalyze RNA production. RNA polymerase II synthesizes mRNAs. The transcription initiation complex for pre-mRNA genes is a multiprotein complex that includes RNA polymerase II and a number of general transcription factors. This multiprotein complex consists of 60–70 polypeptides with a total mass of approximately 3 MDa, almost the size of a eukaryotic ribosome. DNA regulatory elements include pieces of DNA responsible for attracting the transcriptional machinery. The most common promoter elements of eukaryotic genes are promoter-proximal elements and enhancers. The three principle promoter sequences are the TATA box, initiators, and CpG islands. Eukaryotic genes are regulated by the interaction of transcription factors with these cis-acting sequences. Transcription factors are the protein activators and repressors of transcription. Transcription factor activators and repressors consist of two modular domains with distinct functions: a DNA binding domain and an activation/repression domain. The identification of these DNA binding proteins and the regulatory elements they bind was facilitated using a number of techniques such as DNase I footprinting, the electrophoretic mobility shift assay, 5' deletion analysis, linker scanning mutation analysis, and sequence-specific DNA affinity chromatography.

The molecular mechanisms of transcriptional activation and repression include control imparted by chromatin. Since DNA promoter sequences reside within chromatin in the nucleus, modifications of histones, such as acetylation and phosphorylation events, can dramatically influence transcriptional activity. Two large protein complexes, the mediator and the SWI/SNF chromatin remodeling complex, are involved in chromatin structure and preinitiation complex assembly.

The regulation of transcription factors also greatly influences whether or not a specific gene is expressed. The nuclear-receptor superfamily is discussed as it provides a remarkable example of regulation of transcriptional activators. In these receptors, specific domains are responsible for binding hormone ligands which then regulate binding to specific response element sequences present in the promoters of hormone-regulated genes. The chapter also briefly considers regulation of transcript elongation and termination and, lastly, transcription initiation at RNA polymerase I and polymerase III promoters is described.

PART B: *Multiple Choice*

Circle the letter corresponding to the most appropriate terms/phrases. There may be more than one correct answer. Circle all that apply.

11.1 *Overview of Eukaryotic Gene Control and RNA Polymerases*

1. Specific DNA control elements in promoters can

    a. interact with general transcription factors.

    a. interact with repressor proteins.

    b. interact with activator proteins.

    d. remain unavailable because of condensed chromatin.

2. Reporter genes are used by scientists to

    a. express enzymes that are not easily assayed in cell extracts.

    b. express enzymes that are easily assayed in cell extracts.

    c. characterize DNA control elements.

    d. characterize reporter plasmids.

3. The three eukaryotic RNA polymerases can be distinguished by

    a. the types of genes they transcribe.

    b. the number and types of large subunits.

    c. their differential sensitivities to cyclohexamide.

    d. their differential sensitivities to α-amanitin.

11.2 *Regulatory Sequences in Protein-Coding Genes*

4. Which of the following can be identified using a series of promoter linker scanning mutations?

    a. areas of the promoter that are non-essential

    b. areas of the promoter that are essential

    c. the presence of separate transcriptional control regions

    d. spacing constraints on separate transcriptional control regions

5. An enhancer

    a. can be located upstream of a promoter.

    b. can be located downstream of a promoter.

    c. can be located a variable distance from the promoter.

    d. is always located within 1 kb of the promoter.

    e. can be cell-type-specific.

11.3 *Activators and Repressors of Transcription*

6. The fact that a specific protein leaves a "footprint" on a DNA molecule is indicative of

    a. a lack of interaction between the specific protein and DNA.

    b. protection from DNAse by the specific protein.

    c. binding of the specific protein to all types of DNA.

    d. binding of the specific protein to a specific sequence of DNA.

7. The C-terminal activation domain of transcriptional activators is capable of

    a. binding to DNA.

    b. stimulating transcription.

    c. interaction with other transcriptional machinery.

    d. functioning in a fusion with a DNA-binding domain from an unrelated transcriptional activator.

8. Which of the following are NOT found in DNA-binding proteins?

    a. homeodomains

    b. zinc fingers

    c. leucine zippers

    d. CpG islands

9. Transcription factors can

   a. exhibit cooperative binding.

   b. exist as heterodimers.

   c. act to repress transcription of transcription factor genes.

   d. undergo conformational changes which alter activity.

   e. never interact with co-repressors.

11.4 Transcriptional Initiation by RNA Polymerase II

10. Which is the first factor to bind at the promoter of eukaryotic genes?

    a. RNA polymerase

    b. TFIIA

    c. TFIIB

    d. TFIID

    e. TATA-box binding protein

11. Which of the following occurs during Pol II transcription preinitiation complex formation?

    a. TFIIA binds to TFIIB

    b. TFIIB unwinds the DNA

    c. TFIIB contacts both TATA-box binding factor and DNA

    d. DNA bends

11.5 Molecular Mechanisms of Transcription Activation and Repression

12. Chromatin-mediated repression of transcription involves

    a. modification of lysine residues in histones.

    b. large, multiprotein complexes.

    c. acetylation of histone tails.

    d. deacetylation of histone tails.

13. Which of the following does NOT require a DNA helicase activity?

    a. SWI/SNF function

b. Pol II open complex formation

c. transcription factor binding to DNA

d. deacetylation of histone tails

14. The yeast two-hybrid system can be used to identify

    a. proteins that interact with a known or "bait" protein.

    b. proteins that interact with a "fish" domain.

    c. cDNAs that encode interacting proteins.

    d. co-activators and co-repressors.

11.6 Regulation of Transcription Factor Activity

15. Which of the following can occur when steroid hormone binds to its cognate receptor?

    a. conformational changes in the receptor.

    b. translocation of the receptor out of the cell membrane.

    c. binding of the receptor to DNA response elements.

    d. stimulation of transcription.

11.8 Other Eukaryotic Transcription Systems

16. The promoter sequences of RNA polymerase III-transcribed genes are located

    a. within the transcribed sequences of genes.

    b. upstream of the transcribed sequences of genes.

    c. downstream of the transcribed sequences of genes.

    d. within the first intron.

PART C: Reviewing Concepts

11.1 Overview of Eukaryotic Gene Control and RNA Polymerases

17. Compare the structure and function of the three eukaryotic RNA polymerases.

11.2 Regulatory Sequences in Protein-Coding Genes

18. Describe the three types of eukaryotic promoters.

19. How can linker scanning mutation analysis be used to map the location of transcription-control elements?

11.3 Activators and Repressors of Transcription

20. Describe the experimental evidence that shows that transcriptional activators have separable DNA binding and activation domains.

21. What is the functional advantage of heterodimeric transcription factors?

11.4 Transcriptional Initiation by RNA Polymerase II

22. Describe the sequence of events that occur during formation of a transcription initiation complex on a TATA box promoter.

11.5 Molecular Mechanisms of Transcription Activation and Repression

23. Describe the general mechanism of yeast mating-type switching.

24. Describe the mechanism for gene silencing at yeast telomeres.

11.6 Regulation of Transcription Factor Activity

25. Describe how addition of glucocorticoid hormone elicits an increase in transcription of glucocorticoid responsive genes.

11.7 Other Eukaryotic Transcription Systems

26. Describe transcription initiation for RNA polymerase I and III transcribed genes.

PART D: Analyzing Experiments

27. One of the first promoter regions to be analyzed by linker scanning mutations was that of the thymidine kinase (tk) gene of herpes simplex virus. The results of such a study for the –100 region of the gene are shown in Figure 11-1. Each of the boxes in the figure shows the position of a clustered region of 6 to 10 nucleotide substitutions.

    a. Based on these results, draw a map showing the position of specific nucleotide regions important for the efficient transcription of the tk gene.

    b. What type of control element is represented by each of these transcriptionally important regions?

    c. Which of these transcriptionally important regions is likely to contain two neighboring control elements?

28. The NFκB transcription factor binds to an enhancer of the kappa light-chain gene, an enhancer on the human immunodeficiency virus (HIV) genome, and an upstream sequence of the gene encoding the a chain of the interleukin-2 receptor. Binding of NFκB confers transcriptional activity and phorbol ester inducibility to genes controlled by these cis-acting elements. This factor can exist in active and inactive forms. Baeuerle and Baltimore have used electrophoretic mobility shift analysis (EMSA) to investigate the nature of these two forms. EMSA can detect the ability of a transcription factor to retard the mobility of enhancer DNA in a gel.

    In one such experiment, isolated NFκB factor was incubated with $[^{32}P]$-labeled enhancer DNA in the presence or absence of varying amounts of complete cytosol or NFκB-depleted cytosol from pre-B cells. The migration of the labeled DNA in the gel following incubation is shown in Figure 11-2. Free DNA migrates rapidly and is found at the bottom of the gel. The addition of increasing amounts of cytosol or depleted cytosol to the incubation mix are indicated

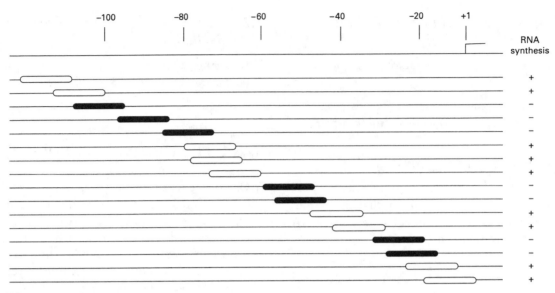

Figure 11-1

by the numbers 1 → 4 at the top of the figure. As controls, incubation is with cytosol or depleted cytosol alone (lanes 1 and 2).

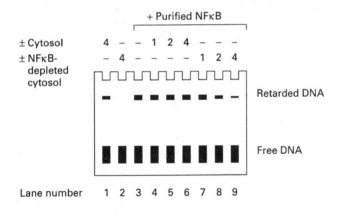

Figure 11-2

a. Based on lanes 1 and 2 of the EMSA pattern, what is the effect of NFκB on enhancer DNA migration?

b. Does cytosol have any apparent effect on the migration of enhancer DNA incubated with purified NFκB?

c. Does NFκB-depleted cytosol have any apparent effect on the migration of enhancer DNA incubated with purified NFκB? If so, does depleted cytosol appear to contain an activator or inhibitor of NFκB activity?

29. A template competition assay for transcription factor specificity includes the following steps: (1) gene A is incubated with limiting amounts of transcription factor, (2) competing gene B is added, and (3) transcription of each gene is assayed by gel electrophoresis of the RNA transcripts. Isolated human transcription factors for RNA polymerase III genes were assayed with this procedure, using a 5S-rRNA gene with an insert (termed long gene) and a wild-type 5S gene. The gel patterns of the reaction products (long rRNA, wild-type rRNA) are shown diagrammatically in Figure 11-3; the components incubated in step 1 are indicated at the top (A, B, and C refer to TFIII A, B, and C, and III to RNA polymerase III). In all reaction mixtures except that shown in gel lane 1, the genes were added sequentially with the long gene added first. In the reaction shown in lane 1, the two genes were added simultaneously.

a. What conclusions regarding the role of TFIIIA, B, and C and RNA polymerase III in the formation of protein-DNA complexes are suggested by these data?

b. If TFIIIA acted catalytically rather than stoichiometrically, would further TFIIIA be required for transcription of the long gene added to a reaction mix containing excess TFIIIB and C, RNA polymerase, wild-type 5S gene, and limiting amounts of TFIIIA?

c. When Roeder and colleagues performed the experiment described in part (b), they found that more TFIIIA was required for transcription of the long gene. In parallel experiments with TFIIIC, similar results were obtained. Based on these data, what is the role of TFIIIA and C in the transcription of 5S-rRNA genes?

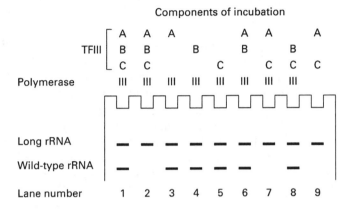

Figure 11-3

Answers

1. a, b, c, d

2. b, c

3. a, d

4. a, b, c

5. a, b, c, e

6. b, d

7. a, b, c, d

8. d

9. a, b, c, d

10. d, e

11. c, d

12. a, b, c

13. c, d

14. a, c, d

15. a, c, d

16. a

17. In eukaryotic cells, three different RNA polymerases catalyze the formation of different RNAs. RNA polymerase I is located in the nucleolus and synthesizes pre-ribosomal RNA (the precursor to 28S, 5.8S, and 18S rRNAs). RNA polymerase II synthesizes messenger RNAs. RNA polymerase III synthesizes tRNAs, 5S rRNAs, and a number of small RNAs. These small RNAs include an RNA involved in RNA splicing and the 7S RNA of the signal recognition particle. All three RNA polymerases contain two large subunits and 12–15 smaller subunits. Some of these subunits are shared among the different polymerases and others are unique.

18. Three types of promoter sequences have been identified in eukaryote genes. One class of eukaryotic genes contains a highly conserved sequence located approximately 25–35 base pairs upstream of the transcription start site known as the TATA box. The TATA box acts similar to a prokaryotic promoter to position RNA polymerase II for transcription initiation. A second class of eukaryotic genes lack a TATA box and instead contain an alternative promoter element called an initiator. The initiator sequence is not highly conserved. A final class of eukaryotic genes contains neither a TATA box nor an initiator element. These genes contain a CG-rich stretch of 20–50 nucleotides (CpG islands) approximately 100 base pairs upstream of the start site.

19. In linker scanning mutation analysis, a region of DNA that contains putative regulatory elements are cloned into a plasmid upstream of a reporter gene. A series of overlapping linker scanning mutations are introduced into the DNA containing the regulatory elements. The mutations are introduced by scrambling the nucleotide sequence in a short stretch of DNA. The effects of these mutations are evaluated by assaying reporter gene expression.

20. Transcription activators were first shown to contain separable DNA binding and activation domains in yeast. A series of Gal4 mutants were constructed to determine their ability to bind to UAS_{GAL}, a regulatory element that contains several Gal4 binding sites, linked to a *lacZ* reporter gene. Deletion of amino acids from the N-terminal end destroyed the ability of Gal4 to bind to UAS_{GAL} and to stimulate expression of *lacZ*. Gal4 mutants lacking amino acids at the C-terminal end were still able to bind UAS_{GAL} but lacked the ability to stimulate expression of *lacZ*. These results localize the DNA binding domain to the N-terminal end and the activation domain to the C-terminal end.

21. Heterodimeric transcription factors increase gene control options due to an increase in the number of combinatorial possibilities. For example, assume that there are three transcription factors that exist as dimers and each one has a different DNA binding specificity. Three different combinations of heterodimers can be formed in addition to the three homodimers making a total of six different dimers. Each of these six dimers could bind to a different DNA sequence and alter transcription.

22. A number of sequential protein binding steps occurs during the formation of a transcription initiation complex. The TATA-box binding protein

(TBP) included within TFIID first binds to a TATA box in the promoter. Once TBP has bound to the TATA box, TFIIB can bind. The next step is the binding of a pre-formed TFIIF and PolII complex. Then TFIIE and TFIIH bind, completing assembly of the transcription initiation complex.

23. The yeast mating-type phenotype (**a** or α) is determined by the presence of a protein termed mating-type factor **a** or α. Only one of these factors is expressed in a haploid cell, even though the genes for both are present in the genome, one at a locus called *HML* and the other at a locus called *HMR*. When at these loci, the genes are transcriptionally silent. Evidence exists that transcriptional silencing is mediated by a condensed chromatin structure that sterically blocks transcription factors from interacting with the DNA. In order to be expressed, the **a** or α coding sequence must be present at the *MAT* locus, which is between *HML* and *HMR*. Thus, just as a tape inserted into a cassette player can be heard, an **a** or α sequence inserted into the *MAT* locus can be transcribed. A haploid yeast cell switches mating type (e.g., α→ **a**) by gene conversion, a process in which the α sequence present at *MAT* is excised and the **a** sequence at the silent *HML* or *HMR* locus is copied into *MAT*. The **a**→ α switch occurs by a similar process in which the **a** sequence at *MAT* is lost and the silent α sequence is inserted into *MAT*. Since the mating-type sequences *HML* and *HMR* are transferred to *MAT* by DNA synthesis, they are not lost from these loci and can be passed onto the next generation of cells

24. The model for silencing at yeast telomeres involves the formation of a large nucleoprotein complex. First multiple copies of the protein Rap1 bind to repeated, nucleosome-free sequences at yeast telomeres. Then a multiprotein complex is formed through protein-protein interactions between the silent information regulator proteins Sir2, Sir3, and Sir4 and the hypoacetylated histones H3 and H4. As a result, the DNA in this stable nucleoprotein complex is largely inaccessible to external proteins and is silenced.

25. In the absence of Glucocorticoid (GC) hormone, the glucocorticoid receptor (GR) is anchored in the cytoplasm by its interaction with inhibitor proteins, including Hsp90. When GC is added to GC responsive cells, GC diffuses across the cell membrane and binds to the ligand binding domain of the GR. This results in release of the inhibitor proteins and translocation of the GR to the nucleus. Once in the nucleus, the ligand-bound GR binds to the glucocorticoid response element contained in the 5' regulatory regions of genes regulated by GC via its DNA-binding domain. Transcriptional co-activators are recruited which acetylate histones and allows for chromatin remodeling and activation of transcription.

26. Transcription initiation by RNA polymerase I is mediated by a core element, which includes the start site of the pre rRNA gene and an upstream control element, located approximately 100 base pairs upstream. Two transcription factors—upstream binding factor (UBF) and selectivity factor 1 (SL1)—bind to and help stabilize the initiation complex. Transcription initiation by RNA polymerase III is mediated by internal promoter elements, termed the A box and B box, present in all tRNA genes. Two transcription factors, TFIIIC and TFIIIB, are required for transcription initiation of tRNA genes. An additional factor (TFIIIA) is required for transcription initiation of 5S rRNA genes.

27a. Only mutations indicated by the black boxes shown in Figure 11-4 affect RNA synthesis. These mutations fall into three regions, all of which are important for efficient transcription of the *tk* gene. These regions are centered at about -20, -50, and -90.

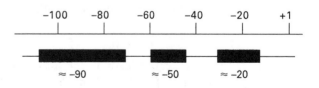

Figure 11-4

27b. Since TATA boxes generally are located within ≈25–35 bp of the start site, the control sequence centered at about -20 ought to be a TATA box.

The other two sequences are likely to be promoter-proximal elements.

27c. The most upstream region is about 35–40 bp in length, whereas the other two regions are each only about 10 bp long. The longer region is the only one that is likely to contain two unresolved control elements.

28a. Removal of NFκB from the cytosol (lane 2) eliminates retardation of enhancer DNA by cytosol (lane 1). Thus NFκB retards enhancer DNA migration.

28b. The amount of NFκB in the cytosol preparation appears to be insufficient to increase the quantity of enhancer DNA retarded by the added purified NFκB (lanes 3–6).

28c. Addition of depleted cytosol to the incubation mixtures containing purified NFκB results in a decrease in the quantity of enhancer DNA that is retarded (lanes 3, 7–9). This observation suggests that depleted cytosol contains an inhibitor of NFκB activity.

29a. In this assay, formation of a stable complex between the long gene and any of the transcription factors during the first incubation would prevent expression of the wild-type gene; thus only long rRNA would be observed in the gel. Lanes 3–5, which reveal both products, indicate that no factor alone forms a stable complex. Lanes 2 and 7 suggest that a stable, bound complex of TFIIIA and C is formed when these transcription factors are incubated with 5S-rRNA genes. Formation of this complex does not require RNA polymerase III (lane 9).

29b. If TFIIIA acted catalytically in the formation of a stable complex of TFIIIC with 5S-rRNA genes, then TFIIIA should be available in the incubation mix and would be able to catalyze the formation of a TFIIIC complex with added DNA. No additional A should be needed.

29c. These data indicate that neither TFIIIA nor TFIIIC acts catalytically. As complex formation is stable and DNA dependent, the most likely role of the transcription factors is to form a stable, DNA-bound intermediate in the formation of a RNA polymerase III transcription-initiation complex.

12 Post-transcriptional Gene Control and Nuclear Transport

PART A: *Chapter Summary*

The regulation of most genes occurs at the first step in gene expression, namely, initiation of transcription. However, once transcription has been initiated, synthesis of the encoded RNA requires that RNA polymerase transcribe the entire gene and not terminate prematurely. As well, the initial **primary transcripts** produced from eukaryotic genes undergo various processing reactions to yield corresponding functional RNAs. These mature, nuclear RNAs are then actively transported to the cytoplasm, as components of ribonucleoproteins.

There are additional opportunities for gene control beyond the regulation of transcription initiation during the multiple steps in the production of RNAs. In the case of a protein-coding gene, the amount of protein expressed can be regulated by controlling the stability of the corresponding mRNA in the cytosol and the rate of its translation. In addition, the cellular locations of some mRNAs are regulated, so that newly synthesized protein is concentrated where it is needed. All of the regulatory mechanisms that control gene expression following transcription initiation are referred to as *post-transcriptional control*. This chapter highlights the various steps in the synthesis of mRNA, tRNA, and rRNA following transcription initiation and the mechanisms by which RNAs and proteins are transported in and out of the nucleus in eukaryotic cells. Relevant examples of how regulation of these steps contributes to the control of gene expression is also presented.

PART B: *Multiple Choice*

Circle the letter corresponding to the most appropriate terms/phrases. There may be more than one correct answer. Circle all that apply.

12.1 *Processing of Eukaryotic Pre-mRNA*

1. Pre-mRNA molecules

    a. exist as free RNA molecules in eukaryotic cells.

    b. are associated with an abundant set of nuclear proteins.

    c. are mostly located in the cytoplasm.

    d. are located in the nucleus.

2. RNA-binding proteins

    a. can be identified by chromatography of UV irradiated nuclear extracts over oligo-dT columns.

    b. can be identified by sequence homology to known RNA-binding domains.

    c. have a conserved structure as seen by X-ray crystallographic analysis.

    d. alter the secondary structure of pre-mRNAs which decreases interactions with other RNAs or proteins.

    e. always remain in the nucleus.

3. Which of the following are associated with most polyadenylation events?:

    a. cleavage of the 5' end of the RNA molecule

    b. an assembly of a large, multi-protein complex

    c. the AAUAAA site upstream of where polyadenylation will occur

    d. a GU-rich or U-rich region near the cleavage site

    e. slow addition of A residues followed by rapid addition of 200 to 250 more A residues

4. Which of the following are required for splicing to occur?

    a. two transesterification reactions

    b. intact, naturally-occurring introns

    c. formation of a lariat-like structure

    d. a branch point G residue

5. What is the evidence that pre-mRNAs are not synthesized unless machinery for processing is properly positioned?

    a. dimeric capping enzyme associates with the C terminal domain (CTD) of RNA polymerase II

    b. polyadenylation factors associate with the CTD of RNA polymerase II

    c. association of splicing factors with the phosphorylated CTD of RNA polymerase II stimulates transcript elongation

    d. none of the above

12.2   Regulation of Pre-mRNA Processing

6. How are the locations of exons and introns within a gene determined?

    a. by isolating expressed sequence tags (ESTs)

    b. by comparing the cDNA and encoded protein sequence for the gene

    c. by comparing the cDNA and genomic DNA sequence for the gene

    d. by comparing the genomic DNA and encoded protein sequence for the gene

7. Which of the following regulatory scenarios of cell-type specific expression of fibronectin isoforms in humans is consistent with information on known splicing factors?

    a. A Tra-like splicing factor expressed in fibroblasts is required for inclusion of EIIIA and EIIIB exons.

    b. A Sxl-like splicing factor is required in hepatocytes for exclusion of EIIIA and EIIIB exons.

    c. A Tra-2-like factor enhances binding of the cleavage/polyadenylation complex in fibroblasts.

    d. A Dsx-like protein is required to inhibit transcription of the fibronectin gene.

8. Which of the following is not involved in alternative splicing in eukaryotes?

    a. RNA binding proteins

    b. hnRNPs

    c. self-splicing group II introns

    d. the splicesosome

9. Differential splicing of the *slo* gene in ciliated neurons results in

    a. hair cells with different response frequencies.

    b. different *slo* isoforms.

    c. *slo* channels that open at different calcium concentrations.

    d. different *slo* mRNAs expressed in different cells.

12.3   Macromolecular Transport Across the Nuclear Envelope

10. The nuclear pore complex

    a. is a large complex made of many different proteins.

    b. is an octagonal, membrane-embedded structure with long filaments.

    c. is required for transport of all molecules in or out of the nucleus.

    d. binds to importin via FG-repeats.

    e. hydrolyzes bound GTP to GDP.

11. When TAT protein fails to bind to the TAR region of the HIV transcript, which of the following is true?

    a. RNA polymerase II terminates transcription

    b. Cyclin T does not interact with Cdk 9

    c. Cdk9 acts on the RNA polymerase II molecule transcribing the HIV genome

    d. the C terminal domain of RNA polymerase II is phosphorylated

12.4   Cytoplasmic Mechanisms of Post-transcriptional Control

12. RNA interference

    a. utilizes *antisense* RNA.

    b. utilizes double-stranded RNA.

    c. is used to silence expression of genes.

    d. utilizes short interfering RNAs.

13. Polyadenylation can promote translation

    a. of the fragile-X gene.

    b. of stored oocyte mRNAs.

    c. by allowing for greater interaction at the 5' end.

    d. by allowing for greater interaction with translation initiation factors.

12.5   Processing of rRNA and tRNA

14. The 28S, 18S and 5.8S rRNAs are

    a. transcribed by RNA polymerase II.

    b. encoded by a single transcription unit.

    c. arranged in tandem arrays.

    d. processed in the cytoplasm.

15. An investigator incubates an unprocessed RNA molecule containing an intron in an extract absent of any protein. Correct splicing of the RNA is observed, but only when guanosine is present. What can the investigator conclude about the RNA molecule?

    a. It contains a ribozyme.

    b. Self-splicing has occurred.

    c. It contains a group II intron.

    d. It contains a group I intron.

PART C: *Reviewing Concepts*

12.1   Processing of Eukaryotic Pre-mRNA

16. You have been asked to isolate mRNA from cultured carrot cells. You know that mRNA constitutes at most a few percent of the total cellular RNA and is heterogeneous in size. Although this isolation appears to be a difficult experimental problem, you are confident that it can be done readily. Describe your approach.

17. Splicing in pre-mRNAs proceeds via two transesterification reactions. Draw a sketch showing the atoms involved in these reactions and the products formed.

18. The spliceosomal splicing cycle involves ordered interactions among a pre-mRNA and several U snRNPs. According to the current model of spliceosomal splicing, which intermediate(s) in the splicing of a pre-mRNA containing one intron should be immunoprecipitated by anti-U5 snRNP? Which additional intermediate(s) should be immunoprecipitated by anti-U1 snRNP?

19. RNAs that are capable of self-splicing have been referred to as *catalytic* RNAs. Discuss how these RNAs are similar to and different from protein enzymes.

12.2   Regulation of Pre-mRNA Processing

20. Differential RNA processing plays an important role in sex determination in *Drosophila*. To a large extent this is due to differential RNA splicing as a consequence of binding of Sex-lethal (Sxl) protein to newly synthesized pre-mRNAs. In *Drosophila*, maleness might be thought of as the default developmental pathway, the pathway taken in the absence of functional Sxl protein. To what extent is this a fair summary of sex determination in *Drosophila*?

12.3   Macromolecular Transport Across the Nuclear Envelope

21. What observations led to the directional model of mRNP transport through the nuclear pore complex?

22. What are nuclear localization signals and why are they needed?

12.4 Cytoplasmic Mechanisms of Post-Transcriptional Control

23. Human and rodent mRNAs on the whole are metabolically fairly stable. The average mRNA half-life is 10 h. Despite this general stability, considerable variation in mRNA half-life is observed. For example, histone mRNA has a half-life of less than 30 minutes, which is much less than that of the typical eukaryotic mRNA. What might be the advantages to organisms of such large differences in mRNA half-life?

12.5 Processing of rRNA and tRNA

24. The 28S, 18S, and 5.8S RNAs in eukaryotic ribosomes are present in equimolar amounts. How is this result assured?

25. What are the main features of splicing in pre-tRNA that distinguish it from splicing in pre-mRNA?

PART D: Analyzing Experiments

26. Transcription of protein-coding genes by eukaryotic RNA polymerase II is a complex process in which termination may occur at any one of a number of sites downstream from the last exon. Early sequencing studies of cDNA indicated that the sequence AAUAAA or in occasional cases AUUAA was located 10–35 nucleotides upstream of the poly-A addition site. Transcription of genes in which this sequence is mutated yields a pre-mRNA that does not undergo polyadenylation and is rapidly degraded. You have inserted an AAUAAA sequence into a human sialyltransferase gene 250 nucleotides upstream from the normal site.

   a. What effect should this insertion have upon the length of the primary transcript?

   b. What effect should this insertion have upon the length of the resulting mRNA?

   c. What effect should this insertion have upon the length of the resulting protein?

   d. What effect should this insertion have upon the function of the resulting protein?

27. One level at which gene expression theoretically can be controlled in eukaryotic cells is the stability of cytoplasmic mRNA. Shapiro and colleagues investigated the balance between the rate of synthesis and degradation of vitellogenin mRNA in a *Xenopus laevis* liver-cell culture system. They found that addition of estrogen, a steroid hormone, to the culture greatly enhanced the synthesis of vitellogenin mRNA and also affected the stability of mRNA in *Xenopus* liver cells. The results of assays for vitellogenin mRNA and poly A–hr containing RNA over time are shown in Figure 12-1.

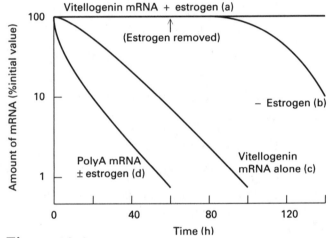

Figure 12-1

   a. What assays could be used to obtain the data in Figure 12-1?

   b. What is the effect of estrogen on the stability of vitellogenin mRNA in *Xenopus* liver cells?

   c. Is the estrogen effect mRNA specific?

   d. Is the estrogen effect reversible? What does this suggest about the nature of the associations that mediate this effect?

28. After comparing a genomic sequence and a cDNA sequence for gene X, a group of investigators hypothesized that gene X contains 2 exons and 1 intron. To monitor splicing of gene X mRNA, a 497 nucleotide, radiolabeled RNA that contains 158 nt of exon 1 and 209 nt of exon 2 separated by a 130 nt intron was prepared. This RNA molecule was incubated with a nuclear extract from HeLa cells for

various times. After incubation, the RNA was purified and subjected to electrophoresis and autoradiography, along with RNA markers (lane M; sizes given in nucleotides) (Figure 12-2).

a. Which bands represent products of the splicing reaction and why? Hint: Non-linear molecules migrate differently than linear molecules. Draw the structure of the products and indicate their sizes if the investigators' hypothesis is correct.

b. Which bands represent intermediates of the splicing reaction? Draw these intermediates and indicate their sizes.

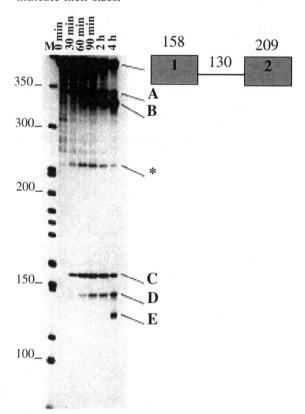

Figure 12-2

Answers

1. b, d

2. a, b, c

3. b, c, d, e

4. a, c

5. c

6. b

7. a, b

8. c

9. a, ,b, c, d

10. a, b, d

11. a

12. b, c, d

13. b, d

14. b, c

15. a, b, d

16. All eukaryotic mRNAs, with the exception of histone mRNAs, have a 3′ poly-A tail, which is about 200 nucleotides long in higher eukaryotes. The poly-A tail is added to pre-mRNA during processing in the nucleus and is not found in other RNA species. Thus mRNA can be readily separated from other RNAs by affinity chromatography with a column to which short strings of thymidylate (oligo-dT) are linked to the matrix.

17. See Figure 12-7 in your textbook. In the first reaction, the phosphoester bond between the 5′ G of the intron and the last (3′) nucleotide in exon 1 is exchanged for a phosphoester bond between the G and the 2′ oxygen of the A at the branch point. In the second reaction, the phosphoester bond between the 3′ G in the intron and the first (5′) nucleotide in exon 2 is exchanged for a phosphoester bond with the last nucleotide in exon 1. No energy is consumed in these reactions, and the intron is released as a circular structure called a lariat.

18. Two different potential intermediates should be immunoprecipitated by anti-U5 snRNP: (1) a structure in the process of joining the two exons together but still containing the intron and (2) a structure that contains the excised intron in lariat form. Both of these structures contain U5 snRNP and other snRNPs. Anti-U1 snRNP should precipitate two additional intermediates: (1) the pre-mRNA with U1 snRNP bound to the 5′ end of the intron and (2) a structure consisting of the pre-mRNA, U1 snRNP, and U2 snRNP bound to the branch site. See MCB, Figures 12-8 and 12-9.

19. Self-splicing RNA is catalytic in the sense that it increases the rate of a reaction as protein enzymes do. However, the substrates of protein enzymes are separate molecules; the intron substrate of a self-splicing RNA is part of the RNA molecule itself. Thus one enzyme molecule carries out its characteristic reaction multiple times on different substrate molecules, whereas self-splicing RNA can cleave itself only once.

20. In *Drosophila,* production of Sxl protein leads to a complicated set of regulated RNA splicing events in which functional Sxl protein is an active participant. Males lack functional sex-lethal protein and, hence what might be thought of as a set of passive default RNA-splicing events happen in males. However, in the end, sex determination in *Drosophila* is the result of an active process of gene repression. In females, the expression of a female-specific Double-sex protein represses transcription of genes required for male sexual development. In males the expression of a male-specific Double-sex protein represses transcription of genes required for female sexual development. Hence, in both *Drosophila* males and females, sexual development requires active repression steps and is not a passive default pathway.

21. From studies on the salivary gland Balbani rings of the insect *Chironomous tentans,* investigators using electron microscopy were able to visualize the transport of fully processed mRNAs and associated proteins (mRNPs) in the form of 50-nm coiled structures. Electron micrographs of sections of salivary gland cells shows that mRNPs move through the nuclear pores to the cytosol. As they uncoil on the cytoplasmic side, they immediately

become associated with ribosomes, indicating that the 5' end leads the way through the complex.

22. A nuclear-localization signal is a domain of amino acids within a protein that allows for the nuclear import of that protein. These domains are required for proteins which eventually function inside the nucleus, as all proteins are synthesized in the cytoplasm. These domains are also needed because most transport of proteins into the nucleus occurs via an active process rather than a passive one.

23. At least two types of general explanations can be proposed for the existence of some short half-life mRNAs in higher eukaryotes. First, the protein encoded by a particular mRNA may have only a short-term physiological role. If the protein itself is only needed for a brief time, then the corresponding mRNA should be unstable, so that unnecessary protein is not produced. Second, the encoded protein may be very tightly coupled to an ordered assembly process in cells. This is the case with histone mRNAs. DNA replication is very closely coupled to the formation of new nucleosomes of which histones form the protein core. Because of the short half-life of the histone mRNAs, anything that interferes with transcription of histone genes will lead in short order to a cessation of DNA replication. There is no point in a cell replicating under conditions that fail to support transcription. Histones as proteins are relatively metabolically stable.

24. The 45S pre-rRNA encodes one copy of each of these mature rRNA molecules. Processing one pre-rRNA molecule thus produces one molecule of 5.8S, 18S, and 28S rRNA, which associate with proteins to form the ribosomal subunits.

25. Splicing of pre-tRNA does not involve spliceosomes. In the first step, an endonuclease-catalyzed reaction excises the intron, which is released as a linear fragment, and a 2',3'-cyclic monophosphate ester forms on the cleaved end of the 5' exon. A multistep reaction that requires the energy derived from hydrolysis of one GTP and one ATP then joins the two exons. In contrast, pre-mRNA splicing occurs in spliceosomes, involves two transesterification reactions, releases the intron as a lariat structure, and does not require GTP. Although these transesterification reactions do not require ATP hydrolysis, it probably is necessary for the rearrangements that occur in the spliceosome.

26a. Transcription by RNA polymerase II terminates at any one of multiple sites 0.5-3 kb downstream from the 3' end of the last exon in the transcript. Since there is little, if any, relationship between termination and poly-A addition signal, insertion of an additional upstream AAUAAA sequence most likely would not affect termination. Thus the primary transcript would be essentially the same length as that produced from the wild-type gene.

26b. The 3' end of a mature mRNA is generated by cleavage and subsequent polyadenylation of the primary transcript (pre-mRNA). The AAUAAA sequence is one of two sequences that signals poly-A modification of pre-mRNA. The other is a GU- or U-rich region located approximately 60–85 nucleotides downstream of the cleavage/poly-A addition site. Both signals are needed for efficient cleavage. Hence, the effect of inserting an extra AAUAAA sequence 250 nucleotides upstream from its normal site would depend on whether a GU- or U-rich region is located an appropriate distance downstream from the inserted sequence. If such a GU- or U-rich region is present, then the primary transcript produced from the altered gene could be cleaved and polyadenylated at the inserted poly-A site, generating an alternative mRNA that is approximately 250 nucleotides shorter than wild-type mRNA. If this second site is not present, then cleavage/polyadenylation could occur only at the wild-type poly-A site, generating an mRNA of normal length.

26c. Once again, there are several possibilities depending on whether a GU- or U-rich region is present downstream from the new AAUAAA sequence, and whether the last exon is greater than 250 nucleotides from the new site. If the primary transcript produced from the altered gene could be cleaved and polyadenylated at the inserted poly-A site, generating an alternative mRNA that is approximately 250 nucleotides shorter than wild-type mRNA, and if this new poly A site did not interrupt the coding region, then the translated protein would be the same length as the normal wild-type protein. If, however, the new poly A site interrupts the coding region, the translated protein will be truncated (i.e., shorter).

26d. If the new poly A site has interrupted the coding region, then the shorter protein may not function normally if the amino acids not present in the truncated protein are important for function.

27a. The data in Figure 12-1 represent the amounts of vitellogenin mRNA and total mRNA (poly A-containing RNA) in the cytoplasm. After the liver cells are disrupted and nuclei removed by low-speed centrifugation, the total mRNA could be determined by affinity chromatography with a poly-T column and vitellogenin mRNA could be determined by Northern blotting with a specific DNA probe.

27b. Comparison of curve (a) and curve (c) indicates that the half-life of vitellogenin mRNA is longer in estrogen-treated than in untreated cells. Thus estrogen appears to increase the stability of vitellogenin mRNA.

27c. The estrogen effect on mRNA stability appears to be specific for vitellogenin mRNA, as the half-life of poly A-containing mRNA is the same in treated and untreated cells (curve d). However, the half-life value for total poly A-containing mRNA is an average for many different mRNA species, some of which are present in small amounts. For this reason, the presence of minor mRNA species that are stabilized by estrogen treatment cannot be proved or disproved from these data.

27d. The stabilization of vitellogenin mRNA by estrogen is reversed when estrogen is removed (curve b). This finding suggests that the stabilization of vitellogenin mRNA more likely results from reversible protein-mRNA interactions than from formation of stable covalent modifications of the mRNA.

28a. The products of the splicing reaction are bands B, D and E (see Figure 12-3). Band B is the 367 nucleotide spliced product. Band D is the 130 nucleotide lariat structure of the intron which was removed by splicing. Band E is the 130 nucleotide "debranched" intron. Note that the lariat

structure migrates slower than expected based on its molecular weight due to its non-linear nature. One can conclude that these are the products of the reaction because they increase in amount throughout the splicing reaction.

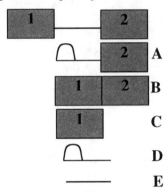

Figure 12-3

28b. The intermediates of the reaction are bands which do NOT increase in a linear fashion throughout the splicing reaction. Band A is the intron plus exon 2 (339 nucleotides). The migration of Band A is slower than expected due to the lariat structure of the intron. Band C is the exon 1 intermediate (158 nucleotides). The (*) band is an aberrant product of the in vitro reaction and can be greatly reduced if capped RNA is included in the reaction.

13

Signaling at the Cell Surface

PART A: *Chapter Summary*

An elaborate intercellular communication network coordinates the growth, differentiation, and metabolism of cells in tissues and organs. This chapter discusses the variety of extracellular signaling molecules, receptors, and signal transduction pathways. Communication by extracellular signals usually involves six general steps: synthesis and release of the signaling molecule by the signaling cell; transport of the signal to the target cell; binding of the signal by a specific receptor, which is thereby activated; initiation of intracellular signal-transduction pathways; change in some cellular process in the target cell; and removal of the signal.

Signaling molecules can act in an endocrine, paracrine, or autocrine fashion, acting on cells at a distance, cells in close proximity, or on the signal-producing cell itself. The binding of a signaling molecule to its receptor is characterized by the dissociation constant, K_d. Some receptors and signal transduction proteins are localized in the cell membrane by clustering in lipid rafts termed caveolae, or by binding to cytosolic adapter proteins containing multiple PDZ domains, which bind to specific sequences in receptors and other proteins. A number of molecules can act as second messengers in intracellular signaling pathways. These second messengers can be soluble molecules, membrane bound molecules or even gases, such as nitric oxide.

A number of G protein-coupled receptors are present in the cell membrane and can activate signaling cascades. Ligand binding to G protein-coupled receptors activates an associated trimeric G protein, leading to the dissociation of the G_α subunit from the $G_{\beta\gamma}$ complex. The G_α-GTP complex can then interact and activate adenylyl cyclase, leading to a cascade involving an increase in cAMP levels, activation of protein kinase A, and phosphorylation of target proteins. The G_α-GTP or the $G_{\beta\gamma}$ subunits can also lead to the opening or closing of ion channels. Ligand binding to G protein-coupled receptors can also activate phospholipase C, which cleaves PIP_2 to DAG and IP_3, leading to DAG activation of protein kinase C and IP_3 mediated opening of Ca^{2+} channels in the ER membrane. The resultant rise in cytosolic Ca^{2+} levels can lead to a number of calmodulin regulated actions. Finally, G protein-coupled receptors can activate transcription factors.

Signal transduction is a complex process that often involves the activation of more than one signaling pathway. It is the coordinated regulation of these multiple, interacting signaling pathways that ultimately leads to the appropriate physiological response and modulation of gene expression.

PART B: Multiple Choice

Circle the letter corresponding to the most appropriate terms/phrases. There may be more than one correct answer. Circle all that apply.

13.1   Signaling Molecules and Cell-Surface Receptors

1. In endocrine signaling, the signaling molecule

   a. acts on target cells distant from the site of synthesis

   b. acts on target cells that are in close proximity to the secreting cell

   c. acts on the same cells that released the signaling molecule

   d. acts on cells that are physically in contact with the secreting cell

2. The dissociation constant K_d for a receptor-ligand complex is

   a. a measure of the affinity of the receptor for the ligand

   b. the concentration of the ligand at which half of the receptors contain bound ligand

   c. a measure of the time it takes for a ligand to be converted to a product

   d. equal to k_{off}/k_{on}

13.2   Intracellular Signal Transduction

3. GTPase switch proteins

   a. are active when bound to GDP but inactive when bound to GTP

   b. are active when bound to GTP but inactive when bound to GDP

   c. are converted from the active to inactive state by a guanine nucleotide exchange factor (GEF)

   d. synthesize GTP from GDP

4. The PDZ domain of some proteins

   a. is a large domain of about 250 amino acids

   b. binds to three residue sequences at the C-terminus of target proteins

   c. contains the amino acid sequence, proline-aspartic acid-zeta-threonine

   e. is important for binding to caveolin in lipid rafts

13.3   G Protein-Coupled Receptors that Activate or Inhibit Adenylyl Cyclase

5. Examples of G protein-coupled receptors include

   a. rhodopsin receptor

   b. β-adrenergic receptor

   c. insulin receptor

   d. acetylcholine receptor

6. Place the events that follow the binding of a ligand to a G protein-coupled receptor that acts on adenylyl cyclase in the correct order:

   a. GTP is hydrolyzed to GDP.

   b. GDP is displaced from G_α

   c. the G_α-GTP complex dissociates from the $G_{\beta\gamma}$ complex.

   d. GTP is bound to G_α

7. During the epinephrine stimulated conversion of glycogen to glucose 1-phosphate, which of the following steps involves protein phosphorylation?

   a. inactivation of glycogen synthase

   b. activation of protein kinase A

   c. activation of adenylyl cyclase

   d. activation of glycogen phosphorylase

8. Which of the following is a mechanism for down regulation of signaling from G protein-coupled receptors?

    a. a change in the K_d of the receptor-hormone complex

    b. proteolysis of the G_α subunit.

    c. hydrolysis of cAMP to 5'-AMP

    d. dephosphorylation of adenylyl cyclase

13.4 G Protein-Coupled Receptors that Regulate Ion Channels

9. Which of the following steps does not occur after acetylcholine binds to the muscarinic acetylcholine receptor in heart muscle cells?

    a. dissociation of $G_{i\alpha}$ subunit from the $G_{\beta\gamma}$ complex

    b. binding of the released $G_{i\alpha}$ subunit to K^+ channels

    c. hydrolysis of GTP to GDP by $G_{i\alpha}$

    d. opening of K^+ channels

10. Which of the following events occurs during light activation of the rhodopsin receptor?

    a. 11-cis-retinal is converted to all-trans-retinal

    b. activation of the trimeric G protein, transducin

    c. activation of cGMP phosphodiesterase

    d. closing of the cGMP-gated ion channel

13.5 G Protein-Coupled Receptors that Activate Phospholipase C

11. Which of the following events occurs following the binding of a ligand to a G protein-coupled receptor that triggers release of Ca^{2+}.

    a. activation of protein kinase A by IP_3

    b. phospholipase cleavage of PIP_2 to IP_3 and DAG

    c. release of DAG into the cytosol

    d. IP_3-mediated opening of Ca^{2+} channels in the ER membrane

12. Which of the following plays a role in the relaxation of smooth muscle cells by nitric oxide?

    a. cGMP

    b. protein kinase A

    c. protein kinase G

    d. guanylyl cyclase

13.6 Activation of Gene Transcription by G Protein-Coupled Receptors

13. Activation of the Tubby transcription factor requires the action of

    a. adenylyl cyclase

    b. phospholipase C

    c. protein kinase A

    d. protein kinase C

14. Which of the following plays a role in epinephrine-induced cardiac hypertrophy?

    a. calmodulin

    b. GPCR-arrestin complex

    c. EGF receptor

    d. MAP kinase cascade

PART C: Reviewing Concepts

13.1 Signaling Molecules and Cell-Surface Receptors

15. Describe the six general steps involved in communication between cells by extracellular signals.

16. How can receptors be cloned by affinity chromatography?

13.2 Intracellular Signal Transduction

17. Describe the structure and function of caveolin-containing lipid rafts.

13.3 G Protein-Coupled Receptors that Activate or Inhibit Adenylyl Cyclase

18. Describe the general model for transduction of the signal from an extracellular molecule to an associated effector protein via a G-protein coupled receptor.

19. Describe the role of β-arrestin in the regulation of β-adrenergic receptors.

13.4 G Protein-Coupled Receptors that Regulate Ion Channels

20. Compare the role of G protein subunits in regulating ion channel function after ligand binding to the muscarinic acetylcholine receptor or rhodopsin.

21. Describe the role of opsin phosphorylation in adaptation of rod cells to changes in ambient light.

13.5 G Protein-Coupled Receptors that Activate Phospholipase C

22. Describe how Ca^{2+} and DAG interact to activate protein kinase C.

23. Describe how Ca^{2+} levels in the ER lumen are returned to normal following the IP_3-mediated rise in cytosolic Ca^{2+} levels.

13.6 Activation of Gene Transcription by G Protein-Coupled Receptors

24. What role does phospholipase C play in the activation of the transcription factor Tubby?

PART D: Analyzing Experiments

25. Many oncogenic mutations occur in normal genes that play important roles in regulating cell growth. The trimeric G proteins, which consist of α, β, and γ subunits, relay signals from ligand-activated receptors to intracellular effector proteins. Landis et al. conducted a search for mutations in the $G_{s\alpha}$ subunit in eight human pituitary tumors. Using PCR and DNA sequencing, they identified mutations at codon 201 (Arg in wild type) and codon 227 (Gln in wild type) in four of eight pituitary tumors as shown in Table 13-1. In addition adenylyl cyclase activity was assayed using membrane homogenates from tumor tissue.

a. What do you conclude about the effect of mutations at amino acid positions 201 or 227 for tumors 5-8? What do you conclude about tumors 1-4?

TABLE 13-1

Tumor number	Codon 201	Codon 227	Basal adenylyl cyclase activity
1	Arg	Gln	13
2	Arg	Gln	6
3	Arg	Gln	16
4	Arg	Gln	43
5	Cys	Gln	170
6	His	Gln	480
7	Cys	Gln	190
8	Arg	Arg	180

b. Based on the adenylyl cyclase activities of tumors 5-8, what do you hypothesize may be the biochemical defect in the $G_{s\alpha}$ protein?

c. In a subsequent paper, Lyons et al surveyed specific pituitary tumors (GH, growth hormone; Prl, prolactin; TSH, thyroid stimulating hormone; ACTH, adrenocorticotropin hormone) for mutations in the Gαs protein (Table 13-2). Of the 18 GH-secreting tumors, 16 had mutations in codon 201 (14 Arg to Cys; 2 Arg to His) and two had mutations in codon 227 (Gln to Arg). Eight of the tumors that contained a mutation in codon 201 or 227 showed elevated adenylyl cyclase activity. What do these results suggest about the role of cAMP in proliferation of the different pituitary cell types?

TABLE 13-2

Pituitary tumor type	Number tested	Number with mutations
GH-secreting	42	18
Prl-secreting	12	0
TSH-secreting	2	0
ACTH-secreting	7	0

26. The transcription regulator Tubby functions as a signal transducer from G protein-coupled receptors. Tubby is localized in the plasma membrane through binding phosphatidylinositol 4,5-bisphosphate (PIP_2). Following hydrolysis by phospholipase C, Tubby is translocated to the nucleus. Thus Tubby contains both a nuclear localization signal and a plasma membrane associated domain. To identify these signals/domains, Santagata et al. constructed fusion proteins consisting of full length, amino terminal (aa 1-247), or carboxyl terminal (aa 248-505) fragments of Tubby fused to green fluorescent protein (GFP). These fusion proteins were transfected into cells and the localization of the GFP-Tubby fusion proteins were examined by fluorescence microscopy.

a. What do you conclude from the results shown in Table 13-3 about the location of the nuclear localization signal and the plasma membrane associated domain? What further experiments could you perform to more precisely localize these signals/domains?

TABLE 13-3

Construct	Cellular localization
Tubby(full length)-GFP	initially plasma membrane, later nucleus and plasma membrane
Tubby(1-247)-GFP	nucleus
Tubby(248-505)-GFP	plasma membrane

b. To determine the amino acid residues that are important for plasma membrane localization, site directed mutagenesis of the Tubby cDNA was performed. For example, in the Y327F mutant, a tyrosine codon (Y in the single amino acid code) is changed to a phenylalanine codon (F) at amino acid position 327. The wild type Tubby-GFP and mutant Tubby-GFP constructs were transfected into cultured cells and the location of Tubby was again determined by fluorescence microscopy. In addition, using an in vitro assay the ability of the wild type and mutant Tubby-GFP proteins to bind to phosphatidylinositol 4,5-bisphosphate (PIP_2) was determined. A "+" indicates binding to PIP_2, whereas a "-" indicates no binding to PIP_2 (Table 13-4). What do you conclude from these results about the amino acids required for plasma membrane localization of Tubby?

TABLE 13-4

Tubby-GFP construct	Amino acid change	Cellular localization	PIP_2 binding
Wild type		plasma membrane/ nucleus	+
Y327F	tyrosine to phenylalanine	plasma membrane	+
K330A	lysine to alanine	nucleus	-
R332A	arginine to alanine	nucleus	-
N310A	asparagine to alanine	nucleus	-
R363A	arginine to alanine	nucleus	-
R290A	arginine to alanine	plasma membrane	+
R397A	arginine to alanine	plasma membrane	+

c. Trimeric G proteins consist of a variety of G_α subunits that act on different effector proteins. In order to determine which G_α proteins are capable of releasing Tubby from the membrane, different constitutively active G_α constructs (indicated by an asterisk) were cotransfected into cells with the Tubby-GFP construct. Again, the localization of the Tubby-GFP fusion protein was determined by fluorescence microscopy (Table 13-5). What do you conclude about the G_α protein subunits required to activate Tubby?

TABLE 13-5

Construct transfected	Cellular localization
$G_{o\alpha}^*$	plasma membrane
$G_{s\alpha}^*$	plasma membrane
$G_{q\alpha}^*$	nucleus
$G_{\beta\gamma}$	plasma membrane

Answers

1. a

2. a, b, d

3. b, c

4. b

5. a, b, d

6. b, d, c, a (in order)

7. a, d

8. a, c

9. b

10. a, b, c

11. b, d

12. a, c, d

13. b

14. b, c, d

15. Communication by extracellular signals usually involves six steps: synthesis and release of the signaling molecule by the signaling cell; transport of the signal to the target cell; binding of the signal by a specific receptor, which is thereby activated; initiation of intracellular signal-transduction pathways; change in some cellular process in the target cell; and removal of the signal.

16. Receptors can be cloned based on their affinity to a known ligand. A ligand for the receptor is first chemically linked to beads used to pack a column. A crude, detergent solubilized cell membrane fraction is then passed through the column of ligand-modified beads. The receptor binds to the ligand, while other proteins wash through. An excess of ligand is then passed through the column causing displacement of the receptor from the beads. The receptor can then be purified away from the ligand.

17. In mammalian cells, certain receptors are clustered in lipid rafts called caveolae. The caveolae are marked by the presence of caveolin, a family of approximately 25 kDa, membrane-spanning proteins. Large oligomers of caveolin form a protein coat on the surface of caveolae. The clustering of receptors and signaling molecules in caveolae may facilitate the interaction between signaling molecules, thus enhancing signal transduction.

18. In the inactive state, i.e., ligand not bound to its receptor, the G_α subunit is bound to GDP and complexed with G_β and G_γ subunits. Upon binding the ligand, the receptor undergoes a conformational change that causes exchange of GTP for GDP bound to G_α. The resulting G_α-GTP complex dissociates from the $G_{\beta\gamma}$ complex and interacts and activates an effector protein.

19. Activated β-adrenergic receptors are phosphorylated at specific serine and tyrosine residues by the β-adrenergic receptor kinase (BARK). β-arrestin binds to phosphorylated receptors and inhibits the ability of the receptor to activate the G_α subunit. β-arrestin also interacts with proteins present in coated pits, which leads to the internalization of β-adrenergic receptors.

20. Binding of ligand to the muscarinic acetylcholine receptor leads to activation of the trimeric G protein and dissociation of the $G_{i\alpha}$-GTP complex from the $G_{\beta\gamma}$ complex. The $G_{\beta\gamma}$ complex then directly binds to and opens the K^+ channel. Binding of ligand to rhodopsin also leads to dissociation of the $G_{t\alpha}$-GTP complex from the $G_{\beta\gamma}$ complex. In this case, the $G_{t\alpha}$-GTP complex activates cGMP phosphodiesterase, which converts cGMP to GMP. The resulting low cytosolic cGMP concentration causes the cGMP-gated ion channel to close. In this pathway, the G protein acts indirectly on the ion channel.

21. In the presence of low light, the opsin part of rhodopsin is activated, which leads to activation of $G_{t\alpha}$. The activated opsin is then phosphorylated by rhodopsin kinase. This phosphorylated opsin is slightly reduced in its ability to activate $G_{t\alpha}$. The ability of activated opsin to activate $G_{t\alpha}$ is inversely proportional to the number of phosphorylated sites and the extent of phosphorylation is directly proportional to the amount of time each opsin molecule spends in the

light activated form. Thus the greater the ambient light levels, the higher the level of opsin phosphorylation and the larger the increase in light level needed to activate $G_{t\alpha}$.

22. After phospholipase cleavage of PIP_2 to IP_3 and DAG, DAG remains associated with the plasma membrane. IP_3 binds to and opens Ca^{2+} channels in the ER membrane. The resultant rise in cytosolic concentration of Ca^{2+} causes protein kinase C, which is present in the cytosol in an inactive state, to migrate to the plasma membrane, where it is activated by membrane-bound DAG.

23. The IP_3-mediated rise in Ca^{2+} levels is only transient, because plasma and ER membrane Ca^{2+} ATPases actively pump Ca^{2+} from the cytosol to the ER lumen and out of the cell. As the stores of Ca^{2+} in the ER are depleted, the IP_3-gated Ca^{2+} channels bind to store-operated TRP channels in the plasma membrane, allowing an influx of Ca^{2+} into the cytosol. The cytosolic Ca^{2+} is then pumped into the ER lumen, restoring Ca^{2+} levels.

24. In resting cells, the transcription factor Tubby is bound tightly to PIP_2 in the plasma membrane. Activated phospholipase C cleaves PIP_2 into DAG and IP_3, causing release of Tubby into the cytosol. Once released into the cytosol, Tubby can translocate to the nucleus, where it can activate transcription of target genes.

25a. For tumors 5, 6, 7 and 8 an amino acid substitution at either codon 201 or 227 results in an increase in basal adenylyl cyclase activity. This result indicates that arginine-201 and glutamine-227 play important roles in regulating adenylyl cyclase activity. In tumors 1, 2, 3, and 4, no mutations at codons 201 or 227 are observed, which demonstrates that pituitary tumors are not always the result of mutations at codons 201 and 227 in the $G_{s\alpha}$ subunit.

25b. For trimeric G proteins in the inactive state, $G_{s\alpha}$ is bound to GDP and complexed with $G_{\beta\gamma}$. Upon activation, GTP is exchanged for GDP and the $G_{s\alpha}$-GTP complex dissociates from the $G_{\beta\gamma}$ complex. The active $G_{s\alpha}$-GTP complex then activates adenylyl cyclase. Hydrolysis of GTP to GDP by the intrinsic GTPase activity of $G_{s\alpha}$ returns $G_{s\alpha}$ to the inactive state. In the case of the mutations in tumors 5-8, a logical hypothesis is that the mutations in codons 201 and 227 inhibit the GTPase activity of $G_{s\alpha}$. As a result the $G_{s\alpha}$-GTP subunit remains active for a longer time, resulting in higher basal adenylyl cyclase activity.

25c. As shown previously in Table 13-1, mutations at codons 201 and 227 are correlated with increases in adenylyl cyclase activity and thus cytosolic cAMP levels. Again, only some pituitary tumors are caused by mutations in $G_{s\alpha}$. These data suggest that $G_{s\alpha}$, adenylyl cyclase, and cAMP play important roles in proliferation of GH-secreting cells but may not for prolactin-, TSH- and ACTH-secreting cells.

26a. Transfection of the Tubby(1-247)-GFP fusion protein shows localization only in the nucleus. This result indicates that the nuclear localization signal is present in the amino terminus of Tubby. Transfection of the Tubby(248-505)-GFP fusion protein shows localization only in the plasma membrane, indicating that the nuclear localization signal is present in the carboxyl terminus of Tubby. The Tubby(full length)-GFP construct initially localizes to the plasma membrane but then translocates to the nucleus. This would be consistent with a protein that contains both a nuclear localization signal and a membrane associated domain. To more precisely localize the nuclear localization signal and the membrane associated domain, smaller amino and carboxyl terminal fragments of Tubby can be fused to GFP and assayed for their cellular localization.

26b. Mutation of the positively charged amino acid residues at positions 330, 332, 310, and 363 to the neutral amino acid alanine led to loss of plasma membrane localization and binding to PIP_2. These results demonstrate that these amino acids likely fold to form a site that binds to PIP_2, which is likely the negatively charged phosphate group of PIP_2. Disruption of Tubby binding to the membrane bound PIP_2 results in the localization of Tubby to the nucleus and not the plasma membrane. Mutation of tyrosine to phenylalanine at position 327 likely does not affect PIP_2 binding because both are hydrophobic amino acids. Mutation of charged amino acids at position 290 and 397 have no effect because they are not part of the PIP$_2$ binding site.

26c. Proteins that are associated with PIP$_2$ are often regulated by the action of phospholipase C, which cleaves PIP$_2$ to DAG and IP$_3$. These results show that not all G$_\alpha$ subunits are capable of causing Tubby to be translocated to the nucleus. Only the activated G$_{q\alpha}$ subunit can induce nuclear translocation of Tubby, which presumably occurs through activation of phospholipase C, cleavage of PIP$_2$ and release of Tubby. Neither the G$_{o\alpha}$* or G$_{s\alpha}$* nor the G$_{\beta\gamma}$ complex can activate phospholipase C and release Tubby from the membrane.

14

Signaling Pathways That Control Gene Activity

PART A: *Chapter Summary*

Binding of extracellular signaling molecules to receptors on the cell surface results in changes in gene expression by activation of particular transcription factors. Such regulated gene expression is important in directing developing cells along particular pathways of differentiation and in coordinating the response of mature cells to environmental changes. Although a cell may express receptors for over 100 signaling molecules, these receptors can be divided into a small number of classes based on structural similarities and the mechanisms by which the signal is transduced. Five classes of cell-surface receptors and the signaling cascade that each elicits are described in this chapter:

1.   Upon binding to their ligands, receptors for the TGFβ superfamily of signaling molecules dimerize and become active serine/threonine kinases. The TGFβ receptor then phosphorylates R-Smad transcription factors, revealing a nuclear localization signal. R-Smads then translocate to the nucleus in a complex where they interact with other transcription factors to elicit changes in gene expression. TGFβ inhibits proliferation of most cells, and loss of TGFβ signaling is common to many cancers.

2.   Cytokine receptors form functional dimers upon binding ligand and activate associated tyrosine kinases called JAKs. JAKs phosphorylate the cytokine receptor, creating binding sites for proteins with SH2 domains, including STAT transcription factors. JAKs then phosphorylate STATs, causing STATs to dimerize and expose nuclear localization signals. STAT dimers then translocate to the nucleus to alter gene expression. Erythropoeitin, a cytokine that induces differentiation of red blood cells, activates the JAK-STAT and several other signaling pathways.

3.   Binding of receptor tyrosine kinases (RTKs) to their ligands induces formation of dimers, which transphosphorylate to form docking sites for proteins containing SH2, PTB or SH3 domains. Dimeric, phosphorylated RTKs bind the adapter protein GRB2, which binds Sos, which in turn binds Ras to promote exchange of GDP for GTP on Ras. Ras•GTP recruits and activates the serine/threonine kinase Raf. Hydrolysis of Ras•GTP to Ras•GDP releases Raf, which phosphorylates and activates the cytosolic kinase MEK. MEK then phosphorylates and activates MAP kinase. MAP kinase then dimerizes and translocates to the nucleus to regulate activity of transcription factors for many genes including genes involved in cell proliferation. Consequently, many cancers are characterized by gain-of-function mutations in some component of RTK-MAP kinase signaling. Both RTKs and cytokines activate other signaling pathways including those that signal through phospholipase C.

4.   Association of tumor necrosis factor α (TNFα) with its receptor results in nuclear localization and activation of the transcription factor NF-_B by inducing the phosphorylation and subsequent degradation of its associated inhibitor, I-_B. NF-_B promotes transcription of more than 150 genes, including those that promote cell survival.

5.   The cell surface receptor Notch binds to its ligand Delta, a transmembrane protein located on an adjacent cell. Upon binding to Delta, Notch is cleaved by the TACE protease. Presenilin 1 then cleaves the cytosolic fragment of Notch, which translocates to the nucleus to regulate the activity of several transcription factors. Both the TNFα/ NF-κB and Notch/Delta signaling pathways are relatively irreversible because they include a proteolytic step. A signaling pathway related to Notch/Delta may be mutated in patients suffering from Alzheimer's disease.

PART B: *Multiple Choice*

Circle the letter corresponding to the most appropriate terms/phrases. There may be more than one correct answer. Circle all that apply.

14.1 *TGFβ Receptors and the Direct Activation of Smads*

1. Arrange the following events in the proper order in which they occur during transduction of the TGFβ signal:

   a. association of Smad3 and Smad4

   b. phosphorylation of TGFβ receptor I by TGFβ receptor II

   c. phosphorylation of Smad3 by TGFβ receptor I

   d. nuclear import of Smad3

   e. binding of TGFβ to TGFβ receptor II

2. Which of the following functions as an inhibitor of TGFβ signal transduction?

   a. R-Smad

   b. I-Smad

   c. Co-Smad

   d. Ski

   e. SnoN

3. Which of the following mutations would promote malignancy in cells whose proliferation is inhibited by TGFβ?

   a. overexpression of Smad2

   b. loss of Smad 4

   c. overexpression of TGFβ receptor I

   d. loss of TGFβ receptor II

14.2 *Cytokine Receptors and the JAK-STAT Pathway*

4. Which of the following becomes active as a kinase when an activation lip residue is phosphorylated?

   a. EpoR

   b. JAK

   c. STAT

   d. Ras

   e. MAPK

5. Which of the following mutations might confer a competitive advantage to an athlete by raising his or her hematocrit?

   a. excess production of erythropoeitin

   b. decreased production of erythropoietin

   c. EpoR that cannot bind STAT5

   d. EpoR that cannot bind SHP1

   e. EpoR that cannot bind JAK2

6. What tertiary structure characterizes all cytokines?

   a. 4 α helices

   b. 4 β sheets

   c. 2 α helices and 2 β sheets

   d. 4 α helices and 4 β sheets

14.3 *Receptor Tyrosine Kinases and Activation of Ras*

7. Which of the following mutations would likely have a similar cancer-promoting effect as the Ras^D mutation?

   a. mutation in Grb2 so that it cannot bind Sos

   b. mutation in Sos so that it binds Ras independent of Grb2

   c. mutation in GAP so that it cannot bind Ras

   d. mutation in EGF receptor so that it binds GRB2 independent of EGF

8. Arrange the following events in the proper order in which they occur during transduction of the EGF signal:

   a. transphosphorylation of the EGF receptor

   b. dissociation of GDP from Ras

   c. dissociation of Ras from Sos

   d. binding of GRB2 to the EGF receptor

   e. binding of GTP to Ras

14.4 *MAP Kinase Pathways*

9. Arrange the following proteins in the proper order in which they become activated in the MAPK signaling pathway:

    a. MAP kinase

    b. MEK

    c. Ras

    d. Raf

    e. RTK

10. Which of the following is a direct substrate of MAP kinase?

    a. TCF

    b. SRF

    c. SRE

    d. p90^{RSK}

    e. none of the above

14.5 *Phosphoinositides as Signal Transducers*

11. Which of the following signaling pathways can be activated by activation of receptor tyrosine kinases?

    a. phospholipase Cγ

    b. PI-3 kinase

    c. MAP kinase

    d. none of the above

12. A loss-of-function mutation in PTEN phosphatase results in:

    a. increased cell proliferation.

    b. decreased cell proliferation.

    c. increased apoptosis.

    d. decreased apoptosis.

14.6 *Pathways That Involve Signal-Induced Protein Cleavage*

13. Arrange the following events in the proper order in which they occur during transduction of the TNF-α signal:

    a. phosphorylation of I-κB

    b. binding of E3 ubiquitin ligase to I-κB

    c. polyubiquitination of I-κB

    d. nuclear localization of NF-κB

    e. activation of TAK1 kinase

14. Arrange the following events in the proper order in which they occur during transduction of the Notch/Delta signaling pathway:

    a. cleavage of Notch by TACE

    b. binding of Delta to Notch

    c. translocation of Notch segment to the nucleus

    d. interaction of Notch with transcription factors

    e. cleavage of Notch by Presenilin 1

14.7 *Down-Modulation of Receptor Signaling*

15. In the developing bone, which of the following molecules is expressed by the osteoblast:

    a. RANK

    b. RANKL

    c. osteoprotegorin

    d. none of the above

PART C: *Reviewing Concepts*

14.1 *TGFβ Receptors and the Direct Activation of Smads*

16. TGFβ receptors fall into three classes, type I, II and III. What is the chemical nature of each, and what is their individual contribution to the transduction of the TGFβ signal.

17. The extracellular matrix of connective tissues is a reservoir of latent TGFβ. Describe how latent TGFβ is converted to the mature active signaling molecule.

14.2 *Cytokine Receptors and the JAK-STAT Pathway*

18. As described in Chapter 14 of the textbook, "if receptors for prolactin or thrombopoietin.... are expressed experimentally in an erythroid progenitor cell, the cell will respond to these cytokines by dividing and differentiating into red blood cells, not into mammary cells or megakaryocytes." Why?

14.3 *Receptor Tyrosine Kinases and Activation of Ras*

19. Many proteins contain phosphotyrosine-binding SH2, SH3 and PTB domains, yet the interaction of these proteins with appropriate target molecules is exquisitely specific. What is the basis for such specificity?

20. Describe the roles of *Sevenless*, *Bride of Sevenless* and *Son of Sevenless* proteins in the signaling pathway that regulates differentiation of the R7 neuron during the development of the compound eye of *Drosophila*.

14.4 *MAP Kinase Pathways*

21. Explain the role that scaffolding proteins such as Ste5 play to ensure that an extracellular signal elicits the proper response through its particular MAP kinase signaling pathway.

14.5 *Phosphoinositides as Signal Transducers*

22. Describe how the binding of PI 3-phosphates promotes the activation of protein kinase B.

14.6 *Pathways That Involve Signal-Induced Protein Cleavage*

23. How does NF-κB function to promote resistance to a bacterial infection?

24. What is regulated intramembrane proteolysis (RIP)? Name two signaling pathways that involve RIP.

14.7 *Down-Modulation of Receptor Signaling*

25. Describe two ways in which the activity of cell surface receptors can be reduced in order to prevent an overly robust response to an extracellular signal.

PART D: *Analyzing Experiments*

26. The TGFβ superfamily of signaling molecules are key regulators of developmental processes in diverse organisms. Genetic studies of developmental mutants of the fruit fly *Drosophila melanogaster* and the roundworm *Caenorhabditis elegans* have led to the identification of many of the components of the TGFβ signaling pathway.

In one such study by Padgett and colleagues, a series of genetic mutants were identified that exhibited the same phenotype as a previously identified *daf-4* mutant. *Daf-4* mutants are defective in egg laying and are unusually small. The *daf-4* gene had been shown to encode a transmembrane receptor for the BMP family of TGFβ-like ligands.

In this study, it was determined that male *daf-4* mutants possessed defects in the sensory rays and copulatory spicules in their tails. Because *daf-4* mutants are also small in size, the investigators screened existing Small mutants (*sma*) for defects in male tail development. Three such mutants, *sma-2*, *sma-3*, and *sma-4*, were identified.

a. Since *daf-4*, *sma-2*, *sma-3* and *sma-4* mutants all exhibit defects in size and tail development, what does this suggest about the signaling pathways in which these gene products reside?

Daf-4 was originally identified in a screen for genes that regulate dauer development in *C. elegans*. Dauers are arrested larva that develop when resources are scarce. Unlike *daf-4*, the other genes that were identified in this screen, *daf-1*, *daf-3*, *daf-5*, *daf-7*, *daf-8* and *daf-14*, do not regulate body size or male tail development.

b. What can be deduced about the relationship among the *sma* genes, *daf-4*, and the other *daf* genes?

To determine whether *daf-4* functioned in the same cells as the *sma* genes, Padgett and colleagues performed the following experiments: Animals were generated that were mosaic for the *daf-4* mutation, that is, only specific, identifiable neurons in the male tail possessed the *daf-4* mutation and exhibited the defective phenotype. Surrounding neurons appeared normal and were likewise wild-type for *daf-4*. Similarly, animals mosaic for *sma-2* exhibited abnormalities in neurons with the *sma-2* mutation but not in adjacent neurons with wild-type *sma-2*.

c. Based on these experiments, do *daf-4* and *sma-2* function in the same cells? Is either the Daf-4 or Sma-2 protein likely to be a diffusible molecule?

The investigators then expressed *daf-4* in *sma* mutants using an exogenous *daf-4* gene regulated by a heat shock inducible promoter. Expression of *daf-4* was able to rescue abnormal phenotypes of *daf-4* mutants but not of *sma-2*, *sma-3* or *sma-4* mutants.

d. What do these experiments indicate about the relationship between the *daf-4* and *sma* genes?

The investigators cloned and sequenced the *sma-2*, *sma-3*, and *sma-4* genes. The predicted amino acid sequences were related to one another and to a *Drosophila* protein called Mothers against dpp (Mad). Eventually this family of proteins came to be known as the Smads (Sma- and Mad-related).

e. Based on your knowledge of the role of Smads in transducing signals initiated by members of the TGFβ superfamily, describe the relationship between Daf-4 and the Sma proteins in the signal transduction pathway that regulates cell size and development of the male tail.

27. Platelet-derived growth factor (PDGF) is a growth-promoting substance that is released from platelets when they adhere to the surface of an injured blood vessel. Binding of PDGF to fibroblasts and other cell types produces extremely diverse effects including an increase in DNA synthesis, changes in ion fluxes, alterations in cell shape, and changes in phospholipid metabolism. These effects are mediated by the PDGF receptor, a receptor tyrosine kinase (RTK). After activation and autophosphorylation of the PDGF receptor, phosphotyrosines in its cytosolic domain can interact with SH2 domains in several different cytosolic proteins, leading to various downstream effects. The cytosolic domain of the PDGF receptor has four binding sites, each containing one or two tyrosine residues; when these are phosphorylated, they can bind PI-3 kinase, GAP, the γ isoform of phospholipase (PLC$_\gamma$) and Syp (a protein phosphatase).

A. Kazlauskas and co-workers were interested in the activation of specific signaling pathways by PDGF. In order to understand which of these multiple binding sites are responsible for activating specific responses, they created a series of mutant-gene-encoding-PDGF receptors in which

tyrosine residues (Y) were replaced by nonphosphorylatable phenylalanine residues (F) in three of these binding sites, as diagrammed in Figure 14-1. These mutant receptors should bind to only one of the four possible cytosolic proteins, and theoretically should activate only one of the four possible signaling pathways. For example, the mutant receptor known as Y40/51 contains phenylalanine residues at positions 771, 1009, and 1021. Such a receptor should interact specifically only with PI-3 kinase, which is known to bind to phosphotyrosine residues at position 40 and 51. The F5 receptor contains phenylalanine replacements for all the phosphorylatable tyrosine residues, and should therefore not interact with any of the intracellular signaling proteins.

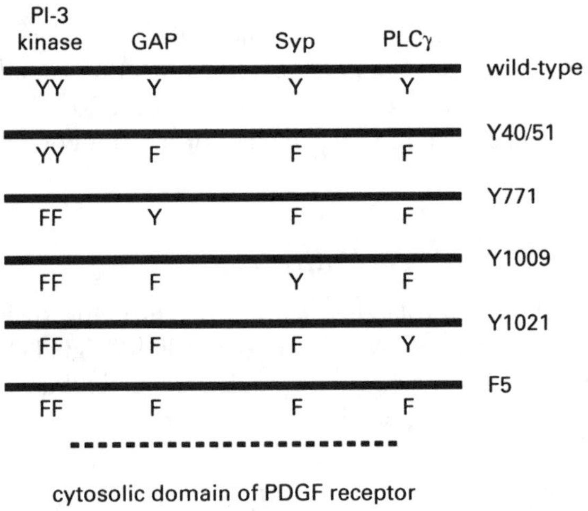

PI-3 kinase	GAP	Syp	PLC$_\gamma$	
YY	Y	Y	Y	wild-type
YY	F	F	F	Y40/51
FF	Y	F	F	Y771
FF	F	Y	F	Y1009
FF	F	F	Y	Y1021
FF	F	F	F	F5

cytosolic domain of PDGF receptor

Figure 14-1

These researchers transfected HepG2 cells, which do not produce an endogenous PDGF receptor, with the genes encoding these mutant receptors. Appropriate controls indicated that the mutant receptors expressed in the transfected cells did indeed associate with the appropriate cytosolic proteins. They then assayed PDGF-induced [³H]-thymidine incorporation into DNA in these transfected cell lines. In this assay, cells were plated at a subconfluent cell density, arrested by serum deprivation, and then stimulated with various concentrations of PDGF. After 18–20 h, the cells were pulsed for 2 h with [³H]-thymidine, and the amount of incorporated radioactivity was measured by scintillation counting of TCA-precipitated nucleic acids. The results, shown in Figure 14-2 are expressed as the percentage of the response stimulated by serum.

a. Based on the results shown in Figure 14-2, which intracellular signaling pathways are involved in initiation of DNA synthesis in response to PDGF?

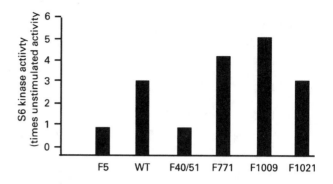

Figure 14-3

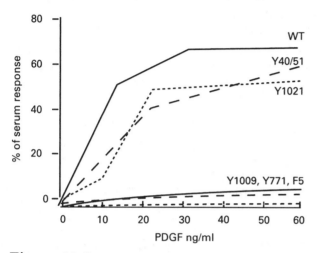

Figure 14-2

Previously, other workers found that activation of a protein known as S6 kinase is important in stimulation of cells to pass through the $G_1 \rightarrow S$ transition in the cell cycle. This protein, which exists in two forms designated αI and αII, is activated by phosphorylation of serine residues, and thus is not a direct substrate for the PDGF receptor tyrosine kinase. Receptor tyrosine kinases activate MAP kinase via the Sos and c-Ras pathway, but it is not known which pathway(s) is coupled to phosphorylation of the S6 kinase. In order to answer this question, Kazlauskas and co-workers generated another series of PDGF receptor mutants. In these mutants, most of the tyrosines were retained, but specific tyrosines were changed to phenylalanines. For example, a mutant receptor with a phenylalanine residue at position 1021 was generated and designated F1021. The researchers incubated quiescent HepG2 cells, which expressed these various mutant PDGF receptors, with PDGF for 45 minutes, and then assayed S6 kinase activity and extent of phosphorylation. Figure 14-3 shows stimulation of S6 kinase activity in response to PDGF in these cells, expressed as multiplied bythe activity in untreated cells.

b. Based on these data, which intracellular signaling pathways are involved in phosphorylation and activation of S6 kinase?

Answers

1. e, b, c, a, d

2. b, d, e

3. b, d

4. b, e

5. a, d

6. a

7. b, c, d

8. a, d, b, e, c

9. e, c, d, b, a

10. a, d

11. a, b, c

12. a, d

13. e, a, b, c, d

14. b, a, e, c, d

15. b, c

16. Chemical crosslinking experiments indicate that TGFb binds to three different types of receptors – type I, II and III. The type III receptor is a proteoglycan called β-glycan that presents TGFβ to the type II receptor. The type II receptor is a constitutively active serine/threonine kinase that, when bound to TGFβ, forms a tetramer consisting of two molecules of the type I and two molecules of the type II receptor. The type II receptor then phosphorylates the type I receptor, activating the type I receptor. The type I receptor, which does not bind TGFβ directly, becomes active as a serine/threonine kinase and transduces the TGFβ signal intracellularly by phosphorylating Smad3 or Smad4.

17. Latent TGFβ consists of two proteolytic fragments, the pro-domain and the mature domain, which are linked by disulfide binds in a complex that contains latent TGFβ-binding protein (LTBP). Upon proteolysis or a conformational change in LTBP, the mature domain is released as a dimer of active TGFβ.

18. Although all cytokine receptors activate similar JAK-STAT signaling pathways, the response of a specific cell to activation of this pathway depends upon the specific set of transcription factors and developmental genes it expresses and its chromatin structure. Thus, an erythroid progenitor cell is programmed to respond to activation of a cytokine receptor by differentiating into a red blood cell rather than another cell type.

19. Although SH2-, SH3-, and PTB-containing proteins possess conserved binding pockets for phosphotyrosine-containing peptides, the detailed structure of these pockets possesses an amino acid sequence that binds to specific consensus sequences of amino acids surrounding the phosphotyrosine.

20. The signaling pathway that regulates the development of the R7 neuron in the *Drosophila* eye is a typical receptor tyrosine kinase (RTK) signaling pathway. *Bride of Sevenless (Boss)* is the ligand that binds the RTK. In this case, the ligand is an integral membrane protein expressed on the surface of adjacent R8 cells. *Sevenless (Sev)* is the RTK, and *Son of Sevenless (Sos)* is the guanine nucleotide-exchange factor that activates Ras. In the tradition of the scientists who study *Drosophila* development, *Sevenless* was named for the phenotype of the mutation – lack of *Sevenless* protein results in the absence of the R7 neuron. Subsequent mutations that resulted in the same phenotype were given clever, related names.

21. Scaffolding proteins such as Ste5 bring the kinase components of a MAP kinase signaling pathway into close proximity such that each kinase will have access only to its appropriate substrate. For example, upon activation of the mating factor signaling pathway, an activated G protein associates with Ste5 and the serine/threonine kinase Ste20. Ste20 phosphorylates, Ste11, which is also bound to Ste5. Ste11, also a serine/threonine kinase, then phosphorylates Ste7, a MEK, which phosphorylates the MAP kinase Fus3. All of these kinases are associated with the Ste5 scaffold.

22. PI 3-phosphates bind the PH domain of protein kinase B. This interaction localizes protein kinase B to the cell surface membrane and releases

inhibition of the catalytic domain. PI 3-phosphates also bind the kinase, PDK1, causing its membrane association as well. PDK1 can then phosphorylate and partially activate protein kinase B.

23. NF-κB stimulates the transcription of numerous genes in multiple cell types that respond to bacterial infection. These include the genes encoding cytokines and chemokines to attract immune cells and fibroblasts to sites of infection, genes encoding receptors in neutrophils that promote their migration to sites of infection, and the gene coding iNOS, an enzyme that produces nitric oxide, a bacterial toxin.

24. Regulated intramembrane proteolysis, (RIP), is signal-induced cleavage of a membrane-bound protein to release a proteolytic fragment into the cytosol to participate in signal transduction. The Notch/Delta signaling pathway involves the cleavage of Notch to release a fragment that translocates to the nucleus to activate transcription factors. The APP protein is likewise cleaved by RIP, releasing a cytosolic fragment of unknown function. Release of an extracellular fragment of APP by the enzyme β-secretase has been implicated in Alzheimer's disease.

25. The activity of cell surface receptors can be down-modulated by receptor mediated endocytosis, which removes the receptor from the cell surface. In many cases, such as with the EGF receptor, the receptor is subsequently degraded in the lysosome, rendering the cell unresponsive to EGF until sufficient new receptor is synthesized, a slow process. Another way that cells down-modulate receptor signaling is to express a decoy receptor that competes for ligand binding but does not transduce the signal. For example, osteoblasts express the soluble decoy receptor osteoprotegerin, which competes for binding of RANKL with the RANK receptor on the surface of osteoclasts.

26a. The data suggest that *daf-4*, *sma-2*, *sma-3* and *sma-4* all reside in the same signaling pathway because mutation in each gene yields a similar phenotype of small animals with male tail defects. However, at this point, the information is just correlative and requires additional genetic and/or

molecular analysis to dissect the signaling pathways.

26b. Because mutation of *daf-4* affects dauer development, body size and mail tail development, it is likely to participate in both the same signaling pathway as the *sma* genes, which regulate size and tail development, and the same signaling pathway as the other *daf* genes, which regulate dauer development. However, the *sma* and other *daf* genes are probably in distinct signaling pathways from each other since mutations in these genes yield different phenotypes.

26c. Yes, these experiments do suggest that *daf-4* and *sma-2* function in the same cells since the same phenotype always localizes only to cells that possess either mutation. Furthermore, neither Daf-4 nor Sma-2 are likely to be diffusible molecules, such as TGFβ or BMP. If either were diffusible, the surrounding unmutated cells in mosaic animals would have been able to supply functional protein to the cells with *daf-4* or *sma-2* mutated, and no abnormal neurons would have been detected.

26d. These data suggest that in a common signaling pathway, Daf-4 protein lies upstream of the *sma* gene products.

26e. Daf-4 encodes a TGFβ-like receptor activated by binding of a BMP ligand. Upon activation, Daf-4 phosphorylates R-Smad proteins, which then bind co-Smads and translocate to the nucleus to regulate transcription of target genes involved in cell size and male tail development. These studies do not determine whether *sma-2*, *sma-3* or *sma-4* encode R-Smads, co-Smads, or both.

27a. These results indicate that PDGF receptors coupled to the PI-3 kinase pathway (Y40/51 curve) and PLCγ pathway (Y1021 curve) initiated near wild-type levels of DNA synthesis. Binding of GAP or Syp did not initiate DNA synthesis. These findings indicate the presence of redundant multiple signaling pathways in PDGF-induced mitogenesis.

27b. All the mutant receptors with partial substitution of phenylalanine for tyrosine exhibit normal activation (Figure 14-3) of S6 kinase except

F40/51. These findings indicate that the tyrosine residues at positions 40 and 51 of the PDGF receptor are required for both phosphorylation and activation of the S6 kinase. Since these are the residues involved in binding PI-3 kinase (see Figure 14-1), phosphorylation and activation of S6 kinase is linked to the PI-3 kinase pathway.

15

Integration of Signals and Gene Controls

PART A: *Chapter Summary*

An elaborate intercellular communication network coordinates the growth, differentiation, and metabolism of cells in tissues and organs. This chapter discusses the integration of multiple signals and responses in the context of development and cell differentiation. With the completion of several different genome sequencing projects, researchers are beginning to think about global responses to signaling. Various techniques are used to measure global signal-induced responses. For example, DNA and protein microarray experiments allow investigators to measure whole genome responses. Results from these types of experiments are yielding new insights into responses from signals.

Cell responses to environmental influences involves second messenger action. Peptides such as insulin and glucagon stimulate specific pathways, and molecules such as oxygen can be sensed by signal transduction pathways in the cell. Developmental events are often triggered by graded regulators that cause different cell responses depending on their concentration Since such regulators are used extensively in *Drosophila* embryo development, a discussion of how different cell types in the fly embryo develop is an area of intense interest. Two major events have been studied: dorsal/ventral and anterior/posterior patterning. To develop a dorsal cell fate, cells utilize two signal systems to alter the nuclear localization of the transcription factor Dorsal. Dorsal turns on transcription of genes required for ventral fate, and represses genes required for a dorsal cell fate. In anterior/posterior patterning, maternal gradients of *bicoid* and *nanos* mRNAs initiate cascades that eventually determine the anterior and posterior ends of the embryo.

Developmental boundaries can also be created by different combinations of transcription factors. Once again, the *Drosophila* development system provides examples of this concept. Further specification of Drosophila cell types requires transcription cascades where one transcription factor activates a gene encoding another transcription factor, which in turn activates the expression of a third transcription factor. Many similar pathways are utilized by other organisms, including plants. For example, a Toll-like signaling pathway is involved in plant disease resistance, and floral development requires several transcription factors. Boundaries can also be induced by extracellular signals, such as the Wingless and Hedgehog secreted proteins. These proteins act through binding to cell surface receptors and stimulate intracellular pathways.

Reciprocal induction and lateral inhibition are developmental events that also require signal transduction pathways. Specific examples of signaling pathways that function during these events are receptor tyrosine kinase signaling in artery and vein development, and Notch signaling during the formation of T cells in humans and vulval cells in worms. Lastly, the integration and control of signals is discussed. The receptor tyrosine kinase-MAP kinase signaling pathway is considered and other specific examples are detailed.

PART B: *Multiple Choice*

Circle the letter corresponding to the most appropriate terms/phrases. There may be more than one correct answer. Circle all that apply.

15.1 *Experimental Approaches for Building a Comprehensive View of Signal-Induced Responses*

1. Which phenomena have contributed to the evolution of gene families?

    a. divergence of genes during evolution

    b. duplication of protein coding genes

    c. signal transduction

    d. in situ hybridization

2. DNA microarray analysis can assess simultaneous expression of multiple genes in an organism if

    a. the entire genome of the organism has been sequenced.

    b. a collection of cDNAs is available.

    c. DNA can be extracted from the organism.

    d. RNA can be extracted from the organism.

3. DNA microarray analysis can identify

    a. sets of regulated genes.

    b. genes regulated by signals.

    c. genes regulated by different developmental states.

    d. genes altered by various diseases.

15.2 *Responses of Cells to Environmental Influences*

4. Which set(s) of signal and second messengers are known to function together?

    a. epinephrine and cGMP

    b. nervous stimulation and calcium

    c. insulin and cAMP

    d. hypoxia and cGMP

5. How is the hypoxia-induced transcription factor (HIF-1) regulated?

    a. by the partial pressure of oxygen

    b. by hemoglobin

    c. by vascular endothelial growth factor

    d. by an unknown oxygen sensor

15.3 *Control of Cell Fates by Graded Amounts of Regulators*

6. Morphogens can induce

    a. graded responses

    b. new gene expression

    c. differentiation events

    d. a change in cell fate

7. Which of the following factors share similarity with epidermal growth factor?

    a. LET-23

    b. LIN-3

    c. VEGF

    d. Delta

15.4 *Boundary Creation by Different Combinations of Transcription Factors*

8. Gap genes are expressed in

    a. single cell stripes.

    b. a seven stripe pattern.

    c. a broad band of cells.

    d. the anterior portion of the embryo only.

9. Gap gene products are

    a. translation factors.

    b. transcription factors.

    c. replication factors.

    d. RNA-binding proteins.

10. Hox genes are important switches that control cell identities by

    a. controlling the transcription of target genes that encode proteins required for various body parts or segments.

    b. stimulating Trithorax and Polycomb gene expression.

    c. by destabilizing *nanos* and *bicoid* RNAs.

    d. by stabilizing *nanos* and *bicoid* RNAs.

15.5 Boundary Creation by Extracellular Signals

11. The segment polarity gene, *wingless,* is expressed

    a. and is secreted after translation.

    b. in the anterior portion of the embryo.

    c. in single-cell wide stripes.

    d. adjacent to Engrailed-producing cells.

12. When Hedgehog protein is absent,

    a. Patched inhibits Smoothened and Fused; Costal-2 and Cubitis interruptus form a complex.

    b. Cubitis interruptus is cleaved and translocates to the nucleus.

    c. Cubitis interruptus represses target gene expression.

    d. inhibition of Smoothened is relieved, and Cubitis interruptus is cleaved.

15.6 Reciprocal Induction and Lateral Inhibition

13. Lateral inhibition is

    a. a process that prevents the duplication of structures at the expense of something not forming.

    b. a process that prevents the omission of structures at the expense of duplicating something.

    c. the process of one cell inhibiting the development of another cell.

    d. the process of one cell sending signals and inducing the development of another cell.

15.7 Integrating and Controlling Signals

14. Which hypotheses describe how diverse cellular responses to a particular signaling molecule result when a signaling pathway is used by many different cell types?

    a. The same sets of genes are turned on in all cell types.

    b. Converging inputs from other pathways modify the response to the signal.

    c. The strength or duration of the signal governs the nature of the response.

    d. The pathway downstream of the receptor is not really the same in different cell types.

15. What evidence supports the idea that limb development depends on integration of extracellular signals?

    a. A cell in the midst of a limb bud receives only one signal.

    b. A bead soaked in fibroblast growth factor 10 can induce limb development if implanted in a place where a limb does not normally develop.

    c. The apical ectodermal ridge secretes fibroblast growth factor 8 and 4 which increases cell division of mesoderm cells.

    d. The apical ectodermal ridge signals only to epidermal cells.

PART C: Reviewing Concepts

15.1 Experimental Approaches for Building a Comprehensive View of Signal-Induced Responses

16. What is the advantage of having an extracellular signal transmitted by a cascade of sequential events?

15.2 Responses of Cells to Environmental Influences

17. Describe the regulation of glycogenolysis by different second messengers in muscle cells.

18. Describe the mechanism for elevating cytosolic Ca^{2+} via the inositol-lipid signaling pathway.

19. In cells transfected with cDNA coding for the insulin receptor, the number of insulin receptors per cell can be much higher than in nontransfected cells. Nonetheless, the physiological response of such transfected cells to insulin often differs very little from that of normal nontransfected cells. Why is this the case?

15.3 Control of Cell Fates by Graded Amounts of Regulators

20. A central concept of developmental biology is that the developmental fate of cells is determined by certain molecules called morphogens. Explain how morphogens are thought to determine different cell fates.

21. Which maternal genes function as anterior and posterior determinants during *Drosophila* embryogenesis, and how do their encoded proteins act to establish the early anterior/posterior patterning of the embryo?

22. What are the molecular controls that result in a gradient of Hunchback (Hb) protein?

15.4 Boundary Creation by Different Combinations of Transcription Factors

23. Predict the phenotypes of *Arabidopsis* mutants defective in both A and B function and B and C function, respectively.

24. Two genes, *def* and *glo*, are required for B function in snapdragon flower identity. What is the phenotype of plants mutant in either *def* or *glo*, and what kind of proteins would def and glo encode? How do such proteins function in floral development?

15.5 Boundary Creation by Extracellular Signals

25. Compared to Wingless, Hedgehog is a non-diffusible inductive signal. How is the diffusion of Hedgehog limited?

15.6 Reciprocal Induction and Lateral Inhibition

26. In lateral interactions between equivalent cells in *C. elegans* that give rise to AC and VU, is there a predetermined pattern for which of the two cells becomes which?

PART D: *Analyzing Experiments*

27. In temperature-shift experiments with *Drosophila* carrying a temperature-sensitive mutation in the *hedgehog* (*hh*) gene, you identify two periods during which *hedgehog* function is critical for development of a normal phenotype: the first occurs 2–10 h after fertilization and the second occurs at pupation. To learn more about regulation of the *hedgehog* gene and the function of its encoded protein, you determine the level of *hedgehog* mRNA by Northern blotting of extracts of wild-type *Drosophila* embryos and adults. Your results are shown in Figure 15-1.

   a. Do the data in Figure 13-8 suggest that *hedgehog* mRNA, like *bicoid* mRNA, has a highly polarized distribution in the egg?

   b. Is Hedgehog protein metabolically stable? Explain.

   c. Would mutations that lengthen or shorten the metabolic stability of Hedgehog protein affect *Drosophila* development?

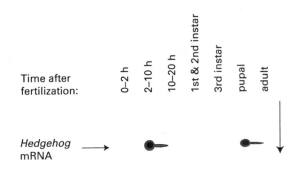

Figure 15-1

28. The genome sequence for the mosquito (*Anopheles gambiae*) has recently become available. Because certain species of *Anopheles* can transmit the Malaria parasite to humans, the development of mosquito embryos is of great interest. A group of researchers have completed a microarray analysis of mosquito embryos. To examine differences in transcription patterns between the dorsal and ventral sides of the embryo, early stage embryos

were divided in half and expression of a group of genes was measured in each half. Table 15-1 highlights some of the results of this experiment.

Genes increased in expression in the dorsal side	Genes increased in expression in the ventral side	Genes not expressed in either half
dpp tolloid short gastrulation zerknullt	rhomboid twist snail	gurken pipe

Table 15-1

Genes expressed in untreated embryos	Genes expressed in treated embryos
dpp tolloid short gastrulation zerknullt twist snail rhomboid dorsal cactus tube pelle	dpp tolloid short gastrulation zerknullt twist dorsal cactus tube pelle

Table 15-2

a. How were the mosquito homologues of genes listed in Table 15-1 identified?

b. Considering the results of the microarray analysis, does dorsal/ventral patterning of the mosquito embryo use pathways similar to *Drosophila*? Why or why not?

c. Why were the mosquito *gurken* and *pipe* genes not expressed in these embryos? Why were the *dorsal*, *tube*, *pelle* and *cactus* genes not differentially expressed?

d. A chemical inhibitor of mosquito development was used to treat mosquito embryos and a microarray analysis was performed on whole treated or untreated embryos. The results are shown in Table 15-2. Considering these results, make an educated guess as to how the chemical inhibitor acts to disrupt mosquito development. Give a rationale for your answer.

29. You have constructed two sets of gene fusions. In one, the *Antennapedia* (*Antp*) Hox gene promoter is fused with CAT, which functions as a reporter gene; in a second, the actin promoter, a strong constitutive promoter, is fused with either the *fushi tarazu* (*ftz*)

pair-rule gene or the *Ultrabithorax* (*Ubx*) Hox gene. The gene fusions are then expressed singularly or in combination in cultured *Drosophila* cells. The results of the specific fusions and their expression/co-expression are summarized in Figure 15-2.

a. Based on these results, are the proteins encoded by *ftz* and *Ubx* positive or negative regulators of *Antp* transcription?

b. In a 4.5 hour-old embryo, *ftz* is expressed in a seven stripe pattern. Reconcile this transcription pattern with the data in Figure 15-2.

c. Based on the results in Figure 15-2, predict the phenotype of a *Ubx* mutant fly.

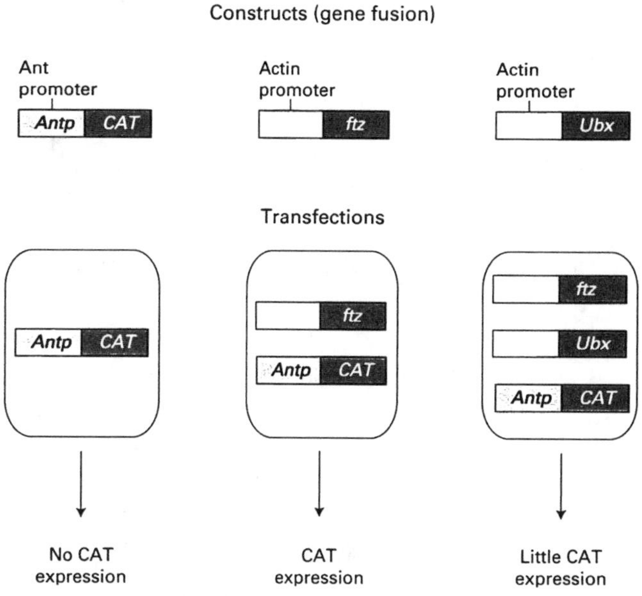

Figure 15-2

Answers

1. a, b

2. b

3. a, b, c, d

4. b, d

5. a, d

6. a, b, c, d

7. b, d

8. c

9. b

10. a

11. a, c

12. a, b, c

13. a, c

14. b, c, d

15. b, c

16. Transduction of an extracellular signal via a cascade of sequential events is advantageous because the signal is thereby amplified. The binding of a few molecules to a receptor can result in the synthesis of a large number of effector molecules because at each step of the cascade the signal is enzymatically amplified.

17. Stimulation of muscle cells by nerve impulses causes an increase in cytosolic Ca^{2+} concentrations. This rise in Ca^{2+} concentration activates glycogen phosphorylase kinase (GPK), which in turn activates glycogen synthase by phosphorylating it. In addition, epinephrine binding to β-adrenergic receptors leads to increased cytosolic concentrations of Ca^{2+} and cAMP. GPK is also activated by a cAMP-dependent protein kinase. In this way, the two second messengers, Ca^{2+} and cAMP, can regulate the same metabolic pathway.

18. The binding of a hormone to its receptor leads to activation of G protein, which in turn activates phospholipase C. Phospholipase C then cleaves phosphatidylinositol 4,5-bisphosphate (PIP_2) to inositol 1,4,5-trisphosphate (IP_3) and diacylglycerol (DAG). The IP_3 diffuses through the cytosol and interacts with IP_3 sensitive Ca^{2+} channels in the membrane of the endoplasmic reticulum causing release of Ca^{2+} stores.

19. Assuming that the insulin receptors in the transfected and nontransfected cells have the same K_D for hormone binding, then at a given insulin concentration, more insulin should bind to each transfected cell than to each nontransfected cell. However, the maximal physiological response of a cell to a hormone may occur when only a fraction of the receptors are occupied by ligand [see MCB, Figure 20-7 (p. 866)]. For this reason, a saturating level of insulin could have the same effect on both transfected and nontransfected cells. In addition, if the physiological response is limited by a second-messenger system or target protein(s), which are similar in both cell types, then the transfected and nontransfected cells would probably exhibit similar responses to insulin.

20. A morphogen can elicit different cellular responses depending on its concentration. Within the embryo, morphogens are distributed along concentration gradients, being present at higher concentrations in some regions than in others. Each morphogen exhibits a finite set of threshold concentrations associated with various cellular responses to it: above the threshold concentration, one response occurs; below the threshold, another response. Morphogens affect the developmental fate of cells by activating or repressing transcription of specific genes or by affecting translation of specific mRNAs. The sensitivity of cells to a particular morphogen may be mediated by the affinity of the morphogen for its binding sites within the genome. This has been demonstrated experimentally in the case of Bicoid protein regulating transcription of the zygotic *hunchback* gene.

21. The anterior determinant is the *bicoid* gene. The *bicoid* mRNA is produced by the mother, deposited in the oocyte, and becomes localized in the anterior portion of the embryo. After Bicoid protein is synthesized, it diffuses posteriorly, establishing an anterior posterior Bicoid concentration gradient. Bicoid protein stimulates transcription of the zygotic *hunchback* gene in the

early embryo, generating a gradient of zygotic *hunchback* mRNA. The posterior determinant is the *nanos* gene. Since the maternal *nanos* mRNA is localized in the posterior region of the embryo, a posterior → anterior gradient of maternal Nanos protein is established following translation of the mRNA. Since Nanos inhibits translation of *hunchback* mRNA, the uniformly distributed, maternally derived, *hunchback* mRNA is not translated in the posterior region of the embryo. Thus, as a result of the opposite gradients of Bicoid and Nanos protein, synthesized from maternal mRNAs, an anterior posterior gradient of Hunchback protein is established.

22. Maternal *Hb* RNA is uniformly distributed throughout the embryo, but these RNAs do not get translated in the posterior region due to repression by *Nanos*. Zygotic *Hb* RNA is under control of the Bicoid protein, which is localized to the anterior region. Both of these controls together ensure that Hb protein is only present at the anterior end of the embryo.

23. An *Arabidopsis* double mutant defective in both A and B function will contain flowers with whorls of carpels, carpels, carpels, and carpels because all information required to make sepals, petals and stamens is lost with loss of A and C function. An *Arabidopsis* double mutant defective in both B and C function will contain flowers with whorls of sepals, sepals, sepals, and sepals, because information required for petals, stamens and carpels is lost with loss of B and C function.

24. Snapdragon flowers mutant in either *def* or *glo* contain four whorls of sepals, sepals, carpels, and carpels only. A good guess is that Def and Glo proteins are MADS family transcription factors that specify petal and stamen identity in conjunction with an A and C gene respectively.

25. Hedgehog is tethered to the cell surface by the attachment of cholesterol to the C-terminus of the 20-kDa N-terminal fragment of Hedgehog precursor.

26. The end outcome of the two equivalent cells to produced AC and VU cells is not set. Either cell could have either fate.

27a. The absence of a hedgehog hybridization band in the lane corresponding to early embryos (0–2 h postfertilization) indicates that hedgehog mRNA is not present in the egg.

27b. The critical periods for hedgehog function identified in temperature-shift experiments with hedgehog mutants coincide with the peaks in hedgehog mRNA production in wild-type flies. This coincidence could only occur if hedgehog protein turned over fairly rapidly, with a half-life of 1 h or less. Thus the protein is metabolically unstable.

27c. Mutations that stabilize hedgehog would increase the amount of protein present and lengthen the duration of its presence within the embryo. Corresponding mutations that destabilized the protein would have the converse effect. In nature, hedgehog, like other developmental proteins, is needed at the right time and likely in the right amount for normal development to occur. Hence, mutations affecting hedgehog stability are likely to affect *Drosophila* development. Hedgehog is involved in establishing segment polarity in the early embryo (2–10 h postfertilization) and is critical for patterning the adult appendages during the pupal stage.

28a. The mosquito homologues were identified by their similarity at both the nucleic acid and amino acid levels to known *Drosophila* genes. Many regulatory genes whose encoded proteins control development and differentiation have been found to be conserved for millions of years.

28b. Yes, from the data one can surmise that mosquito dorsal/ventral patterning is very similar at the gene level to the process in *Drosophila*. The data supports this idea because the gene expression patterns seen in the dorsal and ventral halves of the embryos are similar to what has been found in *Drosophila*.

28c. The *gurken* and *pipe* genes are not expressed in either the dorsal or ventral sides of the embryo because these genes are expressed and function during egg development. Hence, these genes are responsible for early patterning events that are already complete once the egg matures. The *dorsal, tube, pelle* and *cactus* genes are not regulated at the level of transcription. Their differential effects arise through differential Toll signaling present on the dorsal and ventral sides. When Toll signaling is stimulated in the ventral side of the embryo, Tube, and Pelle form a complex which phosphorylates Cactus. Phosphorylation of Cactus then functions to effect Dorsal release and subsequent Dorsal function in the nucleus.

28d. The site of chemical inhibitor action is most likely the snail gene homologue, or one of its regulators. Both untreated and treated embryos express all of the expected genes, except for snail, twist and rhomboid. Since snail functions to turn on twist and rhomboid expression, lack of snail function would give rise to the patterns in Table 15-2.

29a. The Ftz protein is a positive regulator of the *Antp* promoter, whereas the Ubx protein is a negative regulator of the same promoter. This is indicated by the altered levels of reporter-gene expression under control of the *Antp* promoter when different constructs are placed in the same cell.

29b. If Ftz was the only regulator of Antp transcription, Antp would be co-expressed in a seven stripe pattern along with its transcriptional activator, Ftz. However, many genes are involved in the regulation of *Antp*, so the actual pattern is dependent on other genes including repressors such as *Ubx*. This eventually results in a single broad stripe of Antp expression corresponding to the second thoracic segment of the fly.

29c. If a repressor of Antp transcription such as Ubx is missing, then the first abdominal segment (where Ubx normally functions to pattern the embryo) will express the Antp protein. This will result in a homeotic conversion of the first abdominal segment to the third thoracic segment. Since the third thoracic segment specifies the wings, such mutants will have a second pair of wings.

16

Moving Proteins into Membranes and Organelles

PART A: *Chapter Summary*

The typical mammalian cell contains up to 10,000 kinds of proteins; a yeast cell contains about 5000. For normal cell function, all of these proteins must be localized to the proper subcellular compartment. The Na^+/K^+ ATPase pump must, for example, be distributed to the basolateral surface of the intestinal epithelial cell. Water-soluble enzymes such as RNA and DNA must be targeted to the nucleus; still other proteins must be delivered to the endoplasmic reticulum and Golgi apparatus. Many proteins, such as hormones and components of the extracellular matrix, must be directed to the cell surface and secreted.

The process of directing each newly made polypeptide to a particular destination—referred to as protein targeting or sorting—is essential to the organization and functioning of eukaryotic cells. The process occurs at several levels. A small number of proteins, encoded by DNA present in mitochondria and chloroplasts, are synthesized on ribosomes in these organelles and incorporated directly into compartments within these organelles. Most mitochondrial and chloroplast proteins and all proteins of the other organelles, particles, and membranes of a eukaryotic cell are encoded by nuclear DNA. They are synthesized on ribosomes in the cytosol and distributed to their correct destinations by the sequential action of up to several sorting signals. How nuclear-encoded organelle, membrane, and secretory proteins are sorted to their correct destinations are the major subjects of this chapter.

The first sorting event occurs during the initial growth of nascent polypeptide chains on cytosolic ribosomes. Some nascent chains contain, generally at their amino terminus, a specific signal or targeting sequence that directs the ribosomes synthesizing them to the endoplasmic reticulum (ER). Protein synthesis is completed in association with membranes of the rough ER. The completed polypeptide chains may then stay in the ER or move to the Golgi apparatus, where they are sorted for other possible destinations. Proteins synthesized and sorted in this pathway, termed the secretory pathway, include not only those that are secreted from the cell but also enzymes and other resident proteins in the lumen of the ER, Golgi apparatus, and lysosomes as well as integral proteins in the membranes of these organelles and the plasma membrane.

Synthesis of all other nuclear-encoded proteins is completed on "free," non-membrane attached cytosolic ribosomes, and the completed proteins are released into the cytosol. These proteins remain in the cytosol unless they contain a specific signal sequence that directs them to the mitochondrion, chloroplast, peroxisome, or nucleus. Many of these proteins are subsequently further sorted to reach their correct destinations within these organelles; such sorting events depend on multiple signals within the protein. Each sorting event involves binding a signal sequence to one or more receptor proteins on the surface or interior of the organelle.

This chapter details the mechanisms whereby proteins are sorted to the major organelles and compartments of the cell. (The transport of proteins in and out of the nucleus was described earlier in Chapter 12). Despite some variations, the same basic mechanisms govern protein sorting to all the various intracellular organelles. The information to target a particular protein to its destination is encoded within the amino acid sequence of the protein. These sequences are typically 20-50 amino acids in length and are known as signal sequences or uptake-targeting sequences. Organelle-bound, receptor proteins recognize targeting sequences. The receptor recognized protein chain is transferred through the membrane bilayer by some kind of translocation channel. Unidirectional transfer of proteins into an organelle is usually achieved by coupling translocation to an energetically favorable process, e.g., cleavage of ATP.

For each protein-targeting event, four fundamental questions are posed.

1. What is the signal sequence and how is it distinguished from examples of signal sequences?

2. What is the receptor for the signal sequence?

3. What is the structure of the translocation channel? Is the channel so narrow that the protein can pass only in the unfolded state?

4. What is the source of energy that drives unidirectional transfer across the membrane?

PART B: *Multiple Choice*

Circle the letter corresponding to the most appropriate terms/phrases. There may be more than one correct answer. Circle all that apply.

1. Membranes expand by

   a. de novo synthesis

   b. expansion of existing membranes

   c. exocytosis of membranes actively transported from outside the cell

   d. breakdown of endocytic vesicles

   e. breakdown of transport vesicles

2. During protein modification in the secretory pathway, proteins move through organelles in the following order:

   a. mitochondria → peroxisome → lysosome

   b. chlorplast → peroxisome → lysosome

   c. mitochondria → chloroplast → nucleus

   d. ER → Golgi complex → plasma membrane

   e. ER → Golgi complex → lysosome

3. Which of the following statements about translocation of most nascent proteins across the endoplasmic reticulum membrane is true?

   a. Translocation occurs concomitantly with protein synthesis.

   b. Translocation selects a class of ribosomes to be membrane-bound.

   c. Proteins are transported to the lumen of the endoplasmic reticulum by passage through the phospholipids of the membrane.

   d. BiP binds to translocated proteins in the ER lumen, thereby preventing aggregation or misfolding of these proteins.

   e. The N-terminal end of the protein always enters and leaves the membrane first.

4. The ability of very small amounts of protease to release protein fragments from preparations of rough ER was critical in the discovery of which of the following?

   a. mannose 6-phosphate receptor (MPR)

   b. signal peptide

   c. signal peptidase

   d. mRNA

   e. signal recognition receptor (SRP receptor)

5. Which of the following posttranslational modifications of proteins occur in the lumen of the endoplasmic reticulum?

   a. glycosylation

   b. formation of disulfide bonds

   c. conformational folding and formation of quaternary structure

   d. proteolytic cleavage

   e. aldol condensation of sugars

6. Permanent insertion of integral proteins in membranes can be mediated by

   a. a stop-transfer sequence

   b. a single membrane-spanning α-helical domain

   c. a GPI membrane anchor

   d. the Asn-X-Ser/Thr sequence

7. N-linked oligosaccharides in glycoproteins

   a. are characterized by the presence of N-acetylglucosamine

   b. are generally only 2 to 3 saccharide units long

   c. are produced by the sequential addition of sugar residues

   d. are produced from a common precursor

   e. are characterized by the presence of sialic acid

8. Topogenic sequences in proteins are known to be located

   a. at the C-terminus

   b. at the N-terminus

   c. internally

   d. where a mannose is attached

   e. at multiple sites within one protein

9. Order the following choices to indicate the sequence of steps in the synthesis of secretory proteins.

   a. dissociation of SRP from its receptor

   b. translocation across the membrane of the endoplasmic reticulum

   c. initiation of protein synthesis on free ribosomes

   d. binding of SRP to nascent polypeptide

   e. cleavage of signal sequence

   f. binding of SRP, polypeptide, and ribosome to SRP receptor

10. Shared features between translocation of bacterial proteins across the inner membrane and translocation of proteins into the ER of eukaryotic cells include:

   a. the use of ATP to pull the bacterial protein through the translocon

   b. protein transfer through the inner membrane in a translocon composed of proteins that are structurally similar to the eukaryotic cell Sec61 complex

   c. bacterial protein homologs of the SRP and SRP receptor

   d. the presence of a cleavable signal sequence

11. Translocation of bacterial proteins into the extracellular space involves

   a. a unitary, single mechanism

   b. invariably a folding step in the periplasmic space

   c. in some cases a single step in which the protein is translocated through a large complex that spans both membranes

   d. an ATP independent process

12. N-terminal sequences usually function as targeting signals for protein transport into

   a. the nucleus

   b. peroxisomes

   c. chloroplasts

   d. the rough endoplasmic reticulum

   b. mitochondria

13. The energy required for protein transport into mitochondria is provided by

   a. a proton-motive force across the inner membrane

   b. ATP hydrolysis associated with chaperones in the cytoplasm

   c. ATP generated by anaerobic respiration

   d. GTP hydrolysis

   e. ATP hydrolysis in the matrix

14. Translocation of proteins into mitochondria and chloroplasts is characterized by

    a. prefolding of proteins into enzymatically active shapes before import

    b. interaction with a specific receptor

    c. involvement of a mannose-6-phosphate

    d. incorporation into transport vesicles that subsequently fuse with the organelles

    e. direct passage across the lipid bilayer

15. Proteins targeted to nonmatrix regions of mitochondria

    a. contain a matrix-targeting signal if the destination is the outer membrane

    b. utilize a stop-transfer signal if the destination is the outer membrane

    c. always contain a matrix-targeting signal if the destination is the intermembrane space

    d. may require a chaperone for transport

    e. always move first to the matrix and then to their final destination

16. Which of the following proteins is found in the peroxisome matrix?

    a. catalase

    b. cytochrome c

    c. ribulose 1,5-bisphosphate caroxylase (rubisco)

    d. porin

    e. RNA polymerase

PART C: *Reviewing Concepts*

16.1 *Translocation of Secretory Proteins across the ER Membrane*

17. Describe the overall process by which a cell surface glycoprotein such as glycophorin is synthesized and transported to the cell surface.

18. The signal recognition particle (SRP) is involved in regulating the elongation of nascent secretory proteins and targeting them to the endoplasmic reticulum. Describe experiments in which these functions of SRP can be demonstrated.

19. Energy input is required for protein translocation across the ER membrane. What is the source of this energy for most proteins in yeast and mammalian cells?

16.2 *Insertion of Membrane Proteins into the ER Membrane*

20. In some proteins, a signal sequence also functions as a topogenic sequence. Discuss how these dual-function signal sequences differ from monofunctional signal sequences, which simply direct nascent polypeptides to the ER membrane.

21. Describe the process by which a protein becomes glycosylphosphatidylinositol (GPI) anchored.

16.3 *Protein Modifications, Folding, and Quality Control in the ER*

22. Processing of proteins is subject to "quality control" within the ER that prevents improperly folded proteins from reaching the cell surface. What are typical fates of misfolded proteins?

23. What is the amino acid modified during N-linked glycosylation?

24. Explain why eukaryotic cells, rather than bacteria, are the preferred hosts for insertion of cDNA for the production of human secretory proteins of commercial value.

16.4 *Export of Bacterial Proteins*

25. How might the type III secretion apparatus of some pathogenic bacteria be used to inject engineered proteins into eukaryotic cells?

26. The periplasmic space of gram-negative bacteria, like the lumen of the ER, is a site of protein folding. What does the formation of disulfide bonds in this environment indicate about how reducing or oxidizing the periplasmic space is?

16.5 *Sorting of Proteins to Mitochondria and Chloroplasts*

27. Antibodies specific to the C-terminal end of a cytosol-synthesized mitochondrial protein can prevent the precursor protein from being completely translocated into the mitochondrial matrix, although a portion of the precursor is translocated. Explain these observations.

28. What is the salient difference between the energy requirement for translocation of proteins to mitochondria and to chloroplasts?

16.6 *Sorting of Peroxisomal Proteins*

29. Fibroblasts isolated from patients with Zellweger syndrome are incapable of translocating catalase and other lumenal proteins synthesized in the cytosol into peroxisomes. How could you demonstrate that Zellweger fibroblasts can synthesize catalase?

30. One hypothesis that has been advanced to explain the nature and origin of the peroxisome proposes that this organelle is a vestige of the organellar site that arose in primitive, premitochondrial cells to handle oxygen introduced into the atmosphere by photosynthetic bacteria. Since oxygen radicals and some oxygen-containing molecules can be highly toxic to cells, a mechanism for handling them would be necessary for cell survival. How consistent is this hypothesis with the structure and function of peroxisomes?

PART D: *Analyzing Experiments*

31. The light-harvesting complex protein (LHCP) is an example of a chloroplast protein that is synthesized in the cytosol and then imported and subsequently processed to a mature form. It is found in thylakoid membranes of the chloroplast.

    In an experiment designed to explore the forces governing translocation of preLHCP, the in vitro translocation of [^{3}H] preLHCP to isolated chloroplasts was measured. After an appropriate period of incubation, the chloroplasts were lysed and a fraction containing both thylakoid and envelope proteins was prepared and analyzed by SDS-PAGE autoradiography, as depicted in lane 2

of Figure 16-1. Lane 1 is a control incubation of labeled preLHCP in the absence of chloroplasts. Following incubation, samples of the thylakoid membranes were treated with NaOH or a protease and then analyzed, as depicted in lane 3 (NaOH treatment), lane 4 (thermolysin treatment), and lane 5 (trypsin treatment).

a. What does the difference in the migration patterns in lanes 1 and 2 in Figure 16-1 indicate about preLHCP? Why is this a critical part of the experiment?

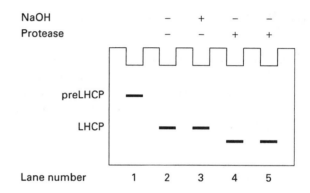

FIGURE 16-1

b. Why were NaOH and proteases used in this experiment?

c. In Figure 16-1, what does the difference in the migration patterns in lanes 4 and 5 compared with the pattern in lane 2 suggest about the location of LHCP in the thylakoid membrane?

d. In other experiments, purified thylakoid membranes, prepared on a sucrose gradient, were incubated with labeled preLHCP in the presence and absence of ATP, stroma, and protease. After an appropriate time, the samples were analyzed by SDS-PAGE autoradiography. The resulting gel profiles are depicted in Figure 16-2. What conclusions can be drawn from these data about the effects of ATP and stroma on binding of preLHCP to thylakoid membranes?

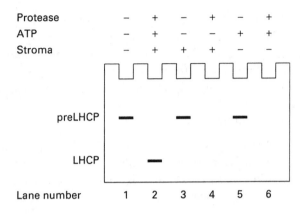

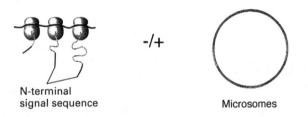

FIGURE 16-2

32. As illlustrated in Figure 16-3, protein synthesis may be done in the presence or absence of microsomal membranes. Under what conditions will protein translocation into the microsomal membranes take place and why?

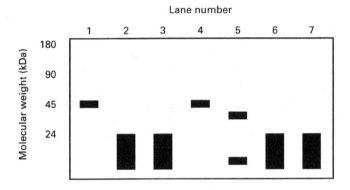

FIGURE 16-3

33. The signal recognition particle (SRP) consists of six polypeptide subunits (9, 14, 19, 54, 68, and 72 kDa), which are organized into three different functional entities surrounding a 7S-RNA molecule. After protein synthesis is initiated in the cytoplasm, the 54-kDa subunit of the SRP binds to the signal sequence shortly after it is synthesized on the ribosome. This SRP-ribosome unit then associates with the SRP receptor (docking protein) on the endoplasmic reticulum where synthesis continues and the newly synthesized protein is translocated across the endoplasmic reticulum.

In experiments designed to sort out the factors necessary to promote translocation, a complete cell-free translational system was incubated with an mRNA encoding a secretory protein of 40-kDa when fully modified, [^{35}S]methionine to monitor protein synthesis, and various preparations containing different factors, as indicated in Table

16-1 (next page). GMP-PNP and AMP-PNP are nonhydrolyzable analogs of GTP and ATP, respectively. After each of the preparations was incubated with the translational system in appropriate buffers, the sample was incubated with a protease (proteinase K). Then all proteins were precipitated, denatured, and separated on an SDS gel. Autoradiography of the gel revealed the pattern shown in Figure 16-4. Each lane of the autoradiogram is labeled with the corresponding preparation number.

FIGURE 16-4

a. What is the significance of the discrete bands in lanes 1, 4, and 5 of the autoradiogram and of the diffuse bands in the other lanes?

c. What can you conclude from this experiment about the factors required for protein translocation and the mechanisms involved in this process?

Table 17-1

Preparation number	Components
1	SRP-ribosome complex + microsomes + ATP + GTP
2	SRP-ribosome complex + microsomes + ATP + GTP + 5mM EDTA
3	SRP-ribosome complex + microsomes + AMP-PNP + GTP
4	SRP-ribosome complex + microsomes + ATP + GMP-PNP
5	SRP-ribosome complex + microsomes + ATP + GTP + more time for incubation
6	SRP-ribosome complex + microsomes
7	SRP-ribosome complex + microsomes + ATP

34. Figure 16-5 schematically depicts proteins containing various types of signal and topogenic sequences. Predict the arrangement of each type of protein shown in the figure with respect to the endoplasmic reticulum membrane and lumen.

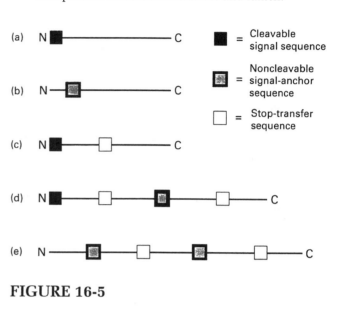

FIGURE 16-5

Answers

1. b

2. d, e

3. a, d

4. e

5. a, b, c, d

6. a, b, c

7. a, b

8. a, b, c, e

9. c, d, f, a, b, e

10. b, c, d

11. c

12. c, d, e

13. a

14. b

15. a, b, d

16. a

17. A membrane protein such as glycophorin is synthesized by ribosomes that associate with the rough endoplasmic reticulum. From there, the protein is transported via vesicles to the *cis* Golgi reticulum (network). In a process termed cisternal progression, or maturation, the *cis* Golgi reticulum matures into the *cis* Golgi, *medial*, and *trans* Golgi. Progression completes with maturation of the Golgi cisterna into the *trans* Golgi reticulum (network). From the *trans* Golgi reticulum, the protein is transported via a vesicular carrier to the plasma membrane. During the whole process the topology of the protein with respect to the membrane is maintained.

18. The functions of SRP can be demonstrated in a series of experiments utilizing a cell-free protein-synthesizing system and mRNA encoding a typical secretory protein. When the mRNA is incubated in the cell-free translational system in the absence of SRP and microsomes, the complete protein with its signal sequence is produced. The addition of SRP to the incubation mixtures causes protein elongation to cease after 70–100 amino acids have been incorporated. When microsomes containing the SRP receptor are also added to the incubations, the block in protein synthesis is relieved and the complete protein minus the signal sequence is extruded into the lumen of the microsomes. The outcome of these experiments can be demonstrated by SDS-polyacrylamide gel electrophoresis combined with protease digestion to test whether the newly synthesized protein is protected from digestion by inclusion in the ER lumen.

19. In mammalian cells, the translation process itself is thought to drive the nascent chain across the ER membrane. In yeast, most secreted proteins are also pushed through the translocon by translation. However, some secreted proteins in yeast translocate post-translationally. Here, the hydrolysis of ATP by an Hsc70 protein, BiP, powers the post-translational import of proteins into the ER lumen.

20. Some integral membrane proteins (e.g., the asialoglycoprotein receptor) contain a single internal hydrophobic sequence of ≈22 residues, which both directs the nascent protein to the ER membrane and embeds/orients the protein in the membrane. Unlike the N-terminal signal sequences in secretory proteins and other membrane proteins, these dual-function sequences, called internal signal-anchor sequences, are not cleaved from the nascent protein. Of course, the situation can be more complicated in the case of membrane proteins that contain multiple topogenic sequences. In these cases, signal anchor sequences and stop-transfer sequences alternate.

21. The protein is initially synthesized as a transmembrane protein with a single stop-transfer membrane-anchor sequence as the transmembrane domain. A short sequence of amino acids in the exoplasmic (luminal) domain, adjacent to the membrane-spanning domain, is recognized by an endoprotease that simultaneously cleaves off the original stop-transfer membrane-anchor sequence and transfers the remainder of the protein to a preformed GPI (glycosylphosphatidlylinositol) anchor in the membrane.

22. Improperly folded proteins in the ER have two common fates. One is to aggregate within the ER and form an almost crystalline aggregate. This is the fate of a misfolded mutant form of α_1-antiprotease. The other fate is to be translocated into the cytosol via the translocon and be degraded in the cytosol by the proteosome.

23. N-linked oligosaccharide is added to the amide nitrogen of asparagine.

24. Many secreted human proteins are stabilized by disulfide bonds and are N-glycosylated. Bacteria are a poor choice for typical production of secreted proteins because they fail to form disulfide bonds properly and do not N-glycosylate proteins. Failure to form the proper disulfide bonds can, for example, result in protein misfolding and precipitation to form inclusion bodies within the bacteria.

25. The type III secretion apparatus in essence acts like a microinjection needle to transfer proteins into animal cells. Designer bacteria such as *Yersinia* cells can be engineered that do not secrete animal harmful proteins. These cells, if engineered to express desirable protein with the proper type III signal sequence, can deliver the engineered protein via the type III mechanism.

26. Disulfide bonds are only formed in a relatively oxidizing environment. Hence the periplasmic space must be an oxidizing environment.

27. Because the precursor protein contains an N-terminal matrix-targeting signal, the N-terminus of the precursor molecule is translocated first. An antibody to the C-terminal end would not prevent passage of the N-terminal end across the mitochondrial membrane. However, the large size of the antibody bound to the C-terminal end would inhibit passage of the entire protein into the organelle, resulting in a translocation intermediate. Such intermediates have been generated in other ways, such as by the addition of methotrexate during the translocation of dihydrofolate reductase. In the presence of the drug, which binds to the protein in its native configuration, translocation is blocked. Experi-

ments such as these indicate that a protein must be in an unfolded state to pass into the interior of the mitochondrion.

28. Energy is required for translocation of proteins to both chloroplasts and mitochondria. Three separate inputs of energy are required for mitochondrial import: ATP hydrolysis in the cytosol, a proton-motive force across the inner mitochondrial membrane, and ATP hydrolysis in the mitochondrial matrix. Protein import into the chloroplast stroma appears to be powered solely by ATP hydrolysis. ATP hydrolysis in all cases appears to be related to the chaperone protein, be it cytosolic or present in the mitochondrial matrix or chloroplast stroma.

29. The ability of Zellweger fibroblasts to synthesize catalase could be demonstrated by carrying out protein synthesis with a cytosolic extract and then adding antibodies to catalase protein to immunoprecipitate newly synthesized catalase in the reaction mixtures.

30. Peroxisomes contain enzymes that degrade amino acids and fatty acids, generating hydrogen peroxide (H_2O_2), which is potentially toxic to cells. The peroxisomal enzyme catalase decomposes H_2O_2 into H_2O. In view of this, the hypothesis that peroxisomes originated to cope with oxygen in primitive cells is very reasonable.

31a. Comparison of lanes 1 and 2 indicates that preLHCP is processed normally in the in vitro system. This is a critical part of the experiment because it is imperative to show that this in vitro system mimics the in vivo situation.

31b. These treatments were used to determine if the mature LHCP is in fact in the thylakoid membrane, where it is found in vivo, or is soluble in the stroma or only partially integrated in the thylakoid membranes. Integral membrane proteins are resistant to extraction with NaOH and to proteolytic digestion, whereas soluble proteins or partially integrated ones would be susceptible to these treatments.

31c. The protease-treated samples (lanes 4 and 5) migrate slightly faster than the untreated sample (lane 2), indicating a decrease in molecular weight. This decrease in molecular weight suggests that part of LHCP is exposed to the stroma and thus is susceptible to partial protease degradation.

31d. The data in Figure 16-2 indicate that both ATP and stroma are necessary for binding of preLHCP to isolated thylakoid membranes; this binding protects the bound protein from degradation by protease (lane 2). However, neither ATP or stroma alone can support binding (lanes 3 and 5); in the presence of either one alone, the unbound preLHCP is susceptible to protease degradation (lanes 4 and 6).

32. Protein translocation will occur only when the microsomes are added to the reaction mixture while protein synthesis is actively occurring. The N-terminal signal sequence acts co-translationally. It interacts with SRP while translation is occurring and is threaded later through the translocon cotranslationally. If the N-terminal signal sequence does not interact rapidly with SRP and then microsomalo membranes, it becomes part of a misfolded protein complex.

33a. Because proteinase K cannot penetrate the microsomal membrane, it can digest only proteins that are not translocated into the lumenal space of the ER. Thus the diffuse bands in Figure 16-4 represent the degradation products of newly synthesized protein that was not translocated during the incubations. The discrete bands represent newly synthesized protein that was translocated into the ER lumen during the incubations; because this protein is enclosed by a membrane, it was protected from the action of the protease.

33b. Comparison of lanes 1 and 6 indicates that ATP or GTP or both is necessary for translocation. However, since translocation did not occur in the presence of ATP alone (lane 7), GTP must be a necessary factor. Comparison of lanes 3 and 4 indicates that hydrolysis of ATP but not of GTP is necessary for translocation, since AMP-PNP in the presence of GTP prevented translocation, whereas GMP-PNP in the presence of ATP did not. Thus

although both ATP and GTP are required for translocation, only ATP is hydrolyzed.

Lane 2 indicates that EDTA inhibits translocation of the nascent protein, suggesting that it "uncouples" the SRP-ribosome complex or prevents ATP hydrolysis, perhaps by chelating Mg^{2+} ions.

Finally, comparison of lanes 1 and 5 suggests that a signal peptidase is present in the translational system. When the incubation time was extended, the peptidase cleaved the 45-kDa nascent protein (lane 1) into the 40-kDa mature protein and a small signal peptide (lane 5).

34. The results are shown diagrammatically in Figure 16-6.

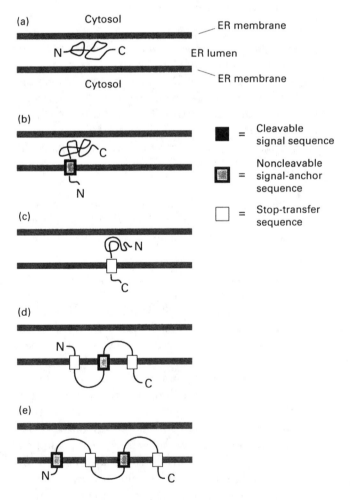

Figure 16-6

17

Vesicular Traffic, Secretion, and Endocytosis

PART A: *Chapter Summary*

Soluble and membrane proteins synthesized on the rough endoplasmic reticulum (ER) move to their final destinations via the secretory pathway. A single unifying principle governs all protein trafficking in the secretory pathway: transport of membrane and soluble proteins from one membrane-bounded compartment to another is mediated by transport vesicles that collect "cargo" proteins in buds arising from the membrane of one compartment and then deliver these cargo proteins to the next compartment by fusing with the membrane of that compartment. As transport vesicles bud from one membrane and fuse with the next, the orientation of membrane components is conserved; the same face of the membrane remains oriented towards the cytosol.

Newly synthesized proteins incorporated into the ER lumen or membrane can be packaged into anterograde (forward-moving) transport vesicles. These vesicles fuse with each other to form a flattened membrane-bounded compartment (cisterna) known as the *cis*-Golgi. From the *cis*-Golgi, certain proteins, mainly ER-localized proteins, are retrieved via a different set of retrograde (backward-moving) transport vesicles. A new cis-Golgi cisterna with its cargo of proteins physically moves from the cis position (nearest the ER) to the trans position (farthest from the ER), successively becoming first a *medial*-Golgi cisterna and then a *trans*-Golgi cisterna. This process, known as cisternal progression, does not involve the budding off and fusion of anterograde transport vesicles. Rather, enzymes and other Golgi-resident proteins are constantly being retrieved from later to earlier Golgi cisternae by retrograde transport vesicles, thereby remaining localized to the *cis*-, *medial*- or *trans*-Golgi cisternae.

Proteins in the secretory pathway that are destined for compartments other than the ER or Golgi eventually reach a complex network of membranes and vesicles termed the *trans*-Golgi network (TGN). Here, a protein can be loaded into one of at least three different kinds of vesicles. After budding from the trans-Golgi network, the first type of vesicle immediately moves to and fuses with the plasma membrane, releasing its contents by exocytosis and topologically conserving the orientation of membrane proteins. Examples of proteins released by such constitutive (or continuous) secretion include collagen by fibroblasts, serum proteins by hepatocytes, and antibodies by activated B lymphocytes. Membrane proteins include receptors and transport proteins. The second type of vesicle to bud from the *trans*-Golgi network, known as secretory vesicles, are stored inside the cell until a signal for exocytosis causes release of their contents at the plasma membrane. Among proteins released by such regulated secretion are peptide hormones from endocrine cells, precursors of digestive enzymes from pancreatic acinar cells, milf proteins from the mammary gland, and neurotransmitters from neurons.

The third type of vesicle that buds from the trans-Golgi network is directed to the lysosome, an organelle responsible for the intracellular degradation of macromolecules, and to lysosome-like storage organelles in certain cells. Secretory proteins destined for lysosomes are first transported by vesicles from the trans-Golgi network to a compartment usually called the late endosome; proteins are then transferred to the lysosome by a mechanism that's not well understood, but may involve direct fusion of the endosome with the lysosomal membrane. Soluble proteins delivered by this pathway include lysosomal digestive enzymes and membrane proteins that pump H^+ from the cytosol into the acidic lumen of the endosome and lysosome. Some of the specific protein processing and sorting events that take place within these organelles depend on their low luminal pH.

The endosome also functions in the endocytic pathway in which vesicles bud from the plasma membrane, bringing membrane protein and their bound ligands into the cell. After being internalized by endocytosis, some proteins are transported to lysosomes, while others are recycled back to the cell surface. Endocytosis is a way for cells to take up nutrients that are in macromolecular form such as cholesterol in the form of lipoprotein particles and iron complexed with the serum protein transferring. Endocytosis can function also as a regulatory mechanism to decrease signaling activity by withdrawing receptors for a particular signaling molecule from the cell surface.

PART B: *Multiple Choice*

Circle the letter corresponding to the most appropriate terms/phrases. There may be more than one correct answer. Circle all that apply.

1. Transport vesicles

   a. can have a clathrin coat

   b. can be coated with a heteromeric COPI and COPII coat

   c. can contain assembly proteins

   d. can shuttle proteins from mitochondria to the endoplasmic reticulum

   e. can fuse with target organelles

2. Secreted proteins

   a. are all present in one type of vesicle

   b. are typically membrane proteins

   c. may be constitutively exocytosed

   d. may require an acidic compartment for maturation

   e. may aggregate within vesicles

3. Assays for following the trafficking of proteins through the secretory pathway in living cells require

   a. sensitive methods to distinguish the very small differences in organelle density caused by passage of a wave of protein transport through the organelle

   b. a way to identify compartments where labeled proteins are located

   c. the use of cell types which are highly dedicated secretory cells

   d. the use of enucleate cells in order to eliminate the nucleus as a site of labeling

   e. a way to label a cohort of secretory proteins

4. VSV G protein has been a very useful tracker protein for both living cell and in vitro assays because

   a. a temperature sensitive mutation in the protein permits reversible accumulation of the protein in the ER

   b. the protein is much smaller than most cellular proteins and therefore can be readily identified by SDS-PAGE

   c. fusion proteins with VSV G retain the temperature sensitive properties of VSV G

   d. VSV infects both animal cells and yeast allowing the full power of yeast genetics to be applied to VSV G transport

   e. the protein is N-glycosylated and hence its trafficking can be inferred from its glycosylation state

5. Examples of cargo proteins within the secretory pathway include:

   a. lysosomal enzymes

   b. cytochrome c oxidase

   c. RNA polymerase

   d. catalase

   e. prochymotrypsinogen

6. Proteins typically included in coated vesicles include:

   a. cargo proteins

   b. cargo protein receptors

   c. small GTPase of the Sar1 or Arf families

   d. v-SNARE molecules

   e. NSF

7. How are cargo protein molecules selectively enriched in transport vesicles?

   a. by interaction with coat proteins/adaptor proteins

   b. by selective aggregation in the *trans* Golgi network

   c. by passive diffusion/bulk flow

   d. by the presence of sequence motifs leading to interactions with transcription factors

8. ARF and Sar1 are active in coat protein recruitment in:

   a. their GDP bound state

   b. their GTP bound state

   c. their nucleotide free state

   d. their Ca^{++} activated state

   e. their soluble, non-membrane associated state

9. All of the following proteins play a role in vesicle formation or vesicle targeting **except**:

   a. rab proteins

   b. coat proteins

   c. v-SNARES

   d. t-SNARES

   e. NSF

10. The sequence Lys-Asp-Glu-Leu (KDEL) is an example of:

    a. a sequence signal for selective inclusion of membrane proteins in COPII vesicles

    b. a sequence signal for interaction with AP2 and inclusion in clathrin/AP2 coated vesicles formed at the cell surface

    c. a sequence signal for interaction with AP1 and inclusion in clathrin/AP1 vesicles formed at the *trans* Golgi network

    d. a sequence signal for inclusion via interaction with the KDEL receptor iin COPI vesicles

    e. a sequence signal for protein aggregation at the trans Golgi network result in inclusion in regulated secretory granules

11. Anterograde transport through the Golgi apparatus occurs via

    a. vesicular transport

    b. cisternal progression

    c. transient tubular connections between Golgi subcompartments

    d. permanent tubular connections between Golgi subcompartments

12. Dynamin is involved in vesicle pinching off at

    a. ER-exit sites

    b. the *medial* Golgi

    c. the *trans* Golgi network

    d. the plasma membrane

    e. nerve endings

13. Proteolytic processing of secretory proteins occurs generally in the the ER and

    a. *cis* Golgi

    b. *medial* Golgi

    c. *trans* Golgi

    d. *trans* Golgi network

    e. post Golgi vesicles/granules

14. An important mechanism for dissociating ligand and receptor within the endocytic pathway is

   a. Ca^{++}-dependent binding

   b. pH-dependent binding

   c. selective proteolysis

   d. GTP hydrolysis

   e. ATP hydrolysis

15. The vesicles within multivesicular endosomes are

   a. formed by inward budding

   b. comparatively rich in endocytized receptor proteins that are to be degraded

   c. formed by mechanisms entirely different than that of retrovirus budding from the plasma membrane

   d. the result of the action of several gene products including Hrs

   e. continuous in three dimensions with all other membranes of multivesicular endosomes

16. What are the role(s) of Ca^{++} in promoting synaptic vesicle fusion with the presynaptic membrane?

   a. recruitment of synaptic vesicles to the presynaptic membrane

   b. loading of acetylcholine into the synaptic vesicle

   c. promoting rapid fusion of docked synaptic vesicles with the presynaptic membrane

   d. interaction with the Ca^{++} sensor synaptotagmin

   e. promoting the dynamin dependent reformation of synaptic vesicles.

PART C: *Reviewing Concepts*

17.1 *Techniques for Studying the Secretory Pathway*

17. Describe how a fusion protein of VSV G and green fluorescent protein (GFP) can be used to track the transport of secretory proteins through the entire secretory pathway.

18. In vitro transport assays always include cytosol as well as purified proteins in the incubation mix. Why is this so?

19. Describe the phenotype of class B *sec* mutants in yeast and explain how any of several different mutations can give a class B phenotype.

17.2 *Molecular Mechanisms of Vesicular Traffic*

20. What are the known types of coat proteins and which small GTPase is required for the requirement of each type to the membrane?

21. What is the role of SNARE proteins in vesicle fusion?

17.3 *Vesicle Trafficking in the Early Stages of the Secretory Pathway*

22. What is the expected subcellular distribution of protein disulfide isomerase (PDI) lacking a C-terminal KDEL sequence?

23. How does the recruitment of proteins into COPII buds program the resulting vesicle for subsequent specific fusion?

24. Explain how strategies for promoting protein folding within the ER might be rational ways to correct some inherited deficiencies in plasma membrane transport proteins.

17.4 *Protein Sorting and Processing in Later Stages of the Secretory Pathway*

25. How does regulated secretion differ from constitutive secretion?

26. When fibroblasts from patients suffering from I cell disease are cultured, they secrete lysosomal enzymes rather than accumulating them within lysosomes. When lysosomal enzymes containing M6P are added to the culture, they are rapidly internalized and accumulate within lysosomes. Why don't the I cell secreted lysosomal enzymes accumulate within cells?

27. Specialization of membranes is a characteristic of mammalian cells. What are the mechanisms by which the apical and basolateral surfaces of certain cells become differentiated?

17.5 Receptor-Mediated Endocytosis and the Sorting of Internalized Proteins

28. Many extracellular proteins are internalized by receptor-mediated endocytosis. What are the molecular signals that trigger uptake of a protein by receptor-mediated endocytosis?

29. The pH of a compartment can be critical for association or dissociation of receptor and ligand. How does the acidic pH of the late endosome result in different fates for the LDL and transferrin receptors?

17.6 Synaptic Vesicle Function and Formation

30. How does the paralysis effect of the *shibire* mutation phenotype in *Drosophila* relate to the underlying molecular mechanism?

31. Most students of molecular cell biology tend to think of nerve endings as being specialized examples with little general relevance. How does the cycling of synaptic vesicle components illustrate general principles of membrane trafficking?

PART D: *Analyzing Experiments*

32. Type I transmembrane proteins are localized to the ER of mammalian cells by virtue of a common signal consisting of two lysine residues at positions −3 and −4 from the C-terminal end of the cytoplasmic domain. A similar dilysine signal has been recently identified in the cytoplasmic domain of a yeast ER membrane protein. In both mammalian and yeast cells, this dilysine sequence appears to function as a retrieval sequence; analysis of the posttranslational modifications of the proteins indicates that they have been retrieved from the Golgi apparatus.

Genetic approaches have been very important to investigations of the machinery involved in the retrieval of type I transmembrane proteins to the ER. In such experiments, a series of yeast temperature-sensitive mutants defective in retrieval were isolated and then characterized for the nature of the defect. The nine temperature-sensitive mutations isolated fell into two complementation classes: *ret1* for retrieval defective 1 (eight examples

all in one allele) and *sec21* (one example, which was a new *sec21* allele termed *sec21-2*). The *sec21* gene was originally identified on the basis of a temperature-sensitive defect in secretion (*sec21-1* allele).

To quantify the effect of mutations on retrieval, the fate of a fusion protein containing the dilysine (diK) retrieval sequence was determined. If this diK-fusion protein is retrieved, it is not exposed to a post-ER protease. After the fusion protein was expressed in yeast cells incubated with a radiolabeled amino acid, the labeled products were resolved by gel electrophoresis and detected by autoradiography. Figure 17-1a (next page) depicts the resulting autoradiograms for wild-type (WT) yeast cells and the *ret1-1, sec21-1,* and *sec21-2* mutants. In the absence of retrieval, the diK-fusion protein is processed by the post-ER protease.

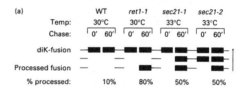

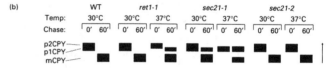

FIGURE 17-1

To quantify the effect of these mutations on secretion, researchers analyzed processing of carboxypeptidase Y (CPY) by a similar approach. CPY exists in two precursor forms termed p1CPY and p2CPY, which are present in the ER and Golgi complex, respectively; the mature form of this enzyme (mCPY) is present in the vacuole. The autoradiograms from this analysis are represented in Figure 17-1b.

a. Based on the data in Figure 17-1a, do all three mutations have a similar effect on processing of the diK-fusion protein?

b. Based on the data in Figure 17-1b, do all three mutations have a similar effect on secretion of CPY to the vacuole?

c. What evidence regarding multiple functional domains in the Ret1 protein and Sec21 protein do these results provide?

d. Ret1p is a subunit of the yeast COP coatomer. In *Saccharomyces cerevisiae*, β-COP, β'-COP, and γ-COP are encoded by the *sec26, sec27,* and *sec21* genes, respectively. Mutations in *sec26* and *sec27* also affect retrieval of the diK-fusion protein. What do these data suggest regarding the role of COP proteins in retrieval and secretion?

33. Donor and acceptor membranes can be used in a test tube assay to establish the biochemical requirements for protein transfer from one Golgi subcompartment to another. In a typical assay, the donor membranes are isolated from cells with a mutation in a specific glycosyltransferase, commonly *N*-acetylglucosamine transferase-1, which is located in the *medial* Golgi. Mutant cells that are infected with vesicular stomatitis virus (VSV) are unable to completely process the virus-encoded G protein, a glycoprotein, en route to the cell surface. Incompletely processed sugar side chains carried by solubilized G protein are sensitive to digestion by endoglycosidase H. Treatment of the sensitive form of the glycoprotein (G_S) with this enzyme yields a product that migrates more rapidly in a polyacrylamide gel than does the resistant form (G_R), thus providing an assay for the two forms.

In one experiment, VSV G protein in mutant donor cells was prelabeled by a brief (pulse) exposure of the cells to [^{3}H]palmitate. VSV G protein is acylated with palmitate in the *cis*-Golgi complex. The cells were then either homogenized immediately or after various chase times in isotope-free media. The donor membranes from the homogenates then were incubated with acceptor membranes from wild-type cells; after 30 min, G protein was solubilized from the in vitro membrane mixtures and assayed for sensitivity to endoglycosidase H. Figure 17-2a illustrates autoradiograms from this assay for two in vivo chase times. Figure 17-2b shows the quantitative effect of chase time on the extent of processing of G protein in vitro.

a. In this experimental protocol, what results would indicate that protein transport between *cis* and the *medial* Golgi has occurred?

b. What is the effect of chase time on the sensitivity of VSV G protein to endoglycosidase H? How does this effect relate to protein transport between Golgi subcompartments?

c. What evidence does this experiment provide regarding whether or not cargo protein transport such as that of VSV-G protein through the Golgi apparatus is unidirectional?

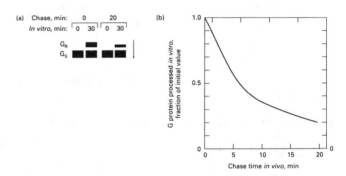

FIGURE 17-2

Answers

1. a, c, e

2. c, e

3. b, e

4. e

5. a, c, e

6. a, b, c, d

7. a, b

8. b

9. e

10. d

11. b

12. d, e

13. e

14. b

15. a, b, d

16. c, d

17. A chimeric protein between ts VSV G and GFP can be constructed by standard recombinant DNA techniques. Curiously, the GFP must be placed C-terminally on the construct, putting the GFP portion in the cytosol. Placement of the GFP on the N-terminus of ts VSV G results in a chimeric protein that never folds properly. The reasons for this are unknown. GFP provides a fluorescent signal that can be detected by fluorescence microscopy. ts VSV G is temperature sensitive in its folding. At elevated temperature, it is retained in the ER due to improper folding. At reduced, permissive temperature, it folds properly and exits the ER. Hence temperature fluctuations can be used to produce the export of a cohort of ts VSV G-GFP from the ER. This can be tracked in living cells by time-lapse fluorescence microscopy. Golgi can be labeled fluorescently in the same cells by using transfection with a second colored resident Golgi protein.

18. Hypothetically, transport between secretory organelles could be reconstituted in a test tube with purified, soluble protein factors. In reality this is not the case. We do not yet know the full range of proteins required. Moreover, even if we did, we do not yet have the ability to purify or produce all the needed soluble proteins. Cytosol provides a complex source of many needed, soluble proteins. Without this complex source, in vitro transport assays would not work.

19. Yeast class B *sec* mutants show as phenotype the accumulation of secretory proteins in the ER. Such a phenotype can be produced by numerous different gene mutations. For example, the COPII coat has four different proteins and Sar1 is the required GTPase for recruitment of COPII to ER membranes. COPII assembly mutations in any of these 5 genes would produce the same class B phenotype. Other such mutations common to ER exit also could be cited.

20. Three classes of coated vesicles have been identified. These are:

    COPII mediating anterograde transport from the ER and requiring Sar1 as the small GTPase involved in coat protein recruitment.

    COPI mediating retrograde transport between the *cis*-Golgi and ER and within the Golgi complex. The small GTPase required is ARF.

    Clathrin mediating anterograde transport from the *trans* Golgi network and endocytosis from the cell surface. The small GTPase is again ARF.

21. The proteins initially synthesized as a SNARE proteins fall into two classes, v-SNAREs (vesicle SNAREs) and t-SNAREs (target SNAREs). By selective v- and t-SNARE pairing, they are involved in determining the specificity of vesicle fusion with target. In a test tube system using liposomes, v- and t-SNAREs alone are sufficient to produce fusion, albeit slowly. Vesicle fusion with target membrane is much more rapid in living cells.

22. Protein disulfide isomerase (PDI) that escapes from the ER is normally brought back to the ER from the *cis*-Golgi reticulum (network) by the

KDEL-receptor. PDI contains a C-terminal KDEL sequence. PDI that is not retrieved will be secreted from the cell. This will lower the ER accumulation of PDI. KDEL-less PDI will not accumulate elsewhere along the secretory pathway as it lacks targeting information for accumulation in a downstream secretory organelle. Rather, it is lost to the extracellular medium. As PDI is lost slowly from the ER, the overall intracellular distribution of PDI will remain predominantly ER.

23. During the formation of COPII buds, specific v-SNARE molecules and rab1 are recruited. These are targeting molecules. The v-SNARE binds with a t-SNARE and the rab1 binds the effector protein, p115, that mediates vesicle docking. Thus budding programs the resulting vesicle for specific targeting,

24. Plasma membrane transporters are secreted membrane proteins that are synthesized on membrane bound ribosomes and fold in the ER. A mutation that results in improper folding will produce ER accumulation rather than normal transport of the protein to the plasma membrane. A drug approach that would stabilize normal protein folding would correct the defect. Cystic fibrosis, which is due to a defect in chloride transport, is an example of such ER accumulation.

25. Regulated secretion occurs only in response to a signal. The proteins to be secreted are stored in special secretory vesicles. Sorting into the regulated secretory pathway likely is controlled by selective protein aggregation. Constitutive secretion appears to occur without any known selectivety properties. In essence, it may be a default pathway.

26. I-cell disease fibroblasts have M6P receptors. They efficiently internalize lysosomal enzymes containing M6P. I-cell patients lack the GlcNAc phosphotransferase that is required for formation of M6P residues on lysosomal enzymes in the *cis*-Golgi. Hence, these proteins can not interact with M6P receptors and the proteins are hence secreted.

27. Membrane proteins are sorted to either the apical or basolateral domains by several different mechanisms. One mechanism for targeting proteins to the appropriate domain of the plasma membrane involves sorting in the *trans* Golgi network. Except for the GPI anchor, which acts as an apical or basolateral targeting signal, no unique sequences have been identified that target proteins to the apical or basolateral domain. Another mechanism operates in hepatocytes where all newly made apical and basolateral proteins are first delivered together from the *trans* Golgi network to the basolateral membrane. From there, both apical and basolateral proteins are endocytosed in the same vesicles. Within endosomes, they are sorted and transported to the appropriate domain. The attachment of integral membrane proteins to the cytoskeleton serves as a retention signal and may assist in the apical-basolateral sorting of some proteins. Hence, depending on cell type, at least three different mechanisms for apical-basolateral sorting are possible.

28. Three different signals that trigger uptake of a protein by receptor-mediated endocytosis have been at least partially characterized. These are the sequence Asn-Pro-X-Tyr (NPXY), Tyr-X-X-ϕ (YXXϕ), and Leu-Leu (LL), an ubiquination signal on some proteins. In all cases, the molecular feature is present in the cytosolic domain of the protein. X = any amino acid, ϕ = bulky hydrophobic residue.

29. At the low pH of the late endosomes, the LDL receptor dissociates from its ligand, low-density lipoprotein (LDL), and is recycled to the cell surface. In this acidic compartment, bound iron is released from ferrotransferrin, but the iron-free transferrin, called apoferritin, remains associated with the transferrin receptor until the complex is cycled back to the cell surface. Apoferritin is released from the receptor at the neutral pH found there.

30. The *shibire* mutation is a mutation in the gene encoding dynamin in *Drosophila*. At the presynaptic membrane of a nerve-muscle synapse, there must be retrieval of exocytosed synaptic vesicle components. This occurs through

endocytosis. Endocytosis is a dynamin dependent process. Without functional dynamin, the vesicle does not pinch off.

31. Neurobiology tends to be treated as a specialized, advanced subject. Membrane trafficking at the synaptic ending might well be considered in this context. However, as emphasized in Chapter 17, there are general principles underlying protein trafficking. In fact, some of these were discovered first or concurrently at the nerve ending.

32a. All three mutations result in processing (i.e., proteolytic cleavage) of the diK-fusion protein, as evidenced by the presence of bands representing lower-molecular-weight cleavage products. There are some variations in the extent of processing among the mutants, but these are relatively minor. These results suggest that all three mutations prevent retrieval of the diK-fusion protein.

32b. The three mutations have divergent effects on secretion of CPY. The *ret1-1* and *sec21-2* mutations have little effect, as evidenced by the strong band corresponding to mCPY at both the permissive and nonpermissive temperatures. In contrast, the *sec21-1* mutation inhibits CPY secretion, as indicated by the weak mCPY band at the nonpermissive temperature. Of these three mutations, *ret1-1* and *sec21-2* were in fact selected on the basis of their effect on protein retrieval, whereas *sec21-1* was selected on the basis of its effect on protein secretion.

32c. The finding that two different alleles of *sec21* differ in phenotype suggests that the Sec21 protein must have at least two different functional domains. Since only one allele of *ret1* has been isolated so far, nothing can be concluded about the possibility of different domains in the Ret1 protein based on the available genetic evidence.

32d. These mutations suggest that COP proteins and the coatomer play a role in both protein retrieval to the ER and protein secretion from the ER. At the time these experiments were done, whether the role in both is direct or indirect was open to question. Today, the generally accepted answer is that COPI mediates directly retrograde transport

which is then required indirectly to support anterograde transport. Components must be recycled for anterograde transport to occur.

33a. Since only VSV G protein labeled with [^{3}H]palmitate is detected in this experiment, all the G protein revealed on the electrophoretograms initially had resided in donor *cis*-Golgi membranes. This G protein is susceptible to endoglycosidase H. The acquisition of resistance to this enzyme during the in vitro incubation is indicated by the appearance of the resistant G_R band. Since this resistance depends on processing in the acceptor *medial*-Golgi membranes, appearance of the G_R band indicates that the labeled G protein is part of a structure that had transformed from the *cis*- to *medial*-Golgi membrane properties.

33b. As the in vivo chase time increases, there is a profound decrease in the proportion of prelabeled G protein that is processed (i.e., acquires endoglycosidase H resistance). This finding suggests that as the chase time is increased, labeled, incompletely processed G protein is found in a subcompartment that is incompetent in the in vitro transport assay.

33c. The results suggest that once G protein has moved from one Golgi subcompartment to another in vivo, it cannot transfer back during the in vitro transport assay for the completion of a given enzymatic processing step on its sugar side chains. In other words, transport through the Golgi is unidirectional. Of course, this conclusion assumes that most of the labeled G protein is present within the Golgi and does not transfer out of the Golgi during the experiment. Experiments to verify this assumption (e.g., by immunolabeling) could be done.

18

Metabolism and Movement of Lipids

PART A: *Chapter Summary*

The cell faces special challenges in metabolizing and transporting lipids, which are poorly soluble in the aqueous interior of cells and in extracellular fluids. Cells use lipids for many purposes, such as: building membranes, storing energy, signaling within and between cells, and covalently modifying proteins. Fatty acids—which are oxidized in mitochondria to release energy for cellular functions—are stored and transported primarily in the form of triglycerides. Fatty acids are also precursors of phospholipids, the structural backbone of cellular membranes. Cholesterol, another important membrane component, is a precursor for steroid hormones and other biologically active lipids that function in cell-cell signaling. Also derived from biosynthesis of precursors of cholesterol are the fat-soluble vitamins.

With the exception of a few specialized cells that store large quantities of lipids, the overwhelming majority of lipids within cells are components of cellular membranes. Discussion of lipid biosynthesis and movement is focused on the major lipids found in cellular membranes and their precursors. In lipid biosynthesis, water-soluble precursors are assembled into membrane-associated intermediates that are then converted into membrane lipid products. Present understanding of intracellular lipid transport is rudimentary. Analysis of of the transport of lipids into, out of, and between cells is far more advanced. Lipid movements by various cell-surface transport proteins and receptors are described.

Finally, the connection between cellular cholesterol metabolism and atherosclerosis, which can lead to cardiovascular disease (i.e., heart attack, stroke), the number one cause of death in Western industrialized societies is examined. Current theories about why large arteries can become clogged with cholesterol-containing deposits and how cells recognize the differences between "good" and "bad" cholesterol are presented. Detailed knowledge of the fundamental cell biology of lipid metabolism has led to the discovery of remarkably effective anti-atherosclerotic drugs.

PART B: *Multiple Choice*

Circle the letter corresponding to the most appropriate terms/phrases. There may be more than one correct answer. Circle all that apply.

1. Precursors of membrane phospholipids are

   a. soluble in the cytosol

   b. incorporated into biosynthetic intermediates

   c. sometimes complexed with fatty acid binding proteins

   d. stored in glycerolosomes

   e. sometimes activated by conjugation with CDP

2. Membrane phospholipids include

   a. cholesterol

   b. plasmologens

   c. sphingolipids

   d. triglycerides

   e. bile acids

3. Membrane lipids are distributed appropriately between the two leaflets of a given membrane by

   a. dockases

   b. trumpases

   c. fatty acid desaturases

   d. flippases

   e. transmembrane tunnelases

4. Cholesterol is

   a. unrelated to any plant lipids

   b. not found in chloroplast membranes

   c. encoded by an intron containing gene on chromosome 23

   d. relatively abundant in the plasma membrane and endocytic membranes

   e. synthesized in the ER

5. Proteins known to be involved in Golgi apparatus independent transport of lipids include:

   a. StAR protein

   b. Neimann-Pick C1 (NPC1) protein

   c. HMG CoA reductase

   d. Acyl:cholesterol acyl transferase (ACAT)

   e. cholesterol 7α-hydroxylase

6. Which of the following is **not** a metabolic derivative of cholesterol?

   a. Dolichol

   b. Vitamin D

   c. Bile acids

   d. v-SNARE molecules

   e. NSF

7. Why was it unexpected that transporter proteins are required for transmembrane fatty acid transport?

   a. Because these molecules are too small for their diffusion to be impeded by membranes.

   b. Because it is the cytosolic fatty acid binding protein-fatty acid complex that is the actual transported complex.

   c. Because these molecules are sufficiently hydrophobic to diffuse across membranes.

   d. Because triglycerides carried by lipoproteins were thought to be the only source of fatty acids in the circulation.

8. ABC proteins

   a. hydrolyze ATP

   b. interact with nuclear receptors such as LXR to stimulate the expression of cholesterol 7α-hydrolase

   c. move phospholipids, cholesterol, and bile acids across the apical surface of liver cells into small ductiles

   d. play an important role in the enterohepatic circulation of biliary lipids

   e. are homologous with SREBP-2

9. Familial hypercholesterolemia (FH) is

   a. an inherited disease in humans

   b. an early on-set disease that typically severely affects infants in the first year or two of their lives

   c. due to mutation in an essential gene

   d. clinically significant in the heterozygous state

   e. a major example that circulatory LDL levels are significant in atherosclerosis

10. In an experiment with normal fibroblasts and ^{125}I-tyrosine labeled LDL, the ratio of cell surface bound LDL to internalized LDL to degraded LDL was 0.2:1:5. These results may be explained on the basis that:

    a. Both internalized LDL and its receptor are degraded rapidly.

    b. LDL receptors mediate repeated rounds of LDL internalization.

    c. Internalized LDL receptors exceed cell surface LDL receptors by a factor of 25-fold.

    d. Internalized LDL receptors exceed cell surface LDL receptors by a factor of 125-fold.

    e. LDL and LDL receptors dissociate from each other upon internalization, the receptor is recycled, and LDL is transported to lysosomes.

11. Antibodies directed against apolipoprotein A should react only with

    a. cholesterol esters

    b. triglycerides

    c. VLDL and LDL

    d. chylomicrons and HDL

12. Selective uptake of cholesterol esters from HDL is mediated by

    a. the receptor SR-BI that also binds LDL and VLDL

    b. a cell surface protease that is selective for the apolipoprotein portion of LDL

    c. the *trans* Golgi network (TGN)

    d. as yet unknown mechanisms for the transfer of cholesterol into the cytosol

    e. nerve endings

13. Negative feedback regulation by cholesterol affects

    a. the activity of HMG-CoA reductase

    b. sodium gradients that drive bile acid import by symporters

    c. stigmasterol production in green plants

    d. the association of Insig-1 with SCAP

    e. the expression of HMG-CoA reductase

14. Three different isoforms of SREBP are known in humans. These isoforms

    a. are redundant with each other and essentially all regulate the same process

    b. permit selective regulation of cholesterol versus triglyceride and phospholipid production

    c. are all important in RIP mediated processes

    d. are all small GTPases

    e. are all ATPases

15. Atherosclerosis

    a. accounts for 75% of deaths due to cardiovascular disease in the United States

    b. is characterized by the progressive deposition of lipids, cells, and extracellular matrix material in the inner layer of the wall of an artery

    c. usually takes decades to develop into overt disease

    d. is protected against by LDL

    e. is dependent upon LDLR uptake of LDL

16. Statins and bile acid sequestrants all act ultimately to reduce plasma LDL-cholesterol by

    a. recruitment of synaptic vesicles to the presynaptic membrane

    b. increasing expression of LDL receptors in the liver

    c. blocking absorption of dietary cholesterol from the intestinal lumen

    d. interaction with the Ca^{++} sensor cholotagmin

    e. promoting the neural suppression of dietary hunger signals

PART C: *Reviewing Concepts*

18.1 *Phospholipids and Sphingolipids: Synthesis and Intracellular Movement*

17. In addition to diacyl glycerophospholipids, animal cells and some anaerobic bacteria contain a different type of glycerol-derived phospholipids. What is this and how does it differ chemically from diacyl glycerophospholipids?

18. How are newly synthesized phospholipids transferred from the cyotosolic leaflet to the exoplasmic leaflet of the ER membrane?

19. Fluorescence quenching can be used as an assay for lipid transfer from one leaflet to the other. Describe how such an assay works.

18.2 *Cholesterol: A Multifunctional Membrane Lipid*

20. Why is cholesterol termed a multifunctional membrane lipid?

21. Cholesterol is often considered from the aspect of disease and, in particular, cardiovascular disease. What evidence points to cholesterol and/or sterols being important normal plasma membrane lipids in both animals and plants?

18.3 *Lipid Movement into and out of Cells*

22. The familial hypercholesterolemia (FH) phenotype can be produced by a number of different mutations in LDL receptor. One such mutation affects receptor numbers at the cell surface. Give examples of different mutations in the gene encoding the LDL receptor that could result in the same phenotype by different means.

23. Seventy per cent of LDL receptors are found in the liver. Yet the classic experiments that established the receptor mediated LDL internalization were all done with fibroblasts from the skin. The same is true for clinical diagnostic studies to assay for LDL receptor levels. Why do such work with skin fibroblasts rather than with liver cells?

24. To what extent are lipoproteins made in the ER interconvertible in the circulation?

25. Cholesterol esters are transferred from HDL to cells without internalization of the HDL. Design a

experiment that provides data which would support this contention.

18.4 *Feedback Regulation of Cellular Lipid Metabolism*

26. What might be the logic to sensing cholesterol levels in the ER membrane via transmembrane domains rather than using a soluble cholesterol receptor or the LDL receptor itself?

27. Why should cholesterol levels be subject to such a multitude of regulatory mechanisms? The normal current concern in Western industrialized societies over cholesterol and cardiovascular disease is generally considered to be one that is not a source of evolutionary selective pressure.

18.5 *The Cell Biology of Atherosclerosis, Heart Attacks, and Strokes*

28. What support does evidence from familial hypercholesterolemia provide that the normal cell surface LDL receptor plays little, if any, direct role in the excessive accumulation of lipids in foam cells?

29. How does reverse cholesterol transport by HDL (good cholesterol) protect against atherosclerosis?

30. Why are balloon stretching, removal of plaques, insertions of metal scaffolds (stents), and grafting of a bypass-vessel only treatments of the symptoms of atherosclerosis rather than the root causes of the disease?

31. Seat belt usage is a legal requirement in most Western industrialized countries. A proposal has been floated in British medical circles to require everyone over 55 years of age to take a similar potentially life-saving measure: a once a day combination pill to lower cholesterol levels and stimulate blood flow. Obviously, this is a public policy question. However, the scientist does have a responsibility to advise on efficacy and risk. Can such a pill be effective and low risk?

PART D: *Analyzing Experiments*

32. Cells mutant in a protease (Pr-, S2P- cell line) that cleaves SREBP to release nSREBP have been isolated. You have used these cells to transfect in two different molecular forms of the protease, S2P-KDEL and S2P-KDAS. You characterize two different properties of these cells. Figure 18-1 shows the distribution by immunofluorescence staining of Insig in these cells cultured in the absence or presence of added sterol (Figure 18-1). The distribution is the same whether wild type cells, mutant cells or transfected cells are examined.

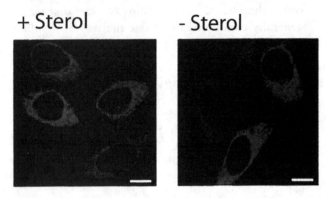

+ Sterol   **- Sterol**

Figure 18-1

a. How can these results be interpreted relative to the possible relocation of Insig in response to sterol levels?

b. How would you expect the distribution of SREBP and SCAP to depend on sterol levels?

   In a second experiment, you test the effect on the SBREP and nSBREP levels of transfecting Pr- cells with either a chimeric S2P protease which have either a KDEL sequence or a KDAS sequence at its C terminus. You incubate cells in the absence or presence of sterols and prepare microsomal and nuclear fractions. These are analyzed for the presence of SREBP forms. The results of probing with anti-SREBP antibodies SDS-PAGE, gels from such an experiment, are shown in Figure 18-2

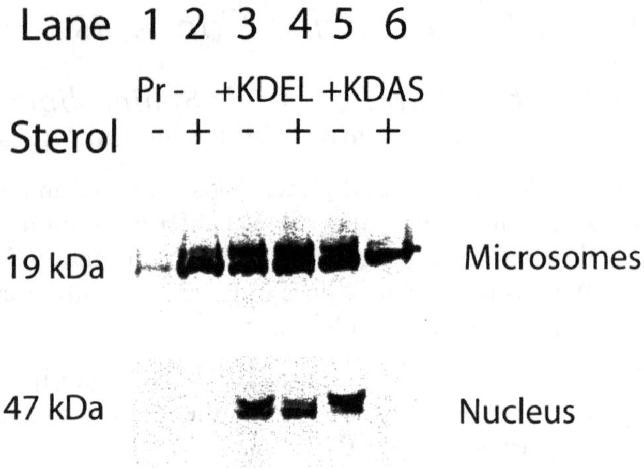

Lane 1 2 3 4 5 6

Pr -   +KDEL +KDAS
Sterol - + - + - +

119 kDa ————————— Microsomes

47 kDa ————— Nucleus

Figure 18-2

c. How can the absence of SREBP in the Lane 1 results be explained?

d. What is the effect of expressing S2P-KDEL on the production of nSREBP? How can this effect be explained?

e. What is the effect of expressing S2P-KDAS on the production of nSREBP? How can this contrasting effect relative to S2P-KDAS be explained?

Answers

1. a, b, c, e

2. b, c

3. d

4. b, d, e

5. a, b

6. a, d, e

7. c

8. a, c, d

9. a, d, e

10. b, e

11. d

12. a, d

13. a, d, e

14. b, c

15. a, b, c

16. b

17. In diacyl glycerophospholipids, the fatty acids are esterified to position 1 and 2 of glycerol. Plasmalogen is the different chemical structure. In plasmalogens, an ether linkage attaches the hydrocarbon chain on carbon 1 of glycerol. Plasmalogens are a reservoir of arachidonate, precursor to the signaling molecules called eicdosanoids. Platelet activating factor is also derived from plasmalogens.

18. The transport of lipids from the cytosolic to the exoplasmic leaflet of the ER is done by a flippase. Flippases are ATPases and members of the ABC super family of small molecule pumps.

19. One example of a fluorescence-quenching assay is cited in Chapter 18. This is the use of dithionite to quench fluorescence from a fluorescent lipid as it is flipped from one membrane leaflet to the other. The quenching reagent is added externally to the vesicles and gains access to the fluorophore only after it is flipped. An alternate approach is fluorescence dequenching, in which fluors are crowded together and quench each other. Transport from one leaflet to the other here results in a decrease in fluorescent lipid crowding and an increase in fluorescence, i.e., dequenching.

20. Cholesterol is termed a multifunction membrane lipid because it has many other functions than being a structural component of the plasma membrane of animals. For example, it is the precursor for the synthesis of vitamin D, bile acids, steroid hormones, cholesterol esters, and modified proteins such as Hedgehog. Cholesterol is secreted as part of bile and incorporated into bile micelles. Cholesterol esters are a normal component of the four major classes of human plasma lipoproteins.

21. Cholesterol is the single most abundant lipid in the mammalian plasma membrane. In fact, it is almost equimolar with all other phospholipids. Between 50 to 90 percent of the cholesterol in most mammalian cells is present in the plasma membrane and related endocytic membranes. The principal yeast sterol (ergosterol) and plant phytosterols (e.g., stigmasterol) differ slightly from that of cholesterol. The sheer abundance of cholesterol in certain membranes is a major indicator that cholesterol is an important normal lipid.

22. The familial hypercholesterolemia (FH) phenotype is the consequence of the lack of functional cell surface LDL receptor proteins. One type of mutation that gives rise to such a phenotype is a nonsense mutation in the LDL receptor gene. A nonsense mutation results in the failure to produce the full-length protein. A truncated protein may fail to fold properly in the endoplasmic reticulum (ER), be secreted from the cell because of the lack of a transmembrane domain, or lack a cytoplasmic domain for interaction with clathrin adapter proteins. Individual amino acid substitutions may affect cholesterol binding or LDL receptor interaction with clathrin adapters. The protein may turnover quickly due to an individual amino acid substitution. Promoter or enhancer mutations may affect the rate of gene transcription. The LDL receptor gene may be deleted. Although not an exhaustive list, it is clear

that the cells may be depleted of functional cell surface LDL receptor by a number of different means.

23. The liver is indeed the major site of LDL receptors in mammals. However, skin fibroblasts do express LDL receptor and are a much more accessible source of cells from patients than are liver cells. Liver biopsies are much more intrusive. One of the principles of medicine is do no harm.

24. To a limited extent, lipoproteins made in the endoplasmic reticulum (ER) can be converted from one to another in the circulation. Lipoproteins are remodeled in the circulation. VLDL secreted from hepatocytes can be converted into LDL. Small pre-HDL particles are generated extracellularly from apoA lipoproteins secreted mainly by liver and intestinal cells, and from small amounts of cholesterol and phospholipids. These can be converted into larger, spherical HDL particles. Chylomicrons are converted into chylomicron remnants. However, one protein species cannot be converted into another in the circulation.

25. At least two different types of experimental approaches can be suggested to test the hypothesis that cholesterol esters from HDL are taken up without internalization of the HDL. One is radiolabeling of components (as indicated in Chapter 18) with assay of degradation of the radiolabeled protein. A second approach is to fluorescently label the protein or to use antibodies to the protein to examine by microscopy whether the apoA protein portion of the HDL has been internalized.

26. Cholesterol, although it has other functions, is a significant membrane lipid. Sensing cholesterol levels with a transmembrane domain in the ER has the advantage that what is being sensed is membrane cholesterol. This is also testing cholesterol levels at an upstream membrane site that is actually the membrane system in which cholesterol is biosynthesized. To sense, through the LDL receptor at the cell surface is perhaps the wrong site because cell surface receptors support a flux of cholesterol and in a membrane where cholesterol is already normally very abundant. A

soluble receptor as sensor simply would not be sensing membrane levels of cholesterol.

27. Cholesterol amounts are indeed highly regulated suggesting that its production and levels are of evolutionary consequence, even if atherosclerosis is not. The normal functions of cholesterol are important. Overproduction of cholesterol can lead to the formation of harmful cholesterol crystals in cells. Cholesterol derived hormone and bile acid levels are important.

28. Familial hypercholesterolemia (FH) patients provide significant evidence that the normal LDL receptor is not involved directly in the accumulation of cholesterol in foam cells. Homozygous FH patients are in essence devoid of functional LDL receptor. Yet they develop foam cells comparatively early in life. The macrophages that become foam cells must use a different receptor for LDL internalization. In fact, it is thought of as a scavenger receptor.

29. Reverse cholesterol transport by HDL (good cholesterol) does indeed protect against atherosclerosis. HDL can remove cholesterol from cells in extrahepatic tissues, including artery walls, and eventually deliver the cholesterol to the liver either directly by selective lipid uptake by the SR-BI receptor or indirectly by transferring its cholesterol to other lipoproteins that will then be internalized by hepatic endocytic receptors. The excess cholesterol can then be secreted into the bile and eventually excreted from the body.

30. Stents and various surgical treatments alleviate the immediate, acute symptoms of atherosclerosis but do not eliminate the causes of atherosclerosis. The causes of atherosclerosis are a delicate balance between a physiological purpose to control infection during an inflammation process and the resulting tissue damage and LDL levels. Drug treatment to control cholesterol levels is an effort to address cause.

31. Requiring a daily cardiovasculature preventative pill is really a question of public policy. From the standpoint of the science, combined statin and bile sequestrant drug strategies are effective in

reducing cholesterol levels. Aspirin taken daily has some cardiovasculature benefit with little risk. Whether there is true risk to a long-term combined medication strategy is an open question. Moreover, what aspects of human behavior that should be regulated by law is an even more open question.

32a. As shown by immunofluorescence in Figure 18-1, the Insig-1 distribution is the same under all conditions in the presence or the absence of added sterols. In all conditions, the distribution is a webby pattern that is characteristic of the endoplasmic reticulum (ER). This indicates that Insig-1 does not relocate towards the ER when sterol levels are low (- Sterol). It may well be in a complex with other components in the ER, but if so that complex does not relocate as an entity with variation in sterol level.

32b. SREBP and SCAP are complexed and are thought to relocate from the ER to the Golgi apparatus at low sterol levels. Therefore, if an antibody against either were used in immunofluorescent studies to assay for the distribution of SREBP or SCAP, the two should show variable location with cholesterol level. At high sterol levels, they should be ER located; at low sterol levels, they should be Golgi located. Of course, degradation in the Golgi might produce an outcome where it is easier to detect loss from ER than accumulation in the Golgi complex.

32c. At low sterol levels SREBP is normally degraded releasing nSREBP. The Pr- cells are deficient or mutant in S2P protease activity. Therefore as shown in Lane 1, there is no nSREBP produced in these cells. The surprising result is that there is almost no detectable SREBP in these cells under these conditions (compare Lanes 1 and 2). This suggests the existence of an unknown and presumably non-specific protease activity that degrades SREBP under low sterol conditions in the absence of S2P.

32d. With expression of S2P-KDEL, production of nSREBP is observed (compare Lanes 3 and 4) under either minus or plus sterol conditions. Presumably because of the KDEL sequence, an ER retrieval sequence, S2P-KDEL is now located in the ER where it can cleave SREBP to nSREBP under any conditions. An antibody to S2P could be used to test this possibility by immunofluorescence or cell fractionation.

32e With the expression of S2P-KDAS, nSREBP is produced in a sterol level responsive manner (compare Lanes 5 and 6). This is exactly the effect expected for the introduction of normal S2P protease into cells. It is normally located in the *cis* Golgi. SREBP normally relocates to cis Golgi only at low sterol levels. Hence, enzyme and substrate are brought together only at low sterol levels. Therefore, nSREBP is produced only at low sterol levels. KDAS is not a functional ER retrieval signal. Localizations could be probed with antibodies by either immunofluorescence, electron microscopy or cell fractionation.

19

Microfilaments and Intermediate Filaments

PART A: *Chapter Summary*

Actin filaments (also known as microfilaments or F-actin) are dynamic polymers assembled from globular actin subunits (G-actin). Actin filaments are involved in generation and alteration of cell shape and in cell locomotion. Muscle is the best-understood example of cell motility, and study of muscle cells has provided much of our knowledge of actin and actin-binding proteins such as myosin. However, non-muscle cells also contain significant amounts of actin and actin-binding proteins. Forces produced by actin polymerization as well as by myosin motors are important for many processes in nonmuscle cells, including cell locomotion and cytokinesis.

Intermediate filaments are less dynamic than actin filaments and are not involved in cell motility, but rather serve an important role in providing structural support to the plasma membrane and to the nucleus.

PART B: *Multiple Choice*

Circle the letter corresponding to the most appropriate terms/phrases. There may be more than one correct answer. Circle all that apply.

19.1 Actin Structures

1. Which of the following describe actin?

    a. Actin is a highly conserved protein.

    b. Multiple actin genes may be found in higher eukaryotes.

    c. Actin is the most abundant protein in a eukaryotic cell.

    d. Individual actin filaments can be observed by electron microscopy.

    e. One end of an actin filament is different from the other.

2. Actin-binding proteins that generate actin filament networks

    a. are long and flexible

    b. have only one actin-binding site

    c. bind only at the ends of actin filaments

    d. are short and inflexible

3. Examples of actin cross-linking proteins include

    a. fimbrin

    b. dystrophin

    c. α-actinin

    d. spectrin

    e. myosin S1

19.2 The Dynamics of Actin Assembly

4. Actin assembly in vitro

   a. consists of three phases

   b. requires addition of actin oligomers termed nuclei

   c. requires ATP hydrolysis

   d. is reversible

   e. requires actin-binding proteins

5. Known functions of actin-binding proteins include

   a. severing of long filaments to generate shorter filaments

   b. inhibition of filament polymerization

   c. bundling of filaments

   d. acceleration of filament polymerization rates

   e. capping of the (+) end

6. Which of the following is true of actin assembly?

   a. ADP-actin can assemble into filaments.

   b. Actin (+) ends grow faster than actin (-) ends.

   c. ATP-actin can assemble into filaments.

   d. Toxins that alter the assembly of actin filaments have been used as tools to study actin in living cells.

19.3 Myosin-Powered Cell Movements

7. All types of myosins

   a. bind and hydrolyze ATP

   b. bind actin filaments

   c. form dimers

   d. contain an associated myosin light chain

   e. move toward the (+) end of an actin filament

8. The contractile ring

   a. is important for cell locomotion

   b. is a permanent structure is all eukaryotic cells

   c. contains myosin II

   d. contains myosin I

   e. is found in adult skeletal muscle cells

9. The structural and functional unit of skeletal muscle is the

   a. myofiber

   b. A band

   c. myofibril

   d. contractile ring

   e. sarcomere

19.4 Cell Locomotion

10. Cell locomotion

    a. requires coordination of motions generated in different regions of the cell

    b. occurs only in cells with a defined polarity

    c. may be controlled by signaling pathways

    d. occurs is a defined sequence of steps

11. Membrane extension involves

    a. bending actin filaments

    b. myosin II

    c. the Arp2/3 nucleation of new filaments

    d. actin depolymerization

    e. actin polymerization

12. Rac, a GTPase superfamily protein,

    a. is related to Ras

    b. may be activated by growth factors

    c. stimulates lamellipodia formation

    d. stimulates filopodia formation

    e. stimulates stress fiber assembly

19.5 Intermediate Filaments

13. Intermediate filaments

    a. are typically 6 nm in diameter

    b. are essential for cell motility

    c. may be used to diagnose the tissue of origin for many cancers

    d. are composed of globular subunits

    e. are formed from tetrameric subunits

14. Keratins

    a. are found in the nucleus

    b. are typically expressed in neuronal cells

    c. contain an acidic and a basic subunit

    d. function in cell adhesion

    e. are the major proteins of hair and horns

15. In the cell, intermediate filament disassembly is triggered by

    a. subunit phosphorylation

    b. subunit dephosphorylation

    c. hydrolysis of ATP bound to the subunits

    d. filament severing proteins

    e. binding of Ca^{2+} to the subunits

PART C: *Reviewing Concepts*

19.1 *Actin Structures*

16. There are four states of actin. What are the four states and what is the significance of each?

17. What structural feature of actin filaments can be visualized by use of filaments "decorated" with myosin S1 fragments?

19.2 *The Dynamics of Actin Assembly*

18. Actin polymerization is accompanied by ATP hydrolysis, but it is thought that ATP hydrolysis is not required for polymerization. What is the experimental basis for this conclusion?

19. What is the effect of gelsolin on actin filaments? How does the cell control gelsolin activity and for what purpose?

19.3 *Myosin-Powered Cell Movements*

20. How is myosin thought to move along actin filaments?

21. Microinjection of an antibody to myosin light-chain kinase inhibits the contraction of vertebrate smooth muscle but not that of vertebrate skeletal muscle. Contraction of both types of muscle, however, is associated with a rise in cytosolic Ca^{2+}. Explain these findings.

19.4 *Cell Locomotion*

22. Cell locomotion depends on actin, but several processes (wound healing, chemotaxis) involve directed cell locomotion. Generally speaking, how is directional information imparted to the actin cytoskeleton?

23. How does actin polymerization lead to movement of a cell's leading edge?

19.5 *Intermediate Filaments*

24. How is the structure of intermediate filament subunits related to the function of the assembled filaments?

25. There are several different types of intermediate filament proteins. What do the different intermediate filament proteins have in common, and how do they differ?

PART D: *Analyzing Experiments*

26. The actin-binding protein cofilin is thought to be involved in the regulation of actin dynamics in cells. To investigate the action of cofilin, in vivo and in vitro experiments were conducted with wild-type cofilin and two cofilin mutants. In both mutants, a different cluster of charge amino acids was substituted with alanines. Mutant 1 had two substitutions and Mutant 2 had three substitutions. To test the effect of wild-type cofilin and the mutant forms of cofilin on actin filaments in vivo, the genes for cofilin and the two cofilin mutants were introduced independently into yeast cells. Yeast cells contain "actin patches" where F-actin structures associate with the cell cortex. The cells containing the introduced genes were then treated with a drug that sequestered G-actin and the effect on actin patches was determined (Figure 19-1a).

    a. What was the purpose of treating the cells with a G-actin sequestering drug?

    b. What effect did wild-type cofilin and the two mutant proteins have on actin patches? What can you infer about the action of cofilin?

    c. To test the effect of wild-type cofilin and the mutant forms of cofilin on actin filaments in vitro, the proteins were expressed, purified, and each was combined with actin filaments. The effect of each protein on the actin filaments was then monitored by light scattered (Figure 19-1b). What do you think the light scattering measures? How does the in vitro data correlate with the in vivo data?

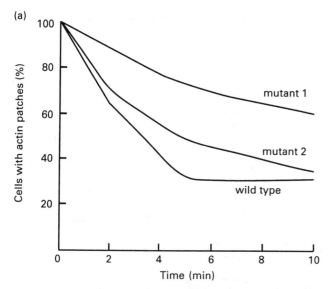

(a)

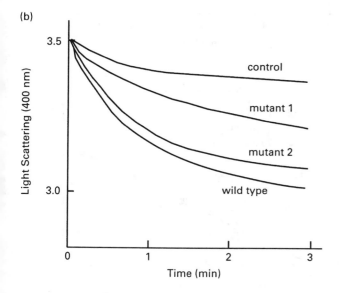

(b)

Figure 19-1

27. Although many of the effects of actin are mediated at the plasma membrane, most known actin-binding proteins are soluble. An exception is ponticulin, a 17-kD membrane glycoprotein from *Dictyostelium discoideum,* which has been shown to traverse the membrane of this cellular slime mold. Membrane preparations from *Dictyostelium* bind F-actin and nucleate actin polymerization at the membrane surface. This process appears to involve the generation of actin trimers at the cytoplasmic surface; these trimers have both the pointed and barbed ends free to elongate.

Ponticulin can be purified by extraction of ameboid cells of *Dictyostelium* with the nonionic detergent Triton X-114 to generate a purified cytoskeletal preparation, and this preparation was then extracted with another detergent, octylglucoside, which removed the ponticulin from the cytoskeletons. F-actin affinity chromatography, high-pressure liquid chromatography (HPLC), and hydrophobic interaction chromatography (HIC), all in the presence of octylglucoside, were used to purify the ponticulin further. SDS-PAGE analysis of ^{125}I-labeled preparations indicated that these preparations contained very minor amounts of contaminating proteins, and no proteins that comigrated with actin.

a. An important question was whether this purified, detergent-soluble preparation retained actin-binding activity. To answer this question, combinations of F-actin and ^{125}I-labeled ponticulin were incubated for 1 h at room temperature, and these reaction mixtures were then centrifuged to pellet the actin filaments. Pellets (P) and supernatants (S) were analyzed by SDS-PAGE and autoradiography. Results are shown in Figure 19-2a. Based on these results, can you conclude that this ponticulin preparation retains actin-binding activity?

b. Membrane proteins solubilized in octylglucoside can be incorporated into membrane vesicles after dilution of the detergent in the presence of excess lipid; this process is called reconstitution. Purified detergent-soluble ponticulin was reconstituted in membranes containing *Dictyostelium* lipids or a synthetic phosphatidylcholine (DMPC). The vesicles were then assayed for the ability to induce polymerization of G-actin. Controls included determination of G-actin polymerization in vesicles reconstituted in the absence of lipid and in lipid-containing vesicles of both types prepared without ponticulin. The results of these assays are shown in Figure 19-2b. What do these data allow you to conclude about the ability of various membrane lipids to support ponticulin-induced actin polymerization?

c. Another possible explanation for the lipid specificity observed in Figure 19-2b is that *Dictyostelium* lipids, but not DMPC or other lipids, promote G-actin polymerization via generation of ponticulin clusters, which provide multivalent actin-binding sites. To test this hypothesis, the nucleation activity of ponticulin reconstituted with *Dictyostelium* lipids at a wide range of lipid:protein ratios was assayed. Polymerization activity was measured as in part (b), yielding data similar to that shown in Figure 19-2b. From these data the polymerization rates at different lipid:protein ratios were calculated and then divided by the rate observed for actin alone, yielding fold increases as shown in Figure 19-2c. Final concentrations of *Dictyostelium* lipids were 1.4 μM, 14 μM, or 70 μM in the assay mixture. Do the data in Figure 19-2c support the hypothesis that membrane-bound ponticulin acts as an oligomer in nucleation of G-actin polymerization? Explain your answer.

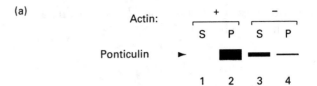

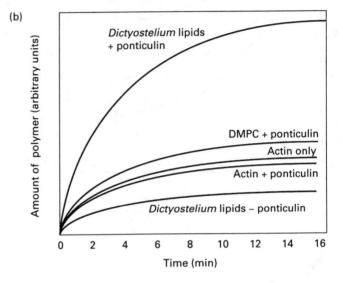

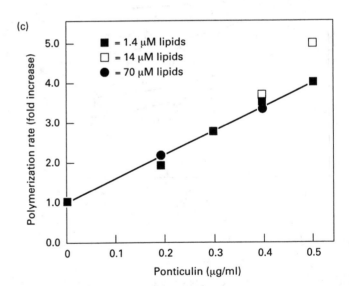

Figure 19-2

Answers

1. a, b, c, d, e

2. a

3. a, b, c, d

4. a, d

5. a, b, c, d, e

6. a, b, c, d

7. a, b

8. c

9. e

10. a, b, c, d

11. a, c, d, e

12. a, b, c, e

13. c, e

14. c, d, e

15. a

16. The four states of actin are ATP-F-Actin, ATP-G-Actin, ADP-F-Actin and ADP-G-Actin. The F-actin designation indicates that the actin subunit is part of a filament while the G-actin designation indicates that the subunit is not part of a filament (i.e. it is unassembled or free in solution/cytosol). The ATP and ADP designations indicate the nucleotide bound to the actin subunit.

17. The observation that all bound S1 fragments point in one direction [toward the (–) end] demonstrates the polarity of actin filaments and provides a method for distinguishing the two ends.

18. G-actin containing ADP or a nonhydrolyzable ATP analog is able to form filaments, so hydrolysis of ATP is not essential for polymerization of actin subunits.

19. Gelsolin severs actin filaments into shorter fragments and caps the (+) ends of the resulting fragments, thereby preventing addition of new subunits. Gelsolin activity is sensitive to Ca^{2+} and

cells can regulate gelsolin's function by regulating Ca^{2+} levels in localized regions of the cell's cortex. Gelsolin appears to play a role in cell locomotion and cytokinesis.

20. Myosin uses the energy of ATP hydrolysis to walk along an actin filament. In the case of myosin II, the motor moves in discrete steps of 5-10 nm, but other myosin motors may move in steps up to 35 nm. For most types of myosin, the step size depends on the length of the neck region – the longer the neck, the longer the step. Some evidence suggests that myosin takes one step per hydrolyzed ATP. The head domain is the region of the motor that interacts with actin and that binds and hydrolyzes ATP. The head domain is thought to go through a cycle of repeating conformational changes that depend on the nucleotide state of the myosin and myosin binding to actin. These conformational changes in the head are transmitted and amplified to the rest of the molecule through the neck as part of the power stroke that generates movement.

21. The mechanism by which a rise in Ca^{2+} triggers contractions differs in skeletal and smooth muscle. In skeletal muscle, binding of Ca^{2+} to troponin C leads to muscle contraction. In comparison, contraction of smooth muscle is triggered by activation of myosin light-chain kinase by Ca^{2+}-calmodulin. Thus only smooth muscle is inhibited by microinjection of antibodies to myosin light-chain kinase.

22. Directional information is delivered to the actin cytoskeleton through signal transduction pathways that begin with a cell surface receptor and may include G proteins and/or Ca^{2+} that ultimately affect the activity of actin-binding proteins and hence actin filament assembly, arrangement and motility.

23. Actin polymerization is thought to provide the force for plasma membrane extension at the leading edge of a moving cell. Pushing of the membrane forward involves the addition of actin subunits to the ends of cross-linked filaments close to the membrane. Since the filament ends abut against the membrane, the most likely mechanism for subunit addition to these ends is described by

the elastic Brownian ratchet model, which postulates that the thermal energy-induced bending of actin filaments close to the membrane briefly creates space for subunit addition. The subsequent straightening of the filaments due to the energy stored in bending provides sufficient force to extend the membrane.

24. Intermediate filaments provide mechanical support to the plasma membrane and the nuclear membrane and are much more stable than actin filaments or microtubules. This stability is derived from the α helical rod structure of the subunits, which overlap along the filament long axis to generate rope-like filaments.

25. All intermediate filament proteins assemble to form 10 nm diameter filaments and all share a common domain structure: globular N-terminal and C-terminal domains connected by a central, α-helical core (or rod) domain. The core domain is conserved among all intermediate filament proteins and mediates dimer and ultimately tetramer formation. The N-terminal and C-terminal domains vary among the different types of intermediate filament proteins and therefore distinguish the different types. There is evidence that the N-terminal domain may play some role in filament assembly, and that both the N-terminal and C-terminal domains are important for filament organization and interaction with other cellular components.

26a. The G-actin sequestering drug was used to eliminate G-actin addition to the cell's actin filaments, and thus permitted examination of cofilin effects on actin disassembly.

26b. Cofilin and the mutant cofilin proteins (to a lesser extent) promoted the loss of actin patches. The results suggest that cofilin promotes the disassembly of actin filaments.

26c. The light is scattered by actin filaments and serves as a measure of the amount of actin filaments present. As the filaments disassemble as shown in Figure 19-1b, the amount of scattered light decreases. The in vitro data correlates well with the in vivo data: both suggest that cofilin promotes actin disassembly.

27a. In the presence of F-actin, the ^{125}I-labeled ponticulin was found in the pellet (lane 2), whereas very little was found in the supernatant (lane 1). In the absence of actin, most of the ponticulin was found in the supernatant (lane 3), but a small amount was found in the pellet (lane 4). This material probably is aggregated ponticulin; such behavior is not unusual for intrinsic membrane proteins even in the presence of detergent. These results indicate that purified soluble ponticulin retained actin-binding activity.

27b. Ponticulin-mediated actin-nucleation activity is dependent on the composition of the lipid bilayer. The rate of actin polymerization in DMPC + ponticulin bilayers was barely elevated over that of actin alone, whereas vesicles containing both *Dictyostelium* lipids and ponticulin stimulated polymerization substantially.

27c. These data do not support the hypothesis. In this assay, the rate of G-actin polymerization would be expected to be linearly dependent upon protein concentration for a monomer, and to be proportional to the square of the protein concentration for a dimer in equilibrium with a monomer. Higher-order dependence would be observed for functional oligomers greater than $n = 2$. The observed linear dependence of the polymerization rate on protein concentration indicates that membrane-bound ponticulin acts as a monomer, not an oligomer, in actin nucleation. This conclusion is also consistent with the observation (data not shown) that ponticulin in *Dictyostelium* lipid vesicles is resistant to chemical cross-linking. Other possible explanations for the lipid specificity indicated in Figure 19-2c are that some particular lipid (or proteolipid) serves as a cofactor for nucleation or provides a specific environment necessary for functional association of ponticulin with actin. These hypotheses have not yet been experimentally tested.

20 Microtubules

PART A: *Chapter Summary*

Microtubules are filaments made of tubulin subunits. These dynamic polymers are important in intracellular organization, vesicle transport, organelle positioning, and in chromosome segregation during mitosis. Microtubules also can be found in cellular appendages such as cilia and flagella, where the bending action of microtubule bundles produces the force to power cell swimming. The various functions of microtubules are orchestrated by many different kinds of microtubule-associated proteins and microtubule-dependent motor proteins.

PART B: *Multiple Choice*

Circle the letter corresponding to the most appropriate terms/phrases. There may be more than one correct answer. Circle all that apply.

20.1 *Microtubule Organization and Dynamics*

1. α- and β-tubulin

   a. are highly conserved

   b. bind GTP

   c. hydrolyze GTP to GDP

   d. are found in all eukaryotes

   e. are related to the bacterial protein FtsZ

2. Microtubule assembly in vitro requires

   a. ATP

   b. a threshold (critical) concentration of tubulin

   c. warming the solution to 37°C

   d. microtubule-associated proteins

   e. GTP

3. Which of the following is true of microtubule assembly?

   a. GDP-tubulin normally assembles into microtubules.

   b. Microtubule (-) ends grow faster than microtubule (+) ends.

   c. Tubulin subunits assemble to form a protofilament.

   d. Microtubules consist of multiple protofilaments.

   e. Microtubules with a GTP cap will shorten.

4. Microtubule-associated proteins

   a. typically co-localize with microtubules inside cells

   b. are divided into two types: those that stabilize microtubules and those that destabilize microtubules

   c. typically co-purify with microtubules

   d. are similar to α- and β-tubulin

5. The drug colchicine acts to

   a. promote microtubule assembly.

   b. inhibit microtubule assembly.

   c. sever microtubules.

   d. block cell division.

   e. cap microtubule ends.

20.2 Kinesin- and Dynein-Powered Movements

6. Kinesin I moves vesicles

    a. from the microtubule (+) end to the (-) end

    b. from the microtubule (-) end to the (+) end

    c. bi-directionally along the same microtubule

    d. along both microtubules and microfilaments

7. All members of the kinesin family of motor proteins

    a. are (+) end-directed motors.

    b. contain associated light chains.

    c. form dimers.

    d. are ATPases.

    e. function in vesicle transport.

8. In the axonemes, the arrangement of 9 outer doublet microtubules, as well as the protofilament number of each outer doublet, is specified by

    a. γ-tubulin

    b. the centrosome

    c. the basal body

    d. nexin

    e. pericentriolar material

9. Axonemal dynein

    a. may contain 2 or 3 head domains

    b. attaches to a B tubule through its base domain

    c. consists of heavy, intermediate and light chains

    d. is a (-) end-directed motor

    e. is an ATPase

10. The dynactin protein complex serves to

    a. link vesicles and chromosomes to cytosolic dynein light chains

    b. link axonemal dynein to B tubules

    c. enhance the processivity of cytosolic dynein

    d. mediate microtubule attachment to centrosomes

    e. attach kinesin to vesicles

20.3 Microtubule Dynamics and Motor Proteins in Mitosis

11. What is the correct order of the stages of mitosis?

    a. anaphase

    b. metaphase

    c. telophase

    d. prophase

    e. prometaphase

12. Which of the following occurs during prophase?

    a. Kinetochores capture microtubule (+) ends.

    b. Chromosomes move poleward.

    c. Chromosomes align on the spindle equator.

    d. Centrosomes separate, and the spindle begins to form.

    e. Chromosome condensation begins.

13. Which of the following contribute to pole separation during anaphase B?

    a. cytosolic dynein

    b. a (-) end-directed kinesin

    c. CENP-E

    d. microtubule depolymerization

    e. microtubule polymerization

14. The average lifetime of microtubules during mitosis is

    a. 5-10 seconds

    b. 60-90 seconds

    c. 10 minutes

    d. 60 minutes

15. MCAK, a kinetochore-associated kinesin,

    a. powers spindle pole separation

    b. prevents spindle pole separation

    c. stabilizes microtubule ends at the cell cortex

    d. promotes disassembly of kinetochore microtubule (+) ends

    e. nucleates microtubules at the spindle poles

PART C: *Reviewing Concepts*

20.1 *Microtubule Organization and Dynamics*

16. Compare α- and β-tubulin in terms of nucleotide binding and hydrolysis.

17. What are the main functions of MTOCs? What are the main components of MTOCs?

18. How have anti-cancer drugs that alter microtubule assembly been used to determine the role of microtubules in cytoplasmic organization?

20.2 *Kinesin- and Dynein-Powered Movements*

19. Various studies on axonal transport have demonstrated the following: (a) Kinesin is a (+) end–directed motor protein; (b) all the microtubules in an individual axon have the same polarity; (c) vesicles can move in both directions simultaneously in an individual axon. How can these seemingly contradictory observations be resolved?

20. Many members of the kinesin family of microtubule motor proteins have been identified. What structural feature makes each member distinct and what is the functional consequence of this difference?

21. Describe the experiment that demonstrated that the outer doublets in axonemes slide during motion of cilia and flagella.

20.3 *Microtubule Dynamics and Motor Proteins in Mitosis*

22. How do microtubule dynamics change as cells proceed from interphase into mitosis?

23. The mitotic spindle is a bipolar structure. How is this structure generated as a cell prepares to divide?

24. What is the difference between a kinetochore and a centromere?

25. During mitosis, some microtubules attain a stable length. Which microtubules exhibit this property? Does such stability mean that these microtubules no longer add or lose tubulin subunits? Explain your answer.

PART D: *Analyzing Experiments*

26. The microtubule motor kinesin exhibits processive movement, meaning that a single motor can move for many micrometers along a microtubule without falling off. This property may be important in vivo in cases where a small vesicle with only a single kinesin bound must travel long distances on the scale of the cell. One model to account for the processive nature of kinesin is termed the hand-over-hand model. A key feature of this model is that one head should always be attached to the microtubule, which would lessen the chance of the motor falling off the microtubule.

a. To test this model, a single headed kinesin was prepared and its motile properties examined. If the hand-over-hand model is correct, would you expect the single headed kinesin to be processive?

b. Figure 20-1 depicts the landing rate of microtubules on a coverslip coated with either two-headed or one-headed kinesin at different motor densities (landing is defined as interacting with and being moved by motors attached to the glass surface). What can you conclude from this data?

c. Another feature of the hand-over-hand model is that one head, upon binding to the microtubule, promotes the release of the other head from the microtubule. If this hypothesis were correct, what would you expect when a one-headed kinesin molecule bound to a microtubule?

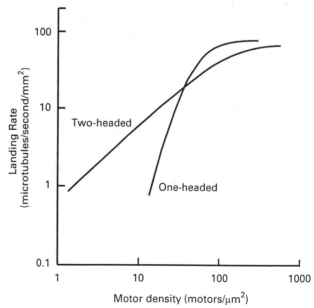

Figure 20-1

27. One approach to characterizing cellular microtubule assembly and disassembly involves injecting fluorescent-labeled tubulin into cells. This fluorescent tubulin is freely incorporated into microtubules in the injected cells, rendering the microtubules visible in a fluorescence microscope. Using a powerful laser, it is possible to irradiate a small region of the microtubules, bleaching the fluorescent dye so that it is no longer fluorescent, and then to measure the recovery of fluorescence in the bleached area.

a. In one experiment, data were collected for cells in interphase and in metaphase as shown in Figure 20-2a. What can you conclude about the relative stability of microtubules in metaphase and interphase cells from this data?

b. What mechanisms are likely to be responsible for the differences in fluorescence recovery shown in Figure 20-2a?

c. In another experiment, cells were microinjected with X-rhodamine tubulin, which labels all microtubules. Areas of the mitotic spindle were then photobleached to determine (1) the rate of movement of the photobleached area and (2) the degree of fluorescence recovery in the photobleached area. One of the first experiments this group performed was to develop a buffer that lysed cells and preserved only kinetochore fiber microtubules. Why was this a critical experiment? What does the effect of this buffer suggest about the differences in microtubules in vivo? What other methods might be useful for preserving kinetochore tubules at the expense of other microtubules?

d. When anaphase cells were labeled as described in (c) and then photobleached, the expected absence of fluorescence was noted. How could you demonstrate that the irradiation does not destroy the kinetochore microtubules at this point?

e. When the position of the bleached spot was monitored in anaphase cells, it had moved 1 μm/100 s toward the pole, but no fluorescence recovery of the photobleached area was noted. What is the interpretation of these findings?

f. Figure 20-2b shows fluorescence recovery after photobleaching results for anaphase and metaphase kinetochore tubules. What is the significance of these data?

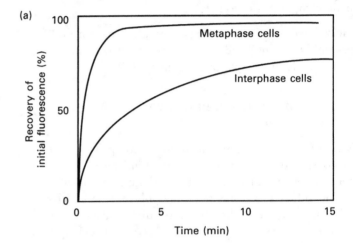

(a)

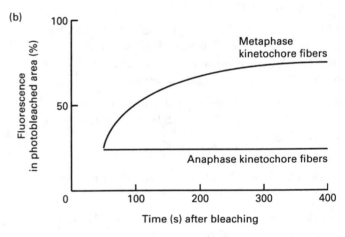

(b)

Figure 20-2

Answers

1. a, b, d, e

2. b, c, e

3. c, d

4. a, b, c

5. b, d, e

6. b

7. d

8. c

9. a, c, d, e

10. a, c, d, e

11. d, e, b, a, c

12. d, e

13. a, e

14. b

15. d

16. Both α- and β-tubulin bind GTP, and this form of the dimer can form microtubules. However, only β-tubulin is able to hydrolyze the bound GTP to GDP (which occurs after assembly into a microtubule) and exchange GDP for GTP (after the subunit disassembles from a microtubule). The GTP bound to α-tubulin is never hydrolyzed or exchanged with a free nucleotide.

17. MTOCs (or centrosomes) serve to nucleate and organize microtubules. Almost all microtubules originate from the MTOC, and the (-) ends of these microtubules remain associated with the MTOC. All MTOCs are composed of pericentriolar material (a lattice of microtubule-associated proteins), and many animal MTOCs contain centrioles. The pericentriolar material is the site of microtubule nucleation, which depends on γ-tubulin ring complexes. It is not clear what function the centrioles serve, but they are not templates for microtubule assembly.

18. Anti-cancer drugs that alter microtubule assembly, particularly those that cause microtubules to disassemble, have been used to determine the role of microtubules in moving and positioning vesicles, the endoplasmic reticulum, the Golgi complex, and mitochondria. Typically, cells were treated with microtubule-depolymerizing drugs and the effect on vesicle movement and/or organelle positioning was observed. Loss of vesicle movement or disruption of organelle structure and/or position following depolymerization of microtubules suggested that microtubules play a role in intracellular transport and cytoplasmic organization.

19. In addition to the (+) end–directed kinesin, individual axons must contain other motor proteins that are (–) end–directed, such as cytosolic dyneins; the latter are responsible for vesicle motion in the opposite direction.

20. Each member of the kinesin family has a unique tail domain. Since the tail domain is responsible for interacting with specific cargo, each member may transport a different cellular component along microtubules.

21. Demembranated axonemes were treated with proteases to digest the cross-linking proteins that held the outer doublet microtubules together. Next, ATP was added to the axonemes, and the axonemal dynein that was still present on the outer doublets converted the energy stored in ATP into sliding force, causing the outer doublets to slide apart. The conclusion of this experiment was that axonemal dynein powers the sliding of outer doublet microtubules relative to each other, but the cross-linking proteins convert this sliding force into bending in the intact axoneme.

22. As cells move from interphase to mitosis, the long microtubules of interphase are replaced by more numerous, shorter, and more dynamic astral and spindle microtubules. The increase in dynamics, apparent in the reduction of a microtubule's average lifetime of 10 minutes in interphase to 60-90 seconds in mitosis, is critical to assembly of the spindle and to its primary function, aligning and separating the cell's DNA.

23. The formation of a bipolar spindle requires centrosome duplication (which occurs prior to M phase) as well as motor proteins that separate the duplicated centrosome to form the two spindle poles. These motors (e.g., BimC) may be present on microtubules in the overlap zone between the poles and act to push the spindles apart, or they may be on the inner surface of the cell membrane and act to pull astral microtubules and hence the poles apart (e.g., cytosolic dynein). It may be true that both pushing and pulling forces are needed to separate the poles.

24. The centromere consists of a specific DNA sequence to which the proteins that form the kinetochore bind. Each sister chromatid has one centromere region and hence one kinetochore. The kinetochore consists of DNA-binding proteins as well as motor proteins and regulatory proteins that allow the kinetochore to capture the (+) ends of spindle microtubules, allowing the sister chromatid pair to attach to the spindle machinery.

25. Both kinetochore and polar microtubules may attain a relatively stable length, at least for short periods of time as mitosis proceeds. Although the length of these microtubules may be constant, tubulin subunit addition and loss still takes place during this time. This assembly/disassembly, particularly evident in kinetochore microtubules, produces a flux, or treadmilling, of tubulin subunits through the microtubules as subunits add at the (+) end and are lost at the (-) end.

26a. If the hand-over-hand model is correct, the one-headed kinesin should not be processive since this protein won't have another head to hold onto the microtubule when the one head releases.

26b. The data in Figure 20-1 demonstrate that two-headed kinesin can support microtubule attachment and movement at lower densities than one-headed kinesin. In addition, the proportional decrease in landing rate as a function of motor density suggests that each two-headed kinesin can support attachment and movement independent of other motors on the coverslip. In comparison, the abrupt decrease in landing rate as a function of motor density for one-headed kinesin suggests that multiple one-headed kinesins are needed to

interact with a microtubule and that once below that threshold density value, there will not be enough motors to ensure attachment and movement.

26c. The one-headed kinesin molecule should bind tightly to the microtubule since there is not another head to promote release.

27a. Recovery of fluorescence in a bleached area results from the addition of new, unbleached dimers. Thus the rate of recovery is proportional to the rate of microtubule depolymerization and repolymerization. The $t_{1/2}$ for fluorescence recovery is a few seconds in metaphase cells, whereas it is a few minutes in interphase cells. This difference indicates that microtubules in interphase cells are more stable than those in metaphase cells.

27b. Since most microtubules in a cell are anchored in the MTOC at their (−) ends, it is likely that changes in the (+) ends account for the observed difference in microtubule stability during metaphase and interphase. According to this mechanism, microtubules in metaphase cells must exhibit a greater rate of disassembly at the (+) end than do those in interphase cells. Another possibility is that microtubule-severing proteins are more prominent in interphase cells, since this would give a different pattern of fluorescence recovery than shrinkage and regrowth.

27c. Because X-rhodamine tubulin labels all microtubules in cells, some method was needed to eliminate all microtubules except the kinetochore fibers, which are the focus of the research. Kinetochore microtubules are more stable than other microtubules in the cell, as evidenced by their persistence in a buffer that causes disassembly of other microtubules. Cold depolymerization or colchicine/nocodazole treatment might induce disassembly of most microtubules but leave the kinetochore tubules intact.

27d. Anti-tubulin immunocytochemical analysis of the cell would reveal antibody labeling in the photobleached zone as well as across the rest of the spindle.

27e. Anaphase microtubules move toward the poles, but exhibit no turnover of the microtubules.

27f. These data indicate that metaphase kinetochore tubules turn over, whereas anaphase kinetochore tubules do not.

21 Regulating the Eukaryotic Cell Cycle

PART A: *Chapter Summary*

The cell division cycle occurs in all eukaryotic cells and is characterized by four phases. These are G_1, S phase—when DNA synthesis occurs—G_2, and M phase (subdivided into prophase, metaphase, anaphase, and telophase), during which the chromosomes are segregated to daughter cells. Transitions from one phase of the cell cycle to the next are catalyzed by heterodimeric protein kinases, consisting of a regulatory subunit called a cyclin and a catalytic subunit, a cyclin-dependent kinase (Cdk); dimers of this type are involved in *all* phases of the cell cycle. The activity of Cdks is influenced by the presence or absence of inhibitors, phosphorylation/dephosphorylation, and the polyubiquitin-targeted degradation of cyclin by proteasomes. Although many substrates for Cdks are unknown, specific cyclin-Cdk complexes are known to phosphorylate components of the DNA replication machinery to regulate S-phase and nuclear laminar and chromosomal proteins to regulate nuclear membrane breakdown and chromatid separation during mitosis. The logic underlying the cycle is that each regulatory event activates one phase of the cell cycle and prepares the cell for the next phase. Degradation of particular molecules ensures that the cycle is unidirectional and irreversible.

Dissection of cell-cycle regulatory mechanisms has involved the use of biochemical, genetic and molecular biological techniques. A variety of organisms have been the subject of investigation, and since many cell-cycle control proteins and regulatory strategies are highly conserved, discoveries made in one model organism have readily fueled research progress in others. Frogs (*Xenopus*) and sea urchins were used to biochemically characterize mitosis-promoting factor (MPF), the first cyclin-Cdk complex to be discovered. Mutations in the fission yeast *S. pombe* were used to study the regulation of the kinase portion of Cdks by phosphorylation events. The budding yeast, *S. cerevisiae*, was particularly useful for investigating the mechanisms that regulate entry into S phase and the accurate replication of chromosomes.

Through the study of analogous pathways in the aforementioned model systems as well as experiments with cultured cells, much has been learned about the regulation of the mammalian cell cycle. In particular, the signaling pathway that growth factors trigger to commit a cell irreversibly to cell-cycle progression by passing the restriction point is well understood. Checkpoints exist in the cell cycle at which signaling pathways trigger cell-cycle arrest in response to threats to genomic integrity including unreplicated DNA, damaged DNA, or improper chromosome alignment or segregation at mitosis. Mutations that bypass restriction point and checkpoint controls are common to most cancers.Meiosis is a special case in which the cell progresses through two M-phases without an intervening S-phase. Several meiosis-specific proteins that regulate this unique cell cycle have recently been identified.

PART B: Multiple Choice

Circle the letter corresponding to the most appropriate terms/phrases. There may be more than one correct answer. Circle all that apply.

21.1 Overview of the Cell Cycle and Its Control

1. Beginning early in G_1, list these events of the cell cycle in the proper order:

    a. degradation of S-phase cyclin-Cdk inhibitor

    b. degradation of securin

    c. degradation of mitotic cyclins

    d. activation of pre-replication complexes

    e. assembly of pre-replication complexes at origins

2. Which of the following are targets of the APC?

    a. securins

    b. S-phase cyclin-Cdk inhibitor

    c. G_1 cyclins

    d. mitotic cyclins

21.2 Biochemical Studies with Oocytes, Eggs and Early Embryos

3. Treatment of *Xenopus* egg extracts with RNase and non-degradable cyclin B results in:

    a. high MPF activity

    b. low MPF activity

    c. M-phase arrest

    d. interphase arrest

4. Which of the following is a ubiquitin ligase?

    a. cyclin B

    b. APC

    c. Cdc14

    d. Cdh1

21.3 Genetic Studies with S. pombe

5. Which of the following *S. pombe* mutants is smaller than normal?

    a. *cdc2⁻*

    b. *cdc2ᴰ*

    c. *wee1⁻*

    d. *cdc25⁻*

    e. *cdc13⁻*

6. Which of the following genes should complement a *cdc2⁻* mutation in *S. pombe*?

    a. *cdc13⁺* from *S. pombe*

    b. *cdc2⁺* from *Xenopus*

    c. *cdh1⁺* from *Xenopus*

    d. *clb5⁺* from *S. cerevisiae*

    e. *cdc28⁺* from *S. cerevisiae*

21.4 Molecular Mechanisms for Regulating Mitotic Events

7. The onset of anaphase depends upon the polyubiquitination and degradation of:

    a. cohesin

    b. condensin

    c. cyclin B

    d. securin

    e. separase

8. Phosphorylation of which of the following by MPF prevents premature cytokinesis?

    a. separase

    b. Ran-GEF

    c. lamin B

    d. lamin C

    e. myosin light chain

21.5 Genetic Studies with S. cerevisiae

9. The translation of which mRNA regulates the length of G_1?

   a. CLB5

   b. CLB6

   c. CLN1

   d. CLN2

   e. CLN3

10. In order for the prereplication complex to assemble at origins of replication, replication initiation factors must be:

    a. polyubiquitinated

    b. degraded

    c. phosphorylated

    d. dephosphorylated

21.6 Cell-Cycle Controls in Mammalian Cells

11. Which mammalian cyclin-CDK plays a role in S phase?

    a. cyclin A-Cdk1

    b. cyclin A-Cdk2

    c. cyclin B-Cdk1

    d. cyclin D-Cdk4

    e. cyclin E-Cdk2

13. Which mammalian cyclin-CDK is inhibited by INK4?

    a. cyclin A-Cdk1

    b. cyclin A-Cdk2

    c. cyclin B-Cdk1

    d. cyclin D-Cdk4

    e. cyclin E-Cdk2

21.7 Checkpoints in Cell-Cycle Regulation

14. Which of the following is a substrate of ATM or ATR?

    a. Cdc25A

    b. Cdc25C

    c. cyclin B

    d. Chk1

    e. Rb

21.8 Meiosis: A Special Type of Cell Division

15. Which of the following proteins is expressed specifically during meiosis?

    a. Clb1

    b. Ime2

    c. Rec8

    d. Scc1

    e. Sic1

PART C: Reviewing Concepts

21.1 Overview of the Cell Cycle and Its Control

16. The diploid number of chromosomes in human is 46. State the number of chromosomes in a somatic cell during G_1, S, G_2 and M. State the number of chromosomes in a mature sperm cell that has completed meiosis.

17. Describe the advantages that each of the following model systems has brought to the study of the cell cycle: a) Xenopus eggs and embryos, b) Schizosaccharomyces pombe, c) Saccharomyces cerevisiae, and d) mammalian cells in tissue culture.

21.2 Biochemical Studies with Oocytes, Eggs and Early Embryos

18. MPF was originally an abbreviation for "maturation-promoting factor" and subsequently came to represent "mitosis-promoting factor." Why are both terms appropriate? What is the molecular composition of MPF?

21.3 *Genetic Studies with S. pombe*

19. Describe how each of the following events affects the three-dimensional structure and catalytic activity of CDK2: a) binding of cyclin A, b) phosphorylation of threonine 161, and c) phosphorylation of threonine 14 and tyrosine 15.

21.4 *Molecular Mechanisms for Regulating Mitotic Events*

20. How does MPF promote the breakdown of the nuclear envelope at mitosis?

21.5 *Genetic Studies with S. cerevisiae*

21. What is the molecular distinction between SPF and MPF in *S. cerevisiae*?

21.6 *Cell-Cycle Controls in Mammalian Cells*

22. What is the evidence that expression of the delayed response genes depends upon the early response gene products?

21.7 *Checkpoints in Cell-Cycle Regulation*

23. Activation of checkpoints ultimately arrests the cell cycle by inhibiting the CDK that drives a particular cell-cycle transition. Name the relevant CDKs and how they are inhibited by the checkpoint pathway that is engaged in response to a) unreplicated DNA and b) damaged DNA.

24. Describe how the Mad2 protein mediates the spindle assembly checkpoint.

21.8 *Meiosis: A Special Type of Cell Division*

25. During sporulation of *S. cerevisiae*, what prevents DNA replication from occurring during meiosis II?

PART D: *Analyzing Experiments*

26. Figure 21-1 illustrates a series of computer simulations of the phenotype of mutants based on a mathematical model of the cell cycle in *S. pombe*. One mutant is *cdc25-*, one is *wee1-* and one is a *cdc25-/wee1-* double mutant. Which is which? Please explain your answer.

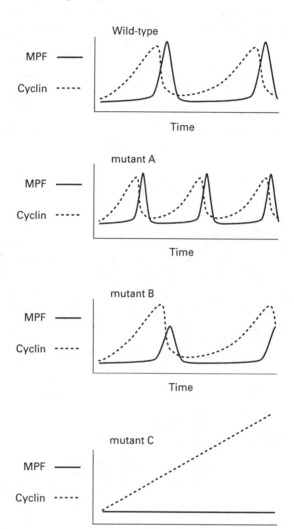

Figure 21-1

27. Our understanding of the pathways that transduce the information that DNA is damaged and result in cell-cycle arrest by inhibition of CDKs has increased dramatically during the past several years. However, many details about these signaling pathways remain to be discovered. A recent study by Cimprich and colleagues has revealed some critical information about the signaling pathway that leads to cell cycle arrest at G$_2$/M in response to UV-damaged DNA.

It was previously known that in *Xenopus* egg extracts supplemented with UV-damaged sperm nuclei, the kinase ATR becomes activated, and ATR in turn phosphorylates and activates the kinase

Chk1. The same signaling pathway was known to become activated when extracts were supplemented with sperm nuclei and aphidicolin, an inhibitor of DNA polymerase. In contrast, egg extracts supplemented with sperm nuclei damaged by ionizing radiation activate the kinase ATM, which in turn activates Chk2 (also known as Cds1 in *Xenopus*). In this study, the investigators wanted to investigate the link between UV-damaged DNA and unreplicated DNA in the ATR/Chk1 signaling pathway.

In the first experiment, Cimprich and colleagues treated sperm nuclei with UV irradiation. Control nuclei were mock-treated. These nuclei were added to egg extracts, and DNA replication was monitored by removal of extracts at the indicated times followed by a brief incubation with radioactively labeled dCTP. Incorporation of radiolabel into DNA was monitored by autoradiography of genomic DNA resolved by electrophoresis. In some samples, caffeine, an inhibitor of ATM and ATR, was added to the extract. The autoradiogram is shown in Figure 21-2.

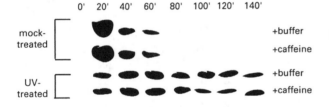

Figure 21-2

a. What does this experiment indicate about the effect of UV-irradiation on DNA replication?

b. What do the effects of caffeine on DNA replication in the extract containing UV-irradiated nuclei indicate about the role of ATM and ATR in regulating DNA replication in these samples?

In the next experiment, extracts containing mock-treated or UV-damaged nuclei were supplemented with geminin, a protein that blocks DNA replication by inhibiting the formation of pre-replication complexes on chromatin. Activation of Chk1 kinase was monitored by Western blotting (Figure 21-3A). Chk1 becomes active when phosphorylated by ATR, resulting in a form of Chk1 that migrates more

slowly when resolved by gel electrophoresis. Entry of these extracts into mitosis was also monitored microscopically by the breakdown of the nuclear envelopes (NEB) of sperm nuclei (Figure 21-3B).

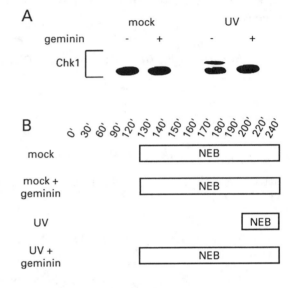

Figure 21-3

c. What is the effect of geminin and UV-damage on the activation of Chk1? On entry into mitosis?

d. What do these experiments reveal about the relationship between UV-damage and DNA replication in the checkpoint-signaling pathway?

Answers

1. e, a, d, b, c

2. a, d

3. a, c

4. b

5. b, c

6. b, e

7. d

8. e

9. e

10. c

11. b, e

12. d

13. d

14. d

15. b, c

16. G_1: 46; S: 46-92 depending on the extent to which DNA replication is complete; G_2: 92; M: 92 at the beginning, 46 in each daughter cell after anaphase and cytokinesis, 23 in a mature sperm cell.

17a. Large numbers of *Xenopus* eggs can be collected and fertilized *in vitro*. These eggs will divide synchronously for multiple cell cycles, providing an excellent model system for biochemical analysis of the cell cycle and for studying the cell cycle in the context of early development. Cytoplasmic cell-free extracts can also be generated from *Xenopus* eggs and these extracts will reproduce the biochemical events of the cell cycle, thereby providing for an easily manipulable system for cell-cycle studies.

17b. The fission yeast *Schizosaccharomyces pombe* is a simple, single-celled eukaryote amenable to genetic analysis and manipulation. The generation of temperature-sensitive mutants that arrest in the cell cycle (cdc mutants) permitted the identification of many key cell cycle genes. Because cell growth and the cell cycle are strictly coordinated in *S. pombe*, mutations that produce cells either too small or too large can be readily identified and used to clone genes for key cell cycle regulators such as *wee1*.

17c. Like *S. pombe*, *Saccharomyces cerevisiae* is a yeast amenable to genetic analysis and manipulation for which many cdc mutants have been identified. Like vertebrate cells, most of the cell cycle regulation in *S. cerevisiae* exists at the G_1/S transition, and thus, much of our understanding about G_1/S control derives from genetic studies with *S. cerevisiae*.

17d. Mammalian cells in tissue culture provide a model to study cell cycle dysregulation in human cells such as cancer cells. Cultured cells can be fused and can also be synchronized in the cell cycle by treatment with various drugs or by serum withdrawal.

18. MPF is a maturation-promoting factor because, when injected into an immature frog oocyte arrested in G_2 of meiosis I, MPF will induce maturation of the oocyte to an egg arrested in metaphase of meiosis II. MPF is also mitosis-promoting factor because, when injected into frog oocytes or other cells, MPF triggers entry from interphase into mitosis. In fact, oocyte maturation is really a cell-cycle phenomenon in which the egg is induced to pass the G_2/M transition. MPF consists of the cyclin-dependent kinase Cdc2 associated with its mitotic cyclin partner, cyclin B or cyclin A.

19a. Binding of cyclin A to CDK2 causes the T-loop in CDK2 to shift from a position that blocks access of substrates to the active site to a position that exposes the active site. Unphosphorylated CDK2 bound to cyclin A possess low kinase activity.

19b. Phosphorylation of threonine 161 in the T-loop causes a conformational change that increases the affinity of CDK2 for its substrates, resulting in a 100-fold increase in the catalytic activity of the cyclin A-CDK2 dimer.

19c. Threonine 14 and tyrosine 15 are located in the ATP-binding region of CDK2. Phosphorylation of these residues places negative charges that prevent binding of ATP by electrostatic repulsion and therefore, inhibits the catalytic activity of cyclin A-CDK2.

20. MPF phosphorylates nuclear lamins A, B and C. This causes the nuclear lamina that support and organize the nuclear membrane to disassemble. MPF also phosphorylates nucleoporins to induce the breakdown of nuclear pore complexes and phosphorylates other integral membrane proteins in the nuclear membrane, decreasing their affinity for chromatin and promoting disassembly of the nuclear membrane.

21. SPF (S-phase promoting factor) and MPF (M-phase promoting factor) are both cyclin-CDK complexes. *S. cerevisiae* possess a single CDK, Cdc28. Therefore, the distinction in substrate specificity and function between SPF and MPF resides in the cyclin subunits. SPF is Cdc28 associated with the G_1 CLN cyclins. MPF is Cdc28 associated with Clbs 1 and 2.

22. When G_0-arrested mammalian cells are treated with serum, early-response genes are transcribed followed by the transcription of delayed-response genes. However, when these cells are treated with serum and inhibitors of protein synthesis, only the early-response genes are transcribed. These data indicate that one or more mRNAs must be translated into protein before delayed-response genes can be transcribed. The most likely candidates for mRNAs that must be translated are the newly synthesized mRNAs encoded by the early-response genes.

23a. Unreplicated DNA triggers a signaling pathway that prevents progression into mitosis by inhibiting cyclin A/B-CDK1. Unreplicated DNA triggers the activation of the ATR kinase, which phosphorylates and activates the Chk1 kinase, which then phosphorylates and inhibits Cdc25C phosphatase. Cdc25C is thus unable to remove inhibitory phosphates from CDK1, and the cell cycle arrests at the G_2/M transition.

23b. DNA damage can arrest the cell cycle by inhibiting the relevant CDKs at several cell-cycle transitions. If DNA damage is detected during G_1, S or G_2,

ATM and/or ATR become phosphorylated and activated, and these kinases phosphorylate and activate p53. (This also occurs indirectly when ATM/ATR phosphorylates Mdm2, a protein binds and destabilizes p53. Phosphorylated Mdm2 cannot bind p53. ATM/ATR also activate Chk1/Chk2, which can directly phosphorylate and stabilize p53.) p53 accumulates and functions as a transcription factor to induce the expression of the CDK inhibitor p21. p21 binds to and inhibits all cyclin-CDK complexes and thus will arrest the cell cycle in response to DNA damage at G_1 by inhibiting cyclin D-CDK4/6, at S by inhibiting cyclin A/E-CDK2 or at G_2/M by inhibiting cyclin A/B-CDK1. During G_1 and S phase, Chk1 or a related kinase, Chk2, phosphorylates Cdc25A, which targets Cdc25A for rapid degradation. In the absence of Cdc25A, cyclin A/E-CDK2 remains phosphorylated at threonine 14 and tyrosine 15, and thus is catalytically inactive. During G_2/M, Chk1 phosphorylates Cdc25C. Cyclin A/B-CDK1 is thereby maintained in its phosphorylated, inactive state.

24. When Mad2 is associated with kinetochores that are not bound by microtubules, it inhibits Cdc20, the specificity factor that targets securin degradation by the APC. Thus, only when Mad2 is inactivated by microtubule attachment to kinetochores can the APC degrade securin, the "glue" that holds sister chromatids together. If the mitotic spindle is not properly formed, then Mad2 remains active, Cdc20 is inactive, securin is not degraded, and sister chromatid separation at anaphase is blocked.

25. During sporulation, G_1 cyclins Cln 1/2/3 that associate with Cdc28 to catalyze phosphorylation of Sic1 and allow entry into S-phase are not expressed. During meiosis I, the kinase Ime2 performs this function. However, Ime2 is not expressed during meiosis II, and thus DNA replication cannot be initiated.

26. Mutant A is *wee1⁻*, mutant B is *cdc25⁻/wee1⁻*, and mutant C is *cdc25⁻*. Wee1 is a tyrosine kinase that phosphorylates the Tyr-15 residue in the Cdc2 subunit of MPF, thereby inactivating it. Cdc25, a phosphatase, has the opposite effect, removing the inhibitory phosphate from Tyr-15. In the absence

of Cdc25, the inactive phosphorylated form accumulates; as a result, no cycling of MPF activity or cyclin B level, and hence no cell division, occurs. In the absence of Wee1, its inhibition of MPF is relieved, so that cycling and cell division are accelerated. Because of the opposite effects of Cdc25 and Wee1, absence of both proteins results in approximately normal (wild-type) cycling of MPF activity and cyclin B levels.

27a. When sperm nuclei are UV-irradiated, the time required for DNA replication is lengthened since radioactive dCTP is incorporated at a slower rate.

27b. Since caffeine inhibits ATM and ATR but does not abrogate the delay to DNA replication induced by UV-irradiation, the mechanism responsible for this delay must be independent of ATM and ATR signaling. This finding is significant because ATR mediates cell cycle arrest in response to unreplicated DNA.

27c. UV damage of DNA activates Chk1, and this activation is abrogated by geminin. UV damage also delays the entry of the extract into mitosis as determined by nuclear envelope breakdown (NEB), and this effect is also abrogated by geminin.

27d. The results of these experiments indicate that the activation of Chk1 and the subsequent delay of mitosis that is induced by UV damage of DNA depends upon DNA replication. The data in this study suggest a model in which UV damage delays DNA replication, thus triggering cell cycle arrest through the DNA replication checkpoint pathway that activates ATR and XChk1. Therefore, there may not be a separate UV-damage checkpoint signaling pathway, but rather, UV-damage activates the DNA replication checkpoint.

22

Cell Birth, Lineage, and Death

PART A: *Chapter Summary*

During the development of a multicellular organism, a single cell must give rise to a large number of cells with multiple phenotypes. During embryogenesis and throughout the lifetime of an adult organism, extraneous, damaged, and deleterious cells must be removed with minimal disruption to the function of other cells that reside in the same tissue. This chapter deals with the lifetime of a cell, from birth, through differentiation along a specific pathway, and, finally, death, in this case by apoptosis.

New cells arise from the division of existing cells. Cell division may be either symmetric, in which the two daughter cells are identical, or asymmetric, in which daughter cells differ, usually differing in their concentration of an mRNA or protein. Stem cells are cells that can both self-renew and give rise to differentiated progeny, providing an example of asymmetric cell division. Most adult tissues possess a population of stem cells, although for most tissues the ability to isolate these cells and/or induce their proliferation remains a challenge to medical biology. Early embryos consist of pluripotent stem cells that have the potential to give rise to all differentiated cell types. As embryos develop, most cells lose potential and give rise to progressively more limited cell lineages. However, recent studies suggest that stem cells in adult tissues may retain their pluripotency, and under proper conditions, can differentiate into cell types other than the ones found in the tissue in which they reside.

Our understanding of how cell division during development gives rise to particular lineages of differentiated cells has been advanced by experiments with the worm, *C elegans*. *C. elegans* are particularly well suited to the study of cell lineage because the adults possess relatively few cells, they are amenable to genetic manipulation, and the patterns of lineage are highly invariant.

The mechanisms by which cells are specified into a particular fate are illustrated by two examples: mating-type specification in yeast and muscle-cell differentiation in animals. In the budding yeast *S. cerevisiae*, a complex interaction of transcription factors encoded by the MAT locus results in positive and negative regulation of target genes. Depending on the complement of transcription factors expressed, cells either express genes that induce differentiation into haploid a cells, haploid α cells or diploid a/α cells. The mechanism by which mating-type switching is limited to the mother cell in *S. cerevisiae* provides an illustration of how asymmetric cell division is achieved. Muscle cell differentiation is also regulated by the interaction of transcription factors with muscle-specific target genes. A related family of transcription factors controls neurogenesis, indicating a common scheme has evolved to regulate the differentiation of multiple cell types.

Both during development and in adult multicellular organisms, many cells die by a program of death called apoptosis. Apoptotic death is responsible for removing unnecessary cells such as extra neurons, damaged cells such as precancerous cells, and deleterious cells such as those that would elicit an autoimmune response. Apoptosis is triggered either by the absence of survival or trophic factors or by extracellular death signals. In both cases, the cell actively participates in it own demise by engaging a signaling pathway that activates a family of proteases called caspases. Caspases activate the destruction of the cell such that the cell shrinks and dies without triggering an inflammatory response and damaging the surrounding tissue.

PART B: *Multiple Choice*

Circle the letter corresponding to the most appropriate terms/phrases. There may be more than one correct answer. Circle all that apply.

22.1   *The Birth of Cells*

1. Which of the following cells is capable of self-renewal?

   a.   differentiated cell

   b.   pluripotent stem cell

   c.   progenitor cell

   d.   totipotent stem cell

   e.   none of the above

2. Which of the following is located in the cell lineage that gives rise to erythrocytes?

   a.   BFU-E

   b.   GM-CFC

   c.   lymphoid stem cell

   d.   myeloid stem cell

   e.   pluripotent hematopoetic stem cell

3. Shoot apical meristems (SAMs) can give rise to:

   a.   floral meristems

   b.   leaves

   c.   shoots

   d.   none of the above

4. Which of the following organs is a mesodermal derivative?

   a.   hair

   b.   heart

   c.   liver

   d.   pancreas

   e.   skeleton

5. In *C. elegans*, loss-of-function mutation in the *let-7* gene will prevent the synthesis of :

   a.   *lin-29* mRNA

   b.   *lin-29* protein

   c.   *lin-41* mRNA

   d.   *lin-41* protein

22.2   *Cell-Type Specification in Yeast*

6. In α cells, α2 functions to:

   a.   activate transcription of α genes

   b.   inhibit transcription of a genes

   c.   activate transcription of haploid-specific genes

   d.   inhibit transcription of haploid-specific genes

7. MCM1 promotes the transcription of _____ when associated with _____.

   a.   haploid-specific genes, α2

   b.   haploid-specific genes, a1

   c.   a-specific genes, a1

   d.   α -specific genes, α 1

   e.   α -specific genes, α 2

8. a factor is produced by _____ cells and binds to receptors on the surface of _____ cells.

   a.   a, a

   b.   α, α

   c.   a, α

   d.   α, a

22.3   *Specification and Differentiation of Muscle*

9. Which of the following is an MRF (muscle-regulatory factor)?

   a.   E2A

   b.   MEF2

   c.   Myf5

   d.   MyoD

   e.   none of the above

10. Muscle cell differentiation is inhibited by expression of :

   a. cyclin D

   b. E2A

   c. Id

   d. MEF2

   e. MyoD

11. Which of the following is a bHLH protein?

   a. Achaete

   b. MyoD

   c. neurogenin

   d. Scute

   e. none of the above

22.4 Regulation of Asymmetric Cell Division

12. Which of the following inhibits mating-type switching in S. cerevisiae?

   a. Ash1

   b. HO endonuclease

   c. MAT locus

   d. Swi/Snf complex

13. Which of the following is localized basally in dividing neuroblasts in Drosophila?

   a. Baz

   b. Miranda complex

   c. Numb complex

   d. Par6

   e. PKC3

22.5 Cell Death and Its Regulation

14. Trks are:

   a. endonucleases.

   b. transmembrane receptors

   c. proteases

   d. protein kinases

15. The vertebrate equivalent of CED-9 is:

   a. Apaf-1.

   b. Bcl-2.

   c. caspase 3.

   d. caspase 9.

PART C: Reviewing Concepts

22.1 The Birth of Cells

16. How is asymmetric cell division critical to the development of most multicellular organisms?

17. Describe the steps required to generate embryonic stem cell cultures from mammalian blastocysts.

22.2 Cell-Type Specification in Yeast

18. Both protein-DNA and protein-protein interactions are important in specifying the characteristics of the three yeast cell types: haploid a, haploid α, and diploid a/α cells. Which interactions are important in regulating cell type-specific genes in a haploid and α haploid cells?

19. How is the production of haploid-specific proteins, as well as a- and α -specific proteins, repressed in diploid cells?

22.3 Specification and Differentiation of Muscle

20. Explain the role of histone acetylases in regulating muscle cell differentiation?

21. In multicellular organisms, development of specific cell types expressing unique proteins occurs in a stepwise fashion and depends upon various developmental genes. Define the two basic steps in cell-type specification and identify the genes involved in each step during myogenesis in mammals and neurogenesis in Drosophila.

22.4 Regulation of Asymmetric Cell Division

22. How is mating-type conversion restricted to the mother cell after mitosis in S. cerevisiae?

23. Explain the role of the mitotic spindle in asymmetric cell division.

22.5 *Cell Death and Its Regulation*

24. Why do you think apoptotic pathways have not been identified in single-celled eurkaryotes such as yeast?

25. What is the phenotype of a loss-of-function *ced-3* mutant in *C. elegans*?

PART D: *Analyzing Experiments*

26. Mating-type conversion is restricted to mother cells after mitosis. The presence of Ash1 protein, an inhibitor of Swi5, is thought to control the activity of Swi5 in promoting HO transcription in mother cells. Researchers have studied the control of Ash1 expression to discern how the presence of Ash1 controls mating-type conversion. The data in Figure 22-1 was derived from experiments utilizing a fluorescently labeled antisense *Ash1* RNA probe which was hybridized to yeast cells undergoing mitosis. Bound probe can be visualized with fluorescent microscopy to detect sense *Ash1* mRNA in wild-type and mutant yeast cells containing a deletion of the *Ash1* gene (ash1Δ). The top panel shows fluorescent detection of *Ash1* mRNA, the middle panel utilizes a fluorescent nuclear stain to indicate the presence of nuclei within the same two yeast cells, and the bottom panel shows the light microscopic image of the same two yeast cells.

a. Is the RNA expression pattern of *Ash1* in wild-type yeast consistent with what is known about mating-type conversion?

b. What is the difference in *Ash1* expression in wild-type vs. ash1Δ mutants?

c. *Ash1* mRNA accumulation was also investigated in various mutant strains of yeast (Figure 22-2). *3′* UTRΔ is a mutant containing a deletion of the 3′ UTR of the *Ash1* gene. Two other mutants, bniΔ and myo4Δ, contain mutations in genes involved in promoting actin-filament formation and myosin respectively. Explain why *Ash1* mRNA accumulates in both mitotic products in the 3'UTR Δ mutant.

d. What can be inferred about the role of the actin filaments and myosin in mating-type conversion?

e. Would **a**-specific or α-specific genes be transcribed in the progeny of the bniΔ mutant cells? Would they be expressed in ash1Δ mutant cells?

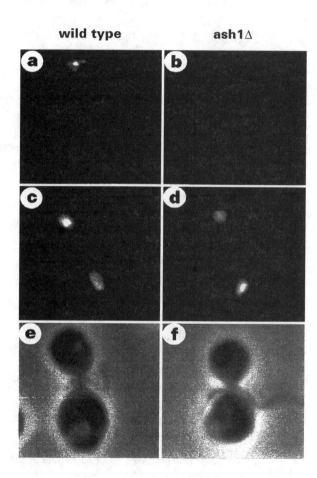

Figure 22-1

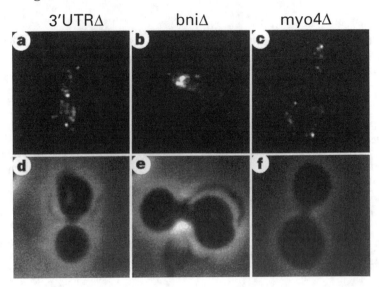

3'UTRΔ   bniΔ   myo4Δ

Figure 22-2

27. DCP-1 is a *Drosophila* caspase. *Drosophila* express other proteins, which are thought to interact with DCP-1. These include DIAP1, Head Involution Defective (HID) and GRIM. These proteins were expressed in yeast. Figure 23-3A shows the effect of expressing either DCP-1 alone or DCP-1 in the presence of DIAP1 alone or plus HID or GRIM. DCP-1 expression is under the control of a galactose-inducible promoter. In glucose grown yeast, DCP-1 is not expressed. The darkened areas in Figure 22-3A are yeast colonies formed over a series of 10-fold dilutions in cell number.

In Figure 22-3B, the results show an affinity column-binding assay for physical interactions between DCP-1, HID, and HIDΔ(1-36) protein with column-bound GST-DIAP1 chimeric protein. The first three proteins were all purified as His6-tagged proteins. HIDΔ(1-36) has a deletion of amino acids 1-36. The results shown are a Western blot to identify possible column binding proteins using specific antibodies. The column matrices are either GST alone or the GST-DIAP1 chimeric protein.

a. What is the effect of DCP-1 expression on yeast colony formation?

b. What is the effect of co-expression of DCP-1 and DIAP1 on yeast colony formation?

c. What is the effect of HID or GRIM on yeast colony formation during co-expression of DCP-1 and DIAP1?

d. Which molecules are apparent inducers of DCP-1 caspase activity and which are inhibitors?

e. What proteins bind to DIAP1?

f. Where within HID is the DIAP1 binding domain?

A

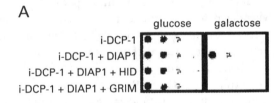

	glucose	galactose
i-DCP-1		
i-DCP-1 + DIAP1		
i-DCP-1 + DIAP1 + HID		
i-DCP-1 + DIAP1 + GRIM		

B

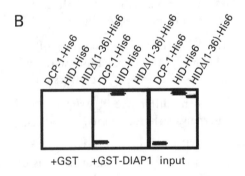

+GST   +GST-DIAP1   input

Figure 22-3

Answers

1. b, d

2. a, d, e

3. a, b, c

4. b, e

5. b

6. b

7. d

8. c

9. c, d

10. a, c

11. a, b, c, d

12. a

13. b, c

14. b, d

15. b

16. Multicellular organisms arise from a single cell, the fertilized egg. Therefore, differentiation of multiple cell types requires that at some stage, cells must give rise to daughter cells that are different from one another. Asymmetric cell division, in which a cellular component is distributed unequally between two daughter cells, provides a mechanism by which a single cell can give rise to cells of two different types.

17. Blastocyst-stage embryos are derived from *in vitro* fertilization or from flushing of oviducts in mice. The inner cell mass is removed by dissection and plated on a layer of non-dividing fibroblast feeder cells. Individual embryonic cells are replated and under careful culture conditions, can be maintained indefinitely as undifferentiated embryonic stem cell populations.

18. In a cells, synthesis of mRNAs encoding a-specific proteins is stimulated by binding of a dimer of MCM1, a constitutive transcription factor, to the P site in the upstream regulatory sequences (URSs) of a-specific genes. In the absence of α1, no activation of α-specific genes occurs. In α cells, transcription of α-specific genes is activated by the simultaneous binding of two proteins to the α-specific URSs, which comprise two sites P and Q. Activation requires binding of dimeric MCM1 to P and binding of α1 protein to Q. Transcription of a-specific genes in α cells is repressed by the cooperative binding of dimeric MCM1 to the P site and dimeric α2 to the flanking α2 sites in a-specific URSs.

19. In diploid cells, transcription of haploid cell-specific genes is repressed by the α2-a1 heterodimer. This heterodimer also represses synthesis of α1 mRNA; in the absence of α1, a-specific genes are not transcribed. Transcription of a-specific genes in diploid cells is repressed by the MCM1-α2 complex by the same mechanism as in α cells.

20. Histone acetylases modify chromatin structure to permit transcriptional activation of certain genes. In myoblasts, MEF2 recruits histone acetylases to activate transcription of genes that induce differentiation to myotubes.

21. The two basic steps in myogenesis, neurogenesis, and other developmental pathways are determination and differentiation. During determination, precursor cells become committed to a particular developmental pathway but do not yet exhibit the characteristics of differentiated cells. During differentiation, committed cells undergo further changes in gene expression resulting in production of cell type-specific proteins. Determination in mammalian myogenesis depends on expression of the *myoD* and *myf5* genes, and determination in *Drosophila* neurogenesis depends on expression of the *achaete* and *scute* genes. The proteins encoded by these genes function in conjunction with general transcription factors (E2A for mammals and Da for *Drosophila*) to activate the mammalian *myogenin* and *Drosophila asense* genes, which are required for differentiation of muscle cells and sensory hairs, respectively.

22. During bud formation in *S. cerevisiae*, the myosin motor protein Myo4p distributes *ASH1* mRNA to the bud. Upon cell division, only the daughter cell

possesses *ASH1* mRNA, which is translated. The Ash1 protein prevents the recruitment of the Swi/Snf complex to the HO gene. HO endonuclease must be synthesized to catalyze the first step in rearranging the *MAT* locus during mating-type conversion. Because the HO gene cannot be transcribed in the daughter cell, mating-type conversion is restricted to the mother cell.

23. During asymmetric cell division, the parent cell is polarized such that a particular factor, generally an mRNA or protein, localizes to a region of the cell. The mitotic spindle must be properly oriented so that this factor will be distributed to one of the daughter cells only. Actin and myosin-related proteins and their regulators function to properly position the mitotic spindle.

24. Apoptosis facilitates the removal of a damaged, deleterious or extraneous cell without disrupting other cells in the surrounding tissue. Because single-celled organisms do not reside in tissues and because the death of a single-celled organism means the organism will not be propagated, apoptosis is not evolutionarily advantageous.

25. CED-3 is a protease that initiates apoptotic death. In the absence of functional CED-3, *C. elegans* develops with 131 extra cells that are normally programmed to die by apoptosis during development.

26a. Yes. Only one progeny cell of the mitotic event (i.e. the daughter cell) is known to express Ash1 protein. The slightly smaller daughter cell in this experiment can be seen to contain *Ash1* mRNA whereas the mother cell does not. This result is consistent with the information that *Ash1* mRNA is directionally transported to the daughter cell.

26b. In wild-type yeast, *Ash1* mRNA becomes segregated to one mitotic product. In Ash1Δ mutants, there is no *Ash1* mRNA produced, so neither mitotic product contains *Ash1* mRNA.

26c. The 3' UTR of *Ash1* mRNA is required for accumulation of *Ash1* mRNA in the daughter cell, probably for the interaction of *Ash1* mRNA with the Myo4p transport protein.

26d. Since mutation in genes required for actin filament formation and myosin production disrupt the normal segregation of *Ash1* mRNA, actin filaments and myosin must be required for the normal segregation of *Ash1*.

26e. Since one does not know the mating phenotype of the parent bniΔ mutant cell before mitosis, it is impossible to predict the phenotype of the progeny afterwards. However, if the parent bniΔ cell originally expressed a-genes, then both products would continue to express a-genes as no mating-type conversion could occur with *Ash1* mRNA present. In *Ash1*Δ mutants, the opposite would occur: mating-type conversion would occur in both progeny cells resulting in a switch from a-specific to α–specific gene expression.

27a. DCP-1 expression (GAL) appears to totally prevent the formation of yeast cell colonies. Since it is a caspase, DCP-1 likely induces apoptotic cell death.

27b. Co-expression of DIAP1 with DCP-1 appears to suppress the killing activity of DCP-1. Colony formation occurs.

27c. HID or GRIM expression appears to counteract the inhibitory effect of DIAP1 on DCP-1 cell killing activity. Colony formation is prevented.

27d. DIAP1 appears to be an inhibitor of DCP-1 caspase activity. HID and GRIM either induce DCP-1 activity or interfere with the ability of DIAP1 to inhibit DCP-1 activity.

27e. DCP-1 and HID bind to GST-DIAP1, i.e., to DIAP1.

27f. As HID did not bind to DIAP1 with the deletion of amino acids 1-36, the HID domain for interacting with DIAP1 is likely to be in the first 36 amino acids of HID.

23   Cancer

PART A: *Chapter Summary*

Cancer is a family of diseases characterized by misregulation of cell proliferation. The cancer phenotype includes progression through the cell cycle independent of growth factors, insensitivity to anti-proliferative signals, resistance to apoptosis, and unlimited replicative potential (immortality). Solid tumors also promote angiogenesis, and metastatic tumors possess the ability to escape the tissue of origin and invade other tissues. Cancer is caused mainly by somatic mutation, although inheritance also predisposes individuals to cancer. Information derived from epidemiologic studies, cell culture experiments and molecular analysis of malignant tumors indicates that multiple mutations are required to transform a normal cell into a cancer cell. Mutation can occur in a wide variety of proteins including growth factors and their receptors, signal transduction proteins, transcription factors, proteins that regulate apoptosis, and cell cycle regulatory proteins.

Both dominant gain-of-function mutations in proto-oncogenes and loss-of-function mutations in tumor suppressor genes are oncogenic. Proto-oncogenes are converted to oncogenes by point mutation or over-expression; overexpression can be caused by gene amplification, translocation, or integration of a slow acting retrovirus. Changes in cell physiology due to oncoproteins include constitutive receptor activity in the absence of ligand, constitutive activity of signal transduction proteins, and loss of regulatory phosphorylation. For example, many oncogenes cause constitutive receptor tyrosine kinase (RTK)-Ras-MAP kinase signaling such that cells progress through the cell cycle even in the absence of growth factor signals.

Loss-of-function mutations in tumor suppressor genes also promote cancer by a variety of mechanisms. The Rb protein normally prevents cells from passing the restriction point and progressing into S-phase in the absence of appropriate growth factors. Like gain-of-function mutations in the RTK-Ras-MAP kinase pathway, loss-of-function mutations in Rb render cell cycle progression growth-factor independent. p53 is another tumor suppressor protein that is mutated in more than half of all cancers. p53 functions both to arrest the cell cycle when DNA is damaged and to induce apoptosis. Loss of p53 function permits cells to accumulate mutations, some of which may be cancer promoting, and to escape programmed cell death. Most loss-of-function mutations in tumor-suppressor genes are recessive, and thus both alleles of these genes must be inactivated in order for cancer to develop. If an individual inherits one defective copy of a tumor-suppressor gene, the chances of developing cancer due to inactivation of the second allele are much greater. Such is the case with hereditary retinoblastoma in which individuals inherit one nonfunctional copy of the *RB* gene and typically develop bilateral eye tumors early in life.

Normal chromosomes are required for regulated DNA synthesis and cell division. Because the development of cancer requires multiple mutations, individuals with inherited defects in DNA repair are sensitive to environmental factors and susceptible to a variety of cancers. Chromosomal abnormalities such as aneuploidy, translocations, and duplications are common in many malignancies. Gain-of-function mutations in telomerase, the enzyme that repairs the ends of chromosomes, promotes immortality in cancer cells.

PART B: Multiple Choice

Circle the letter corresponding to the most appropriate terms/phrases. There may be more than one correct answer. Circle all that apply.

23.1   Tumor Cells and the Onset of Cancer

1. Which of the following properties is characteristic of a benign tumor?

    a. well differentiated

    b. localized to tissue of origin

    c. metastatic

    d. promotes angiogenesis

    e. surrounded by a fibrous capsule

2. Loss-of-function mutations in which of the following genes typifies colorectal carcinomas?

    a. *myc*

    b. *K-ras*

    c. *p53*

    d. *APC*

3. Which property is shared by malignant tumor cells and transformed cells in culture?

    a. reduced growth factor requirement

    b. attachment-dependent growth

    c. loss of actin microfilaments

    d. altered morphology

23.2   The Genetic Basis of Cancer

4. Which of the following are proto-oncogenes?

    a. *fos*

    b. *myc*

    c. *p53*

    d. *ras*

    e. *Rb*

5. Which of the following can lead to loss-of-heterozygosity in a tumor-suppressor gene?

    a. deletion of the normal copy

    b. nondysjunction during mitosis

    c. somatic mutation of the normal copy

    d. mitotic recombination between a wild-type and mutant allele

6. Tumor-suppressor genes encode proteins with which of the following functions?

    a. growth-promoting

    b. growth-arresting

    c. pro-apoptotic

    d. anti-apoptotic

    e. DNA repair

23.3   Oncogenic Mutations in Growth-Promoting Proteins

7. Which of the following oncoproteins possesses protein kinase activity?

    a. E5

    b. ErbB

    c. Myc

    d. Ras

    e. Sis

8. Which of the following are transcription factors?

    a. Myc

    b. Neu

    c. Ras

    d. Sis

    e. Trk

9. A gain-of-function mutation in which the following would have the same effect as a gain-of-function mutation in Ras?

 a. Crk

 b. GAP

 c. Myc

 d. NF1

 e. none of the above

23.4 *Mutations Causing Loss of Growth-Inhibiting and Cell-Cycle Controls*

10. Which of the following is a consequence of p53 stabilization?

 a. stabilization of Cdc25A

 b. activation of ATM

 c. blocked apoptosis

 d. transcription of p21

 e. DNA repair

11. Which of the following is pro-apoptotic?

 a. Bax

 b. Bcl-2

 c. p53

 d. protein kinase B

 e. PTEN

12. Human papilloma virus encodes proteins that inhibit the function of:

 a. Bcl-2

 b. Myc

 c. p53

 d. PDGF receptor

 e. Rb

23.5 *The Role of Carcinogens and DNA Repair in Cancer*

13. Which DNA repair system is affected in patients with xeroderma pigmentosum?

 a. repair by homologous recombination

 b. mismatch repair

 c. nucleotide excision repair

 d. base excision repair

14. The following events take place during the repair of double-strand DNA breaks by homologous recombination. Place them in the proper order.

 a. activation of exonucleases

 b. joining of complementary homologous strands by Rad51

 c. activation of ATM kinase

 d. DNA polymerase activity

 e. ligase activity

15. A gain-of-function mutation in telomerase causes which of the following?

 a. apoptosis

 b. immortalization

 c. error-prone repair

 d. nucleotide excision repair

PART C: *Reviewing Concepts*

23.1 *Tumor Cells and the Onset of Cancer*

16. Describe the paradoxical roles that the extracellular matrix plays in both promoting and hindering the metastasis of tumor cells of epithelial origin.

17. How are stem cells predisposed to give rise to cancer cells?

23.2 *The Genetic Basis of Cancer*

18. In referring to the war on cancer, J. Michael Bishop, a Nobel laureate for his work with oncogenes, quoted a line from the Pogo comic strip: "We have met the enemy and he is us." Why is this quotation appropriate?

19. Distinguish between the methods of transformation by transducing retroviruses versus slow-acting retroviruses.

20. How might microarray technology become a powerful new weapon in the war on cancer?

23.3 *Oncogenic Mutations in Growth-Promoting Proteins*

21. What is the Philadelphia chromosome and how does it contribute to chronic myelogenous leukemia?

23.4 *Mutations Causing Loss of Growth-Inhibiting and Cell-Cycle Controls*

22. To leave the quiescent state and enter the cell cycle, mammalian cells must pass the restriction point by activating the transcription factor E2F. Describe the mechanisms by which the Rb and Swi/Snf proteins inhibit E2F function to maintain cells in quiescence.

23. What is aneuploidy? Why is aneuploidy a common feature of cancer cells?

23.5 *The Role of Carcinogens and DNA Repair in Cancer*

24. What mutational event has been linked to lung cancers caused by cigarette smoke?

25. Why is end-joining repair of double-strand breaks more error-prone than repair by homologous recombination?

PART D: *Analyzing Experiments*

26. Loss-of-function mutations in p53 typify the majority of human cancers. Discovery of the roles for p53 in the DNA damage checkpoint and in promoting apoptosis helps to explain its potency as a tumor suppressor protein. Recent studies suggest that p53 may possess an additional anti-tumor function: preventing telomerase activity. In one such

study by Ogawa and colleagues, a human pancreatic cell line called MIA PaCa-2, which lacks functional p53, was infected with replication-defective adenoviruses encoding p53, p21 or a control protein (lacZ). Western analysis of these cells for p21 and p53 is shown in Figure 23-1.

Figure 23-1

a. Why do cells infected with adenovirus encoding p53 (Ad-p53) also express p21? Why do uninfected cells fail to express endogenous p21?

Control and adenovirus-infected MIA PaCa-2 cells were then tested for telomerase activity. The results of the assay are shown in Table 23-1.

	pre-treatment	un-infected	Ad-lacZ	Ad-p21	Ad-p53
telomerase activity (%)	100	104	100	97	14

Table 23-1

b. What is the significance of telomerase activity in the pretreatment and uninfected MIA PaCa-2 cells? How does telomerase contribute to the malignant phenotype?

c. What can be concluded by the persistence of telomerase activity in cells infected with Ad-p21?

The investigators then performed reverse-transcriptase PCR (RT-PCR) analysis of hTERT mRNA levels in control and adenovirus-infected cells. hTERT codes for telomerase in humans. The results are shown in Figure 23-2.

Figure 23-2

d. What does this experiment indicate about the mechanism by which p53 regulates telomerase activity? Is this mechanism consistent with known activities of p53?

Thanks to Brian Wroble for help with this question.

27. Acoustic neuroma is a human tumor derived from cells that surround the vestibulocochlear nerve. Although most cases of acoustic neuroma occur as unilateral, apparently non-inherited tumors, bilateral tumors are characteristic of an inherited form of this cancer. In order to understand the basis for the inherited disorder, DNA from leukocytes (normal cells) and tumor cells of patients with bilateral tumors were examined by Southern blotting with probes from known locations on human chromosomes. Hybridization of a labeled probe with two bands in the patient's DNA digests indicates heterozygosity in the patient's restriction fragments. The data from this study are shown in Table 23-2.

a. Which chromosomal probes reveal differences in the leukocyte and tumor-DNA from patients? What is the most likely cause of these differences?

b. What do these data suggest about the mechanism of oncogenesis in familial acoustic neuroma?

c. In the tumor DNA from fourteen patients, no loss of heterozygosity was detected with the chromosome 22 probes. How might these data be explained?

Probe	Chromosomal location of probe	Number of patients with two leukocyte fragments that hybridize with probe	Number of patients with two leukocyte fragments that also have two tumor fragments that hybridize with probe
A	1	5	5
C	4	9	9
D	4	3	3
E	10	3	3
F	11	10	10
J	12	5	5
N	13	2	2
O	14	11	11
P	17	10	10
Q	18	9	9
R	19	2	2
U	22	6	3
V	22	12	8
W	22	4	3

Table 23-2

Answers

1. a, b, e

2. c, d

3. a, c, d

4. a, b, d

5. a, b, c, d

6. b, c, e

7. b

8. a

9. a

10. d, e

11. a, c, e

12. c, d, e

13. c

14. c, a, b, d, e

15. b

16. In order for tumor cells to metastasize, they must escape their tissue of origin, enter the circulation or lymphatics, and repopulate a distal tissue. To do so, cancer cells must break through any extracellular matrices between the tissue of origin and the circulation. Epithelial tumors must escape the basal lamina. Metastatic tumors typically express proteins such plasminogen activator that promote degradation of the basal lamina. However, to migrate into and out of various tissues, metastatic cells must utilize extracellular matrices as "roadways" along which to migrate. To do so, they often express cell surface proteins such as integrins that promote interaction with the basal lamina and other matrices.

17. Unlike most differentiated cells, stem cells can divide throughout the lifetime of the organism, allowing for the multiple mutations required for malignancy to accumulate in a single cell lineage. Stem cells express telomerase and thus, already possess one of the requirements for the malignant phenotype, immortality.

18. The Pogo quotation is apt because many human cancers result from alterations in normal cellular genes or metabolic processes. Even retroviral oncogenes are derived from normal cellular proto-oncogenes. Once this was understood, researchers realized that retroviral oncogenes provided a model system for studying the structure and function of cancer-causing genes and their encoded proteins.

19. Transducing retroviruses encode an oncogene derived from a cellular proto-oncogene and can induce the formation of a tumor within days. Slow-acting retroviruses lack an oncogene and cause cancer only when they integrate near a proto-oncogene within the host genome and promote its constitutive activation. Such a random event can take months or years to occur.

20. Microarray technology should allow comprehensive analysis of all of the genes whose expression is altered in a particular tumor. This information will contribute to the understanding of how cancers differ from one another. For example, why are some types of breast cancer responsive to a particular chemotherapy while others are resistant? Microarray analysis may eventually allow a cocktail of chemotherapeutics to be tailored to best treat cancer in an individual depending upon the profile of alterations in gene expression in each case.

21. The Philadelphia chromosome results from a translocation that fuses part of the *c-abl* gene, which encodes a protein tyrosine kinase, with part of the *bcr* gene. The resulting chimeric *bcr-abl* gene is an oncogene that encodes a constitutively active kinase that phosphorylates and activates growth-promoting proteins such as JAK2 kinase and STAT5. In white blood cells, the *bcr-abl* gene causes abnormal proliferation. Mutation in a second gene, such as p53, is required for the full malignant phenotype.

22. Unphosphorylated Rb binds E2F, preventing E2F from activating the transcription of genes required for DNA synthesis. Swi/Snf are chromatin-remodeling proteins that bind to the E2F genes and

prevent their transcription. Rb plays a role in recruiting Swi/Snf to the E2F genes.

23. Aneuploidy is the presence of an aberrant number of chromosomes in a cell. In a cancer cell, typically there are too many chromosomes since a loss of chromosomes is usually lethal to a cell. Cancer cells lack one or more cell cycle checkpoints since a loss of checkpoints permits a cell to accumulate an adequate number of cancer-promoting mutations. These cells are predisposed to aneuploidy, especially when spindle assembly or chromosome-segregation checkpoints are disrupted.

24. Lung cancer caused by cigarette smoke is typically characterized by loss-of-function point mutations in *p53*. This mutation occurs because the carcinogen benzo(*a*)pyrene, which is found in cigarettes, is activated by cytochrome P-450 enzymes in the liver to become a mutagen that converts guanine to thymine bases, including several guanines within the coding region of the *p53* gene.

25. Because homologous recombination utilizes the undamaged DNA from the sister chromatid as template for repair, damage can be repaired without loss of genetic fidelity. In contrast, in end-joining repair of double-strand breaks, nonhomologous fragments of DNA are ligated together. Even if these fragments derive from the same chromosome, several bases are lost at the site of rejoining, resulting in mutational alteration. Sometimes DNA fragments from distinct chromosomes are joined, resulting in a translocational mutation, which will occasionally generate an oncogene such as *bcr/abl*.

26a. p53 functions as a transcription factor to regulate the expression of multiple genes, including p21. Therefore, in cells transfected with p53, p21 expression is also induced, whereas uninfected cells, which lack p53, also lack p21, most likely because its transcription cannot be induced by p53.

26b. Telomerase confers unlimited replicative potential to cells, also referred to as immortality, by preventing the shortening of telomeres that normally occurs during each round of DNA replication and eventually leads to cell senescence. Immortality is one of the hallmarks of cancer cells.

26c. Since cells expressing exogenous p53, but not p21, lose telomerase activity, it can be concluded that p53 is not exerting its anti-telomerase effects indirectly through p21. Rather, p53 must be acting either directly on telomerase expression or activity or indirectly by modifying the expression of another gene.

26d. Because exogenous p53 affects the levels of hTERT mRNA, p53 is most likely affecting telomerase activity by acting as a transcription factor to repress hTERT expression. The role of p53 as a transcription factor is well established and this study is consistent with such a role. Remember that transcription factors can both activate (as with p21) or repress (as with hTERT) the expression of target genes.

27a. Probes U, V, and W reveal that the DNA of tumor cells from some patients has only one restriction fragment that hybridizes with markers on chromosome 22, whereas the leukocyte DNA from the same patients has two different restriction fragments hybridizing with these markers. This finding suggests that the tumor cells of these individuals have lost at least some of the DNA from one of their copies of chromosome 22.

27b. These data suggest that a deletion on chromosome 22 is oncogenic. Acoustic neuroma, like retinoblastoma, appears to be brought about by a relatively rare somatic cell mutation, making the cell homozygous recessive. (The inherited defect is the presence of one recessive allele, indicated by the observed heterozygosity in leukocytes.) The fact that the loss of the second allele causes cancer suggests that the wild-type gene functions as a tumor suppressor gene.

27c. The lack of heterozygosity in some patients has several possible explanations: 1. The tumors in these patients may have been induced by some mechanism that does not involve the loss of genes on chromosome 22. 2. These tumor cells may have the same defective allele as those of other patients, but the change that occurred in the normal allele of the tumor-associated gene on chromosome 22 may not have altered the size of the restriction fragment, e.g. a point mutation. 3. A deletion may have occurred in the normal allele but it did not include the DNA in the restriction fragments that hybridized to the probes.

2

Chemical Foundations

Review the Concepts

1. Less energy is required to form noncovalent bonds than covalent bonds, and the bonds that stick the gecko's feet to the smooth surface need to be formed and broken many times as the animal moves. Since van der Waals interactions are so weak, there must be many points of contact (a large surface area) yielding multiple van der Waals interactions between the septae and the smooth surface.

2a. These are likely to be hydrophilic amino acids, and in particular, negatively charged amino acids (aspartate and glutamate) which would have an affinity for K^+ via ionic bonds.

2b. Like the phospholipid bilayer itself, this portion of the protein is likely to be amphipathic, with hydrophobic amino acids in contact with the fatty acyl chains and hydrophilic amino acids in contact with the hydrophilic heads.

2c,d. Since both the cytosol and extracellular space are aqueous environments, hydrophilic amino acids would contact these fluids.

3. The structure of the peptide is:

Note: this is the linear structure and does not portray the three-dimensional configuration.

At pH = 7.0, the net charge is –1, due to the negative charge on the carboxyl residue of glutamate (E). After phosphorylation by a tyrosine kinase, two additional negative charges (due to attachment of phosphate residues to tyrosines (Y)) would be added. Thus, the net charge would be –3. The most likely source of phosphate is ATP since the attachment of inorganic phosphate (P_i) to tyrosine is highly energetically unfavorable, but when coupled to the hydrolysis of the high-energy phosphoanhydride bond of ATP, the overall reaction is energetically favorable.

4. Disulfide bonds are formed between two cysteine residues. The formation of disulfide bonds increases the order, and therefore decreases the entropy (ΔS becomes a greater negative).

5. Stereoisomers are compounds that have the same molecular formula but are mirror images of one another. Many organic molecules can exist as stereoisomers due to two different possible orientations around an asymmetric carbon atom (e.g. amino acids). Because stereoisomers differ in their three dimensional orientation and because biological molecules interact with one another based on precise molecular complementarity, stereoisomers often react with different molecules, or react differently with the same molecules. Therefore, they may have very distinct physiologic effects in the cell.

6. The compound is guanosine triphosphate (GTP). Although the guanine base is found in both DNA and RNA, the sugar is a ribose sugar because of the 2' hydroxyl group. Therefore, GTP is a component of RNA only. GTP is an important intracellular signaling molecule.

7. At least three properties contribute to this structural diversity. First, monosaccharides can be joined to one another at any of several hydroxyl groups. Second, the C-1 linkage can have either an α or a β configuration. Third, extensive branching of carbohydrate chains is possible.

8. In the acidic pH of a lysosome, ammonia is converted to ammonium ion. Ammonium ion is unable to traverse the membrane because of its positive charge and is trapped within the lysosome. The accumulation of ammonium ion decreases the concentration of protons within lysosomes and therefore elevates lysosomal pH. At neutral pH, ammonia has little, if any, tendency to protonate to ammonium ion and, thus has no effect on cytosolic pH.

9. $K_{eq} = [LR]/[L][R]$

    Since 90% of L binds R, the concentration of LR at equilibrium is $0.9(1 \times 10^{-3} M) = 9 \times 10^{-4}$ M. The concentration of free L at equilibrium is the 10% of L that remains unbound, 1×10^{-4} M. The concentration of R at equilibrium is $(5 \times 10^{-2}$ M$) - (9 \times 10^{-4}$ M$) = 4.91 \times 10^{-2}$ M. Therefore, $[LR]/[L][R] = 9 \times 10^{-4}$ M$/ ((1 \times 10^{-4}$ M$) (4.91 \times 10^{-2}$ M$)) = 183.3$ M^{-1}.

    The equilibrium constant is unaffected by the presence of an enzyme.

    $K_d = 1/ K_{eq} = 5.4 \times 10^{-3}$ M.

10. The pH of cytosol is 7.2, which is the pK$_a$ for the acid-base reaction, $H_2PO_4^- \Leftrightarrow H_2PO_4^{2-} + H^+$. Therefore, phosphoric acid will exist as a mixture of $H_2PO_4^-$ and $H_2PO_4^{2-}$. Because the pK$_a$ of phosphoric acid equals the optimal pH of cytosol, phosphoric acid serves as a biological buffer to help the cell maintain a constant pH.

11. $\Delta G = \Delta G^{o'} + RT\ln$ [products]/[reactants]

    For this reaction, $\Delta G = -1000$ cal/mol $+ [1.987$ cal/(degree$\bullet$mol) $\times$ (298 degrees) $\times \ln (0.01$ M$/ (0.01$ M $\times 0.01$ M$))]$.

    $\Delta G = -1000$ cal/mol $+ 2727$ cal/mol $= 1727$ cal/mol

    To make this reaction energetically favorable, one could increase the concentration of reactants relative to products such that the term $RT\ln$ [products]/[reactants] becomes smaller than 1000 cal/mol. One might also couple this reaction to an energetically favorable reaction.

3

Protein Structure and Function

Review the Concepts

1. The primary structure of a protein is the linear arrangement or sequence of amino acids. The secondary structure of a protein is the various spatial arrangements that result from folding localized regions of the polypeptide chain. The tertiary structure of a protein is the overall conformation of the polypeptide chain, its three dimensional structure. Secondary structures, which include examples such as the alpha helix and the beta sheet, are held together by hydrogen bonds. In contrast, the tertiary structure is primarily stabilized by hydrophobic interactions between non-polar side chains of the amino acids and hydrogen bonds between polar side chains. The quaternary structure describes the number and relative positions of the subunits in a multimeric protein.

2. Chaperones and chaperonins are proteins that play a key role in the proper folding of newly synthesized proteins. Chaperones bind to and stabilize unfolded or partially folded proteins and thereby prevent these proteins from aggregating and being degraded. The major chaperone protein in all organisms is the protein Hsp70 and its homologues. When bound to ATP, a hydrophobic region on Hsp70 binds to a hydrophobic region on the unfolded protein. Hydrolysis of the ATP induces a conformational change in Hsp70 that allows the bound protein to be folded. Chaperonins, which form large macromolecular assemblies, directly facilitate the folding of proteins. A partially folded or misfolded protein binds to the inner wall of the cylindrical chaperonin assembly and is folded into its native conformation. In an ATP-dependent step, the chaperonin assembly undergoes a conformational change that releases the folded protein.

3. Ubiquitin is a 76 amino acid protein that serves as a molecular tag for proteins destined for degradation. Ubiquitination of a protein involves an enzyme-catalyzed transfer of a single ubiquitin molecule to the lysine side chain of a target protein. This ubiquitination step is repeated many times, resulting in a long chain of ubiquitin molecules. The resulting polyubiquitin chain is recognized by the proteasome, which is a large, cylindrical, multisubunit complex that proteolytically cleaves ubiquitin-tagged proteins into short peptides and free ubiquitin molecules.

4. An enzyme increases the rate of a reaction that is already energetically favorable by lowering the activation energy required for the reaction. The reaction occurs within the active site of the enzyme, which consists of a number of amino acid side chains brought into close proximity by the native folding of the enzyme. The active site serves two functions: to recognize and bind the substrate and to catalyze the reaction once the substrate is bound. For an enzyme-catalyzed reaction, the V_{max} represents the maximal velocity of a reaction at saturating substrate concentrations and the K_m represents the substrate concentration that yields a half-maximal velocity ($1/2 \ V_{max}$). For an enzyme-catalyzed reaction, the K_m is a measure of the affinity of a substrate for its enzyme. A substrate with a low K_m has a high affinity for the enzyme, and thus substrate B ($K_m=0.01$mM) would have a higher affinity for enzyme X than substrate A ($K_m=0.4$ mM).

5. Motor proteins possess three general properties: 1) the ability to transduce a source of energy, either ATP or an ion gradient, into linear or rotary movement, 2) the ability to bind and translocate along a cytoskeletal filament, nucleic acid strand, or protein complex, and 3) net movement in a given direction. One cycle of myosin movement relative to an actin filament requires a number of biochemical steps. Step 1 involves binding ATP to myosin, resulting in the freeing of actin. During step 2, the myosin head hydrolyzes the

ATP causing a conformational change in the head that moves it to a new position closer to the (+) end of the actin filament. There, it rebinds to the actin filament. In step 3, the phosphate dissociates and the myosin head undergoes a second conformational change, the power stroke, which restores myosin to its rigor conformation. Because myosin is bound to actin, this second conformational change exerts a force that causes myosin to move relative to the actin filament.

6. Cooperativity, or allostery, refers to any change in the tertiary or quaternary structure of a protein induced by the binding of a ligand that affects the binding of subsequent ligand molecules. In this way, a multisubunit protein can respond more efficiently to small changes in ligand concentration compared to a protein that does not show cooperativity. The activity of many proteins is regulated by the reversible addition/removal of phosphate groups to specific serine, threonine, and tyrosine residues. Protein kinases catalyze phosphorylation (addition of phosphate groups), while protein phosphatases catalyze dephosphorylation (removal of phosphate groups). Phosphorylation/dephosphorylation changes the charge on a protein, which typically leads to a conformational change and a resulting increase or decrease in activity. Some proteins are synthesized as inactive propeptides, which must be enzymatically cleaved to release an active protein.

7. Proteins can be separated by mass by centrifuging them through a solution of increasing density, called a density gradient. In this separation technique, known as rate-zonal centrifugation, proteins of larger mass generally migrate faster than proteins of smaller mass. However, this is not always true because the shape of the protein also influences the migration rate. Gel electrophoresis can also separate proteins based on their mass. In this technique, proteins are separated through a polyacrylamide gel matrix in response to an electric field. Because the migration of proteins through a polyacrylamide gel is also influenced by shape of proteins, the ionic detergent sodium dodecyl sulfate is added to denature proteins and force proteins into similar conformations. During rate zonal centrifugation,

a protein of larger mass (transferrin) will sediment faster during centrifugation; whereas a protein of smaller mass (lysozyme) will migrate faster during electrophoresis.

8. Gel filtration, ion exchange, and affinity chromatography typically involve the use of a bead consisting of polyacrylamide, dextran or agarose packed into a column. In gel filtration chromatography, the protein solution flows around the spherical beads and interacts with depressions that cover the surface of the beads. Small proteins can penetrate these depressions more readily than larger proteins and thus spend more time in the column and elute later from the column; whereas larger proteins do not interact with these depressions and elute first from the column. In ion exchange chromatography, proteins are separated based on their charge. The beads in the column are covered with amino or carboxyl groups that carry a positive or negative charge, respectively. Positively charged proteins will bind to negatively charged beads, and negatively charged proteins will bind to positively charged beads. In affinity chromatography, ligand molecules that bind to the protein of interest are covalently attached to beads in a column. The protein solution is passed over the beads and only those proteins that bind to the ligand attached to the beads will be retained, while other proteins are washed out. The bound protein can later be eluted from the column using an excess of ligand or by changing the salt concentration or pH.

9. Proteins can be made radioactive by the incorporation of radioactively-labeled amino acids during protein synthesis. Methionine or cysteine labeled with sulfur-35 are two commonly used radioactive amino acids, although many others have also been used. The radioactively labeled proteins can be detected using a technique known as autoradiography. In one example of autoradiography, cells are labeled with a radioactive compound and then overlaid with a photographic emulsion sensitive to radiation. The presence of radioactive proteins will be revealed as deposits of silver grains after the emulsion is developed. A Western blot is a method for detecting proteins that combines the resolving power of gel electrophoresis, the specificity of

antibodies, and the sensitivity of enzyme assays. In this method, proteins are first separated by size using gel electrophoresis. The proteins are then transferred onto a nylon filter. A specific protein is then detected using an antibody specific for the protein of interest (primary antibody) and an enzyme-antibody conjugate (secondary antibody) that recognizes the primary antibody. The presence of this protein-primary antibody-enzyme-conjugated-secondary antibody complex is detected using an assay specific for the conjugated enzyme.

10. X-ray crystallography can be used to determine the 3-dimensional structure of proteins. In this technique, x-rays are passed through a protein crystal. The diffraction pattern generated when atoms in the protein scatter the x-rays is a characteristic pattern that can be interpreted into defined structures. Cryoelectron microscopy involves the rapid freezing of a protein sample and examination with a cryoelectron microscope. A low dose of electrons is used to generate a scatter pattern which can be used to reconstruct the protein's structure. In nuclear magnetic resonance (NMR) spectroscopy, a protein solution is placed in a magnetic field and the effects of different radio frequencies on the spin of different atoms are measured. Based on the magnitude of the effect of one atom on an adjacent atom, the distances between residues can be calculated to generate a 3-dimensional structure.

c. Proteins 1 and 6 are strictly nuclear proteins. Proteins 2, 3, and 4 are strictly cytoplasmic proteins. Protein 5 is present in the nucleus and cytoplasm. Protein 7 is a cytoplasmic protein that migrates to the nucleus in response to drug treatment.

d. Protein 1 is a nuclear protein whose level declines in response to drug treatment.

Protein 2 is a cytoplasmic phosphoprotein whose level increases in response to drug treatment.

Protein 3 is a cytoplasmic phosphoprotein whose level does not change in response to drug treatment.

Protein 4 is a cytoplasmic protein whose level increases in response to drug treatment.

Protein 5 is a nuclear and cytoplasmic protein whose level does not change in response to drug treatment.

Protein 6 is a nuclear phosphoprotein whose level does not change in response to drug treatment.

Protein 7 is a cytoplasmic protein which migrates from the cytoplasm to the nucleus, in response to drug treatment.

Analyze the Data

a. Proteins 3, 5, 6, and 7 do not change in response to the drug. Protein 1 declines in response to the drug. Proteins 2 and 4 are induced in response to the drug.

b. Proteins 2, 3 and 6 are phosphoproteins; the others are not. Protein 2 is a phosphoprotein normally present in cells whose level increases in response to the drug. Protein 3 is not a phosphoprotein in control cells and is phosphorylated in response to the drug. Protein 6 is a phosphoprotein whose level does not change in response to the drug.

4

Basic Molecular Genetic Mechanisms

Review the Concepts

1. Watson-Crick base pairs are interactions between a larger purine and a smaller pyrimidine base in DNA. These interactions result in primarily G-C and A-T base pairing in DNA. They are important because these interactions allow for the antiparallel, complementary strands of DNA.

2. The atoms on the edges of each base within the minor groove of DNA at the binding site for the TATA box binding protein form a binding surface (see Figure 4-5 in MCB 5e). This binding surface is recognized by the TATA box binding protein. In most instances this DNA binding surface is located near the consensus sequence TATAA in the promoter.

3. At 90° C, the double-stranded DNA template will denature and the strands will separate. As the temperature slowly drops below the Tm of the plasmid DNA, the single-stranded oligonucleotide primer will anneal to its complementary sequence on the plasmid template. The resulting duplex contains a short double-stranded stretch with a free 3' OH which can be used by DNA polymerase enzyme in sequencing reactions.

4. RNA is less stable chemically than DNA because of the presence of a hydroxyl group on C-2 in the ribose moieties in the backbone. Additionally, cytosine (found in both RNA and DNA) may be deaminated to give uracil. If this occurs in DNA, which does not normally contain uracil, the incorrect base is recognized and repaired by cellular enzymes. In contrast, if this deamination occurs in RNA, which normally contains uracil, the base substitution is not corrected. Thus the presence of deoxyribose and thymine make DNA more stable and less subject to spontaneous changes in nucleotide sequence than RNA. These properties might explain the use of DNA as a long-term information-storage molecule.

5. In prokaryotes, many protein-coding genes are clustered in operons where transcription proceeds from a single promoter that gives rise to one mRNA encoding multiple proteins with related functions. In contrast, eukaryotes do not have operons but do transcribe intron sequences that must be spliced out of mature mRNAs. Eukaryotic mRNAs also differ from their prokaryote counterparts in that they contain a 5' cap and 3' poly (A) tail.

6. A simple explanation is that the larger, membrane-spanning domain-containing protein and the small, secreted protein are encoded by the same gene which is differentially spliced. Specifically, the final exon of the gene could contain the information for the membrane-spanning domain, and in the smaller, secreted protein, this exon could be omitted during splicing.

7. In the presence of glucose the *lac* operon is turned off. The *lac* repressor is synthesized and bound to the operator, which prevents binding of RNA polymerase to the promoter and subsequent activation of the *lac* operon. When *E. coli* cells are shifted into a lactose-containing medium in the absence of glucose, the lactose binds to the *lac* repressor and prevents its binding to the operator. In the absence of *lac* repressor binding to the operator, RNA polymerase can bind to the promoter and turn on the *lac* operon.

8. In two-component regulatory systems, one protein acts as a sensor and the other protein is a response regulator. In *E. coli* the proteins PhoR (the sensor) and PhoB (the response regulator) regulate transcription of target genes in response to the free phosphate concentration. PhoR is a transmembrane protein, whose periplasmic domain binds phosphate and whose cytoplasmic domain has protein kinase activity. Under low phosphate conditions, PhoR undergoes a conformational

change that activates the protein kinase activity and results in the transfer of a phosphate group to Pho B. The phosphorylated form of PhoB then induces transcription from several genes that help the cell respond to low phosphate conditions.

9. In prokaryotes, the eight nucleotide Shine-Dalgarno sequence located near the AUG start codon binds to specific sequences in the 16S rRNA allowing for positioning of the 30S ribosomal subunit near the start site of translation. In eukaryotes, recognition of the start site involves other factors such as eIF4, eIF3 proteins, and Kozak sequences near the start site of the mRNA. eIF4 recognizes and binds to the 5′ cap structure on eukaryotic mRNAs, and eIF3 proteins are part of the preinitiation complex which is thought to scan along the mRNA, most often stopping at the first AUG. The Kozak sequences facilitate the preinitiation complex in choosing the proper start site. Poliovirus initiates translation of its transcripts utilizing host cell machinery so it is identical to the eukaryotic host except for the presence of internal ribosome entry sites (IRES)s, which are internal AUG sites.

10. Even if all proteins are removed from the 23S rRNA, it is capable of catalyzing the formation of peptide bonds. X-ray crystallographic studies support this by revealing that no proteins lie near the site of peptide bond synthesis in the crystal structure of the large subunit.

11. Since poly (A)-binding protein I is involved in increasing the efficiency of translation, a mutant in poly (A)-binding protein I would have less efficient translation. Polyribosomes from such a mutant would not contain circular structures of mRNAs during translation because lack of the poly (A)-binding protein I would eliminate the 3' binding site for eIF4G.

12. DNA synthesis is discontinuous because the double helix consists of two antiparallel strands and DNA polymerase can synthesize DNA only in the 5′ to 3′ direction. Thus one strand is synthesized continuously at the growing fork, but the other strand is synthesized utilizing Okazaki fragments that are joined by DNA ligase.

13. The gene encoding the reverse transcriptase enzyme is unique in retroviruses. These viruses contain RNA as their genetic material; a DNA copy of the viral RNA is made during infection and reverse transcriptase catalyzes this reaction. Non retroviruses have DNA-containing genomes and thus, do not require DNA synthesis using RNA as the template, or starting material.

Analyze the Data

a. From translation of the first synthetic mRNA, you can conclude that the amino acid lysine is specified by a group of A residues. By examining the results using the second and third synthetic mRNAs, you can conclude that a triplet codon is required, rather than a mono, bi, or quadruplet, etc. codon. Specifically, using the second mRNA, an alternating threonine-histidine peptide is derived. This could result if an odd number of bases specified an amino acid, but not if an even number of bases specified the code. The third mRNA can be translated in three different frames resulting in three different polypeptides composed of a single type of amino acid. From this, you can conclude that a triplet codon is required. Therefore, AAA must specify lysine, ACA must specify threonine, and CAC must specify histidine.

b. The anticodon loop of a tRNA specifying threonine would have the sequence UGU.

c. Since this microbe contains four bases and a triplet codon genetic code, one could conclude that the genetic code of this organism is degenerate in that more than one codon probably specifies the same amino acid.

5

Biomembranes and Cell Architecture

Review the Concepts

1. The spontaneous assembly of phospholipid molecules into a lipid bilayer creates a sheetlike structure that is two molecules thick. Each layer is arranged so that the polar head groups of the phospholipids are exposed to the aqueous environment on one side of the bilayer and the hydrocarbon tails associate with the tails of the other layer to create a hydrophobic core. In cross section, the bilayer structure thus consists of a hydrophobic core bordered by polar head groups. When stained with osmium tetroxide, which binds strongly to polar head groups, and viewed in cross section, the bilayer looks like a railroad track with a light center bounded on each side by a thin dark line.

2. The three main types of lipid molecules in biomembranes are phosphoglycerides, sphingolipids, and steroids. All are amphipathic molecules having a polar head group and a hydrophobic tail, but the three types differ in chemical structure, abundance and function.

3. Lipid bilayers are considered to be two-dimensional fluids because lipid molecules (and proteins if present) are able to rotate along their long axes and move laterally within each leaflet. Such movements are driven by thermal energy, and may be quantified by measuring fluorescence recovery after photobleaching, the (FRAP) technique. In this technique, specific membrane lipids or proteins are labeled with a fluorescent reagent, and then a laser is used to irreversibly bleach a small area of the membrane surface. The extent and rate at which fluorescence recovers in the bleached area, as fluorescent molecules diffuse back into the bleach zone and bleached molecules diffuse outward, can be measured. The extent of recovery is proportional to the fraction of labeled molecules that are mobile, and the rate of recovery is used to calculate a diffusion coefficient, which is a measure of the molecule's rate of diffusion within the bilayer. The degree of fluidity depends on factors such as temperature, the length and saturation of the fatty acid chain portion of phospholipids, and the presence/absence of specific lipids such as cholesterol.

4. The amphipathic nature of phospholipid molecules (a hydrophilic head and hydrophobic tail) allows these molecules to self-assemble spontaneously into closed bilayer structures when in an aqueous environment. The phospholipid bilayer provides a barrier with selective permeability that restricts the movement of hydrophilic molecules and macromolecules in and out of the compartment. The different types of proteins present on the two faces of the bilayer contribute to the distinctive functions of each compartment's interior and exterior, and control the movement of selected hydrophilic molecules and macromolecules across the bilayer.

5. Membrane-associated proteins may be classified as integral membrane proteins, lipid-anchored membrane proteins or peripheral membrane proteins. Integral membrane proteins pass through the lipid bilayer and are therefore composed of three domains: a cytosolic domain exposed on the cytosolic face of the bilayer, a exoplasmic domain exposed on the exoplasmic face of the bilayer, and a membrane-spanning domain, which passes through the bilayer and connects the cytosolic and exoplasmic domains. Lipid-anchored membrane proteins have one or more covalently-attached lipid molecules, which embed in one leaflet of the membrane and thereby anchor the protein to one face of the bilayer. Peripheral proteins associate with the lipid bilayer through interactions with either integral membrane proteins or with phospholipid heads on one face of the bilayer.

194

6. Since biomembranes form closed compartments, one face of the bilayer is automatically exposed to the interior of the compartment while the other is exposed to the exterior of the compartment. Each face therefore interacts with different environments and perform different functions. The different functions are in turn directly dependent on the specific molecular composition of each face. For example, different types of phospholipids and lipid-anchored membrane proteins are typically present on the two faces. In addition, different domains of integral proteins are exposed on each face of the bilayer. Finally, in the case of the plasma membrane, the lipids and proteins of the exoplasmic face are often modified with carbohydrates.

7. The major organelles of the eukaryotic cell are the nucleus, the endoplasmic reticulum, the Golgi complex, mitochondria, chloroplasts (plants), endosomes, and lysosomes (animals) or vacuoles (plants). The nucleus encloses the cell's DNA, contains the machinery for RNA synthesis, and physically separates the process of RNA synthesis from that of protein synthesis. The endoplasmic reticulum is responsible for the synthesis of lipids, membrane proteins, and secreted proteins. The Golgi complex processes and sorts secreted and membrane proteins. Mitochondria are the principal sites of ATP synthesis in aerobic cells. Chloroplasts are the site of photosynthesis, which produces ATP and carbohydrates in plant cells. Endosomes take up soluble material from the extracellular environment and lysosomes (in animal cells) degrade much of the material internalized in endosomes for use in biosynthetic reactions. In plant cells, vacuoles function like lysosomes in that they degrade material, but they also serve as storage sites for water, ions, and nutrients, and generate the turgor pressure that drives plant cell elongation.

8. The multiple membranes of mitochondria and chloroplasts act to create additional compartments with specialized functions within these organelles. The polarized stack of compartments that form the Golgi apparatus is associated with the assembly line organization of enzymes that modify many ER-derived products.

9. Specific types of cells in suspension may be isolated by a fluorescence-activated cell sorter (FACS) machine in which cells previously "tagged" with a fluorescent-labeled antibody are separated from cells not recognized by the antibody. The scientist selects an antibody specific for the cell type desired. Specific organelles are generally separated by centrifugation of lysed cells. A series of centrifugations of successive supernatant fractions at increasingly higher speeds and corresponding higher forces serve to separate cellular organelles from one another on the basis of size and mass (larger, heavier cell components pellet at lower speeds). This is often combined with density-gradient separations to purify specific organelles on the basis of their buoyant density.

10. Integral and lipid-anchored membrane proteins require the use of detergents during the isolation process as the membrane must be disrupted, or solubilized, in order for these proteins to be purified. Detergents are amphipathic molecules that can intercalate into lipid membranes and fragment the bilayer by forming mixed micelles of detergents, phospholipids and intergral/lipid-anchored membrane proteins. Peripheral membrane proteins, in comparison, may be isolated with the use of high salt solutions, which disrupt the protein-protein or protein-lipid interactions that hold peripheral membrane proteins to the bilayer but have no effect on the fundamental structure of the lipid bilayer.

11. Actin filaments (microfilaments) are composed of monomeric actin protein subunits assembled into a twisted, two-stranded polymer. Actin filaments provide structural support, particularly to the plasma membrane, and are important for certain types of cell motility. Microtubules are composed of α- and β-tubulin heterodimers assembled into a hollow, tube-like cylinder. Microtubules provide structural support, are involved in certain types of cell

motility, and help generate cell polarity. Intermediate filaments are formed from a family of related proteins such as keratin or lamin. The subunits assemble to create a strong, rope-like polymer that, depending on the specific protein, may provide support for the nuclear membrane or for cell adhesion.

12. Cytoskeletal filaments are typically organized into bundles or networks within the cell. These structures are created and maintained by various cross-linking proteins.

13. Electron microscopy has better resolution than light microscopy, but many light-microscopy techniques allow observation and manipulation of living cells.

14. Chemical stains are required for visualizing cells and tissues with the basic light microscope because most cellular material does not absorb visible light and therefore cells are essentially invisible in a light microscope. The chemical stains that may be used to absorb light and thereby generate a visible image usually bind to a certain class of molecules rather than a specific molecule within that class. For example, certain stains may reveal where proteins are in a cell but not where a specific protein is located. This limitation can be overcome by fluorescence microscopy, in which a fluorescent molecule may be either directly or indirectly attached to a molecule of interest then viewed by an appropriately equipped microscope. Only light emitted by the sample will form an image, so the location of the fluorescence indicates the location of the molecule of interest. Confocal scanning microscopy and deconvolution microscopy build on the ability of fluorescence microscopy by using either optical (confocal scanning) or computational (deconvolution) techniques to remove out-of-focus fluorescence and thereby produce much sharper images. As a result, these techniques facilitate optical sectioning of thick specimens as opposed to physical sectioning and associated techniques that may alter the specimen.

15. Certain electron microscopy methods rely on the use of metal to coat the specimen. The metal coating acts as a replica of the specimen, and the replica rather than the specimen itself is viewed in the electron microscope. Methods that use this approach include metal shadowing, freeze fracturing and freeze etching. Metal shadowing allows visualization of viruses, cytoskeletal fibers, and even individual proteins, while freeze fracturing and freeze etching allow visualization of membrane leaflets and internal cellular structures.

Analyze the Data

a.

Organelle	Marker molecule	Enriched fraction (no.)
Lysosomes	Acid phosphatase	11
Peroxisomes	Catalase	7
Mitochondria	Cytochrome oxidase	3
Plasma membrane	Amino acid permease	15
Rough endoplasmic reticulum	Ribosomal RNA	5
Smooth endoplasmic reticulum	Cytidylyl transferase	12

b. The rough endoplasmic reticulum is more dense than the smooth endoplasmic reticulum, since it is found in a gradient fraction with a higher sucrose concentration (more dense solution).

c. The plasma membrane represents the least dense fraction because it has been fragmented into small pieces that re-seal to form small vesicle-like structures.

d. Addition of a detergent to the homogenate would eliminate the basis for equilibrium density-gradient centrifugation, as each organelle membrane would be solubilized by the detergent. Subjecting a detergent-treated homogenate to equilibrium density-gradient centrifugation would most likely produce a single peak at relatively low percent sucrose that would contain all marker molecules.

6

Integrating Cells Into Tissues

Review the Concepts

1. The diversity of adhesive molecules has arisen from 1) duplication of a common ancestor gene followed by divergent evolution producing multiple genes encoding related isoforms; and 2) alternative splicing of a single gene to yield many mRNAs, each encoding a distinct isoform.

2. Homophilic interactions are those between like cell types, e.g., epithelial cells with epithelial cells. One approach to demonstrating homophilic cell interactions experimentally is to use L cell lines transfected with E-cadherin and P-cadherin. L cells adhere poorly to each other and express no cadherins. When transfected with E- and P-cadherin, L cells adhere tightly to E-cadherin-positive cell to E-cadherin-positive cell and P-cadherin-positive cell to P-cadherin-positive cell. Cadherins directly cause homotypic interactions among cells.

3. Tight junctions help to hold cells together in tissues, and control the flow of solutes between cells in an epithelial sheet. Several things can happen to cells when tight junctions do not function. In hereditary hypomagnesemia, defects in tight junctions prevent the normal flow of magnesium through tight junctions in the kidney. Low blood magnesium levels result and this can lead to convulsions. Altering tight junctions in hair cells of the cochlea of the inner ear can result in deafness.

4. Collagen is a major component of the extracellular matrix in animal cells. Collagen is a protein that has a trimeric structure with rodlike and globular domains that form a 2-dimensional network. Collagen is synthesized in its precursor form by ribosomes attached to the endoplasmic reticulum. These pro-α chains undergo a series of covalent modifications and get folded into a triple helical procollagen molecule. The folded procollagen is transported through the Golgi and the chains are secreted to the outside of the cell. Once outside the cell, peptidases cleave the N- and C-terminal propeptides. The triple helices are then able to form larger structures called collagen fibrils.

5. The RGD sequence on fibronectin mediates binding to integrin proteins. If RGD-containing peptides were added to a layer of fibroblasts grown on a fibronectin substrate in tissue culture, the RGD peptides would compete with fibronectin for binding to the integrins present in the fibroblast extracellular matrix. As a result, the fibroblasts would likely lose adherence to the fibronectin substrate.

6. Fibronectin contains fibrin-binding domains. Since fibrin is a major constituent of blood clots, this domain allows fibronectin to recruit blood clots.

7. Structural studies have shown that integrin exists in both a non-active, low-affinity or "bent" form and an active, high affinity or "straight" form. In outside-in signaling, molecules of the ECM can bind to the extracellular portion of inactive integrin and induce conformational changes that lead to the straightening of the intracellular, cytoplasmic tails of integrin. The straightening of the cytoplasmic tails can stimulate intracellular components such as the cytoskeleton and parts of signaling pathways. This structure also facilitates inside-out signaling. For example, when the metabolic state of the cell is altered, adapter proteins inside the cell can interact with the cytoplasmic tails of integrin and cause straightening from the inside. This would result in new ECM interactions on the outside of the cell.

8. The dystrophin gene, which is defective in Duchenne muscular dystrophy, is an adapter protein that binds to cytoskeletal components such as actin, and to the cell adhesion molecule dystroglycan. Normally, dystrophin and dystroglycan function in an important part of the

signaling relay linking the extracellular matrix on the outside of the muscle cell to the cytoskeleton and signaling components inside of the muscle cell. When any of these components is defective, the muscle cells do not develop and function properly and muscular dystrophy results.

9. A cell strain is a lineage of cells originating from a primary culture taken from an organism. Since these cells are not transformed, they have a limited lifespan in culture. In contrast, a cell line is made of transformed cells and so these cells can divide indefinitely in culture. Such cells are said to be immortal. A clone results when a single cell is cultured and gives rise to genetically identical progeny cells.

10. Normal B lymphocyte cells can produce a single type of antibody molecule. However, such cells have a finite lifespan in culture. Researchers use cell fusion of B lymphocytes and immortalized myeloma cells to create immortalized, antibody-secreting cells. Such cells, called hybridoma cells, retain characteristics of both parent cells which allows for production of a single type, or monoclonal, antibody.

Analyze the Data

a. Cells transfected with wild-type E-cadherin aggregate more than untransfected cells because the increase in E-cadherin allows for more cell-cell interactions via the ECM.

b. Mutant A behaves almost identically to the wild-type E-cadherin in the aggregation assay, therefore this mutation does not change the function of E-cadherin as far as homophilic interactions are concerned. Expression of mutant B, however, does not result in increased aggregation, so this particular mutation does alter the adhesive qualities of E-cadherin.

c. The monoclonal antibody specific for E-cadherin blocks the resulting aggregation because it binds to E-cadherin and blocks homophilic interactions. In contrast, the non-specific antibody does not specifically interact with E-cadherin, and does not block homophilic interactions.

d. Since cadherins require calcium for function, lowering the calcium in the media during the assay would lower the aggregation ability.

7

Transport of Ions and Small Molecules Across Cell Membranes

Review the Concepts

1. Ethanol is the much more membrane permeant of the two at neutral pH. It has no acidic or basic group and is uncharged at a pH of 7.0. The carboxyl group of acetic acid is predominantly dissociated at this pH and hence acetic acid exists predominantly as the negatively charged acetate anion. It is non-permeant. At a pH of 1.0, ethanol remains uncharged and membrane permeant. The carboxyl group of acetic acid is now predominantly non-dissociated and uncharged. Hence, acetic acid is now membrane permeant. Any difference in permeability is very small.

2. Uniporters are slower than channels because they mediate a more complicated process. The transported substrate both binds to the uniporter and elicits a conformational change in the transporter. A uniporter transports one substrate molecule at a time. In contrast, channel proteins form a protein-lined passageway through which multiple water molecules or ions move simultaneously, single file, at a rapid rate.

3. The three classes of transporters are uniporters, symporters, and antiporters. Both symporters and antiporters are capable of moving organic molecules against an electrochemical gradient by coupling an energetically unfavorable movement to the energetically favorable movement of a small inorganic ion. The ΔG for bicarbonate has two terms, a concentration term and an electrical term, because bicarbonate is an anion. Glucose is neutrally charged and hence its ΔG for transport has only a concentration term. Unlike pumps, neither symporters nor antiporters hydrolyze ATP or any other molecule during transport. Hence, these cotransporters are better referred to as examples of secondary active transporters rather than as actual active transporters. The term active

transporter is restricted to the ATP pumps where ATP is hydrolyzed in the transport process.

4. Uniporters mediate substrate-specific facilitated diffusion. Uptake of glucose by GLUT1 exhibits Michaelis-Menten kinetics and approaches saturation as the concentration of glucose is increased. For any given substrate, GLUT1 displays a characteristic K_m at which concentration GLUT1 is transporting the substrate at 50% of V_{max}. Determination of the rate of erythrocyte (GLUT1) transport of substrate versus concentration allows the determination of K_m for glucose versus galactose versus mannose. The K_m for glucose will be lowest, indicating that GLUT1 is glucose-specific, not galactose- or mannose-specific.

5. The four classes of ATP-powered pumps are: P-class, V-class, F-class and ABC superfamily. Only the ABC superfamily members transport small organic molecules. All other classes pump cations or protons. The initial discovery of ABC superfamily pumps came from the discovery of multidrug resistance to chemotherapy and the realization that ultimately this was due to transport proteins, i.e., ABC superfamily pumps. Today, the natural substrates of ABC superfamily pumps are thought to be small phospholipids, cholesterol and other small molecules.

6. Total genome information allows the identification of the complete set of open reading frames within an organism. These may then be compared by sequence homology to known example transporters. For example, GLUT1 was characterized from erythrocyte plasma membranes. By homology, there are 12 proteins in the human genome that are highly similar in sequence. As the complete mouse genome is known, a similar homology grouping of proteins can be made. Several of the human GLUT family members remain "orphan" transporters as their natural substrate and physiology remains

unknown. In other words, they are members by homology, but that alone is not sufficient to establish their actual physiological roles. For example, GLUT5 transports fructose. Establishment of actual roles requires "hunt and peck" biochemical and physiological experiments. For example, Northern blotting of a DNA probe to mRNA from various tissues will establish in which tissue(s) a given GLUT family member is present, but nevertheless does not establish actual substrate or physiological importance. In conclusion, a series of different kinds of experiments are needed.

7. The P-class Ca^{2+} ATPase of the sarcoplasmic reticulum (SR) pumps Ca^{2+} ions from the cytosol into the SR. Hence, it lowers the cytosolic Ca^{2+} concentration and induces muscle relaxation. Selective inhibition of this Ca^{2+} ATPase will prolong the period of elevated cytosolic Ca^{2+} associated with muscle contraction and hence prolong the length and/or strength of muscle contraction.

8. Membrane potential refers to the voltage gradient across a biological membrane. The generation of this voltage gradient involves three fundamental elements: a membrane to separate charge, a Na^+/K^+ ATPase to achieve charge separation across the membrane, and nongated K^+ channels to selectively conduct current. The major ionic movement across the plasma membrane is that of K^+ from inside to outside the cell. Movement of K^+ outward, powered by the K^+ concentration gradient generated by Na^+/K^+ ATPase, leaves an excess of negative charges on the inside and creates an excess of positive charges on the outside of the membrane. Thus an inside-negative membrane potential is generated. These potassium channels are referred to as resting K^+ channels. This is because these channels, although they alternate between an open and closed state, are not affected by membrane potential or by small signaling molecules. Their opening and closing is non-regulated. Hence, the channels are called nongated. K^+ channels achieve selectivity for K^+, versus, say, Na^+, through coordination of the nonhydrated ion with carbonyl groups carried by amino acids within the channel protein. The ion enters the channel as a hydrated ion, the water

of hydration is exchanged for interaction with carbonyl residues within the channel, and then as the ion exits the channel it is rehydrated. Within the confines of the channel protein structure, Na^+ unlike K^+ is too small to replace fully the interactions of water with those with amino acid carried carbonyl groups. Because of this, the energetic situation is highly unfavorable for Na^+ versus K^+.

9. Expression of a channel protein in a normally nonexpressing cell permits the patch clamp assessment of channel properties. Typically, the cell used is a frog oocyte. Frog oocytes do not normally express plasma membrane channel proteins. Channel protein expression may be induced by microinjection of in vitro transcribed mRNA encoding the protein. Frog oocytes are large and hence technically easier to inject and to patch clamp than other cells.

10. Plant cells, unlike animal cells, are surrounded by a cell wall. This cell wall is relatively stiff and rigid. The hyperosmotic situation within the plant vacuole that typically constitutes most of the volume of the plant cell is resisted by the rigid cell wall and the cell does not burst. Overall, a plant cell is considered to have a turgor pressure because of the hyperosmotic vacuole. The Na^+/K^+ ATPase is key to animal cells avoiding osmotic lysis. Animal cells have a slow inward leakage of ions. In the absence of a countervailing export, this would result in osmotic lysis of the cells even under isotonic conditions. The main countervailing export is the net transport of cations by Na^+/K^+ ATPase (3 Na^+ ions out for 2 K^+ in).

11. The Na^+/K^+ ATPase located on the basolateral surface of intestinal epithelial cells uses energy from ATP to establish Na^+ and K^+ ion gradients across the intestinal epithelial cell plasma membrane. Cotransporters couple the energetically unfavorable movement of glucose and amino acids into epithelial cells to the energetically favorable movement of Na^+ into these cells. The accumulation of gucose and amino acids here is an important example of secondary active transport. Tight junctions are essential for the process because they seal the

interstitial space between cells and hence allow the transport proteins in the apical and basolateral membranes of the epithelial cell to be effective. Effective transport could not be achieved through a leaky cell layer. The coordinated transport of glucose and Na+ ions across the intestinal epithelium creates a transepithelial osmotic gradient. This forces the movement of water from the intestinal lumen across the cell layer and hence promotes water absorption from sport drinks.

12. An action potential has three phases: depolarization, in which the local negative membrane potential goes to a positive membrane potential; repolarization, in which the membrane potential goes from positive to negative; and hyperpolarization, in which the resting negative membrane potential is exceeded. Depolarization corresponds to the opening of voltage-gated Na^+ channels and a resulting influx of Na^+. Repolarization corresponds to the opening of voltage-gated K^+ channels. Hyperpolarization corresponds to a period of closure and inactivation of voltage-gated Na^+ channels. The Na^+ channels involved in the propagation of an action potential are referred to as voltage-gated channels because they only open in response to a threshold potential.

13. Myelination is the development of a myelin sheath about a nerve axon. The myelin sheath is an outgrowth of neighboring glial (Schwann) cell plasma membrane about the axon that repeatedly warps itself around the neural extension until all the cytosol between the layers of membrane is forced out. The remaining membrane is the compact myelin sheath. The myelin sheath serves as an insulator about the axon and hence speeds the rate of action potential propagation by 10-100-fold.

The myelin sheath surrounding an axon is formed from many glial cells. Between each region of myelination is a gap, the node of Ranvier. The voltage-gated Na^+ channels that generate the action potential are all located in the nodes. The action potential spreads passively through the axonal cytosol to the next node. This produces a situation where the action potential in effect jumps from node to node. If the nodes are located too far apart, for example, a 10-fold increase in

spacing, then the passive spread of the action potential may become too slow to jump from node to node.

14. Both the accumulation of neurotransmitters in synaptic vesicles and the accumulation of sucrose in the lumen of the plant vacuole are mediated by H^+-linked antiporters. In both cases, the proton moves down an electrochemical gradient to power the inward movement of the small organic molecule against a concentration gradient. Acetylcholine (as a released neurotransmitter) is degraded by acetylcholine esterase after its release into the synaptic cleft. Decreased acetylcholine esterase activity at the nerve-muscle synapse will have the effect of prolonging signaling and hence prolonging muscle contraction.

15. The dendrite is the neuron extension that receives signals at synapses and the axon is the neuron extension that transmits signals to other neurons or muscle cells. Typically, each neuron has multiple dendrites that radiate out from the cell body and only one axon that extends out from the cell body. At the synapse, the dendrite may have either excitatory or inhibitory receptors. Activation of these receptors results in either a small depolarization or small hyperpolarization of the plasma membrane. These depolarizations move down the dendrite to the cell body and then to the axon hillock. When the sum of the various small depolarizations and hyperpolari-zations at the axon hillock reaches a threshold potential, an action potential is triggered. The action potential then moves down the axon.

Analyze the Data

a. As shown in the top figure, there is a time dependent increase in glutamate (Glu) transport in the vesicles prepared from transfected cells but not in those from wild type cells. Note that Glu transport reaches an essentially maximal value after 2 min. In vesicles prepared from wild-type cells, the Glu transport fluctuates and shows no pattern of time dependent increase. By inspection of the bottom figure, the apparent K_m for Glu transport is approximately 1 mM. This is 10-100-fold higher than that expected for a plasma membrane excitatory amino acid transporter.

b. Addition of L-glu, D-glu or valinomycin plus nigericin (V + N) all depress BNP1 transport of labeled L-glutamate. In contrast, Asp (aspartate) and Gln (glutamine) do not depress BNP1 transport; transport remains at about 100. L-glu is the normal stereoisomer of glutamate while D-glu is not. Overall, these results indicate that BNP1 is an L-glu selective transporter whose activity is dependent on the membrane electrical gradient.

c. Clearly, there is a biphasic dependence on chloride concentration for BNP1 transport in isolated vesicles. Glu transport increases as chloride concentration is increased to 4-5 mM and then decreases at higher chloride concentrations. The control, wild-type vesicles show no change in transport properties with variation in chloride concentration. How then might the BNP1 chloride dependence be explained? The normal cytosolic concentration of chloride in vertebrate cells is 4 mM (Table 7-2, MCB 5e). One plausible explanation is that the glutamate transporter BNP1 is an antiporter that transports glutamate and chloride in opposite directions. If so, it might well be evolved to work best at physiological concentrations of chloride anion. An alternate explanation is that BNP1 has an anion binding site whose extent of saturation with chloride affects its transport properties. Based on the experimental data presented, it is not possible to know what the actual explanation is.

d. Logically, mutations modeled on the C. elegans EAT-4 protein ought to be important in the mammalian homolog. A direct approach to testing this is to prepare modified BNP1 encoding cDNAs based on the C. elegans mutations, express these in PC12 cells, isolate vesicle populations from these new transfectants and then characterize their Glu transport properties.

Cellular Energetics

8

Review the Concepts

1. The pmf is generated by a voltage and chemical (proton) gradient across the inner membrane of mitochondria and the thylakoid membrane of chloroplasts. Like ATP, the pmf is a form of stored energy, and the energy stored in the pmf may be converted to ATP by the action of ATP synthase.

2. The unique properties of the mitochondrial inner membrane include the presence of membrane invaginations (termed cristae), a higher than normal protein concentration, and an abundance of the lipid cardiolipin. The cristae increase the surface area of the inner membrane, thereby increasing the total amount of membrane and hence electron transport chain components, ATP synthase molecules, and transporters of reagents and products of the citric acid cycle and/or oxidative phosphorylation are all increased. The higher protein concentration, mostly proteins involved in electron transport and ATP synthesis, further increases the capacity of mitochondria to synthesize ATP. Finally, cardiolipin enhances the barrier properties of the inner membrane by reducing the membrane's permeability to protons.

3. Glycolysis does not require oxygen, but the citric acid cycle and the electron transport chain do require oxygen to function. In the case of the citric acid cycle, oxygen is not directly involved in any reaction, but the cycle will come to a halt as NAD^+ and FAD levels drop in the absence of oxygen. For electron transport, oxygen is required as an electron acceptor. In the absence of oxygen, certain eukaryotic organisms (facultative anaerobes) as well as certain cells (mammalian skeletal muscles during prolonged contraction) can produce limited amounts of ATP by glycolysis (a process known as fermentation).

4. Electrons produced by glycolysis are delivered to the electron transport chain via electron shuttles in the mitochondrial inner membrane. The most common shuttle is the malate-aspartate shuttle, which utilizes two different antiporters and various intermediate carriers to remove electrons from NADH in the cytosol and add them to NAD^+ in the matrix. The steps in this process are as follows. First, cytosolic malate dehydrogenase transfers electrons from cytosolic NADH to oxaloacetate to form malate. Second, the malate/α-ketoglutarate antiporter transports malate into the matrix in exchange for α-ketoglutarate. Third, mitochondrial malate dehydrogenase converts malate to oxaloacetate, reducing matrix NAD^+ to NADH in the process. Fourth, oxaloacetate is converted to the amino acid aspartate by addition of an amino group from the amino acid glutamate (the "rest" of glutamate is converted to α-ketoglutarate). Fifth, the glutamate/aspartate antiporter exports aspartate to the cytosol in exchange for glutamate. Finally, aspartate is converted to oxaloacetate in the cytosol to complete the cycle. If a mutation inactivated the malate-aspartate shuttle, initially there would be a small reduction in the efficiency of oxidative phosphorylation as there would be a small reduction in the electrons delivered to the electron transport chain per glucose molecule. Longer term, glycolysis would be inhibited as the availability of NAD^+ for glycolytic reactions would become limiting.

5. Prosthetic groups are small nonpeptide organic molecules or metal ions that are tightly associated with a protein or protein complex. Several types of heme, an iron-containing prosthetic group, are associated with the cytochromes. The various cytochromes in the electron transport chain contain heme prosthetic groups with different axial ligands, and as a result, each cytochrome has a different reduction potential, so that electrons can move only in sequential order through the electron carriers.

6. The underlying reason for the difference in ATP yield for electrons donated by $FADH_2$ and NADH is that the electrons carried in $FADH_2$ have less potential energy (43.4 kcal/mol) than the electrons carried in NADH (52.6 kcal/mol). Thus, $FADH_2$ transfers electrons to the respiratory chain at a later point than NADH does, resulting in the translocation of fewer protons, a smaller change in pH, and fewer synthesized ATP molecules.

7. Aerobic bacteria carry out oxidative phosphorylation by the same processes that occur in mitochondria (and are simpler and easier to work with than mitochondria). Glycolysis and the citric acid cycle take place in the bacterial cell cytosol, while electron transport components are localized to the bacterial plasma membrane. Since electron transport takes place at the plasma membrane, the pmf is generated across the plasma membrane. In addition to using the pmf to synthesize ATP, aerobic bacteria also use the pmf to power uptake of certain nutrients and cell swimming.

8. In addition to providing energy to power ATP synthesis, the pmf also provides the energy used by several active transport proteins to move substrates into the mitochondria and products out of the mitochondria. The OH^- gradient, which results from generation of the pmf by electron transport, is used to move HPO_4^{2-} into the matrix, and the voltage gradient contribution of the pmf drives exchange of ADP for ATP.

9. The Q cycle functions to double the number of protons transported per electron pair moving through a specific complex of the electron transport chain, and thereby maximizes the pmf across a membrane. In mitochondria, the specific complex is the $CoQH_2$-cytochrome c reductase complex, while in chloroplasts it is the cytochrome bf complex and in purple bacteria it is the cytochrome bc_1 complex. Using mitochondria as an example, the Q cycle is believed to function as follows: $CoQH_2$ arrives at the Q_o site on the intermembrane space side of the $CoQH_2$-cytochrome c reductase complex; it delivers two electrons to the complex, and releases two protons into the intermembrane space. Next, one electron is transported directly to cytochrome c while the

other partially reduces a CoQ molecule bound to the Q_i site on the inner side of the complex, forming a CoQ semiquinone anion. CoQ dissociates from the Q_o site and is replaced by another $CoQH_2$, which delivers two more electrons and releases two protons to the intermembrane space. As before, one electron is transferred to cytochrome c, but the other combines with the CoQ semiquinone anion at the Q_i site to produce $CoQH_2$, thus regenerating one $CoQH_2$. In sum, the net result of the Q cycle is that four protons are transported to the intermembrane space for every two electrons moving through the $CoQH_2$-cytochrome c reductase complex.

10. $$6CO_2 + 6H_2O \rightarrow 6O_2 + C_6H_{12}O_6$$

O_2-generating photosynthesis uses the energy of absorbed light to create, via electron donation to quinone, the powerful oxidant P^+ form of the reaction center chlorophyll. This, in turn, acts to remove electrons from H_2O, a poor electron donor. The electrons are then passed along an electron transport chain, and the stored energy is converted to other forms for subsequent use in ATP synthesis and carbon fixation. The O_2 is not used in subsequent reactions in this pathway and thus is a by-product of the removal of electrons from H_2O.

11. Photosynthesis consists of four stages. During Stage 1, which occurs in the thylakoid membrane, light is absorbed by the reaction center chlorophyll, a charge separation is generated, and electrons are removed from water, forming oxygen. During Stage 2, electrons are transported via carriers in the thylakoid membrane to the ultimate electron donor, $NADP^+$, reducing it to NADPH, and protons are pumped from the stroma into the thylakoid lumen, producing a proton gradient across the thylakoid membrane. During Stage 3, protons move down their electrochemical gradient across the thylakoid membrane through F_0F_1 complexes and power ATP synthesis. Finally, during Stage 4, the ATP and NADPH generated in the earlier stages is used to drive CO_2 fixation and carbohydrate synthesis. CO_2 fixation occurs in the stroma and

carbohydrate (sucrose) synthesis occurs in the cytosol.

12. Chlorophyll *a* is present in both reaction centers and antenna. Additionally, antennas contain either chlorophyll *b* (vascular plants) or carotenoids (plants and photosynthetic bacteria). Antennas capture light energy and transmit it to the reaction center, where the primary reactions of photosynthesis occur. The primary evidence that these pigments are involved in photosynthesis is that the absorption spectrum of these pigments is similar to the action spectra of photosynthesis.

13. Photosynthesis in green and purple bacteria does not generate oxygen because these bacteria have only one photosystem, which cannot produce oxygen. These organisms still utilize photosynthesis to produce ATP by utilizing cyclic electron flow to produce a pmf (but no oxygen or reduced coenzymes), which can be utilized by F_0F_1 complexes. Alternatively, this photosystem can exhibit linear, noncyclic electron flow, which will generate both a pmf and NADH. For linear electron flow, hydrogen gas (H_2) or hydrogen sulfide (H_2S) rather than H_2O donates electrons, so no oxygen is formed.

14. PSI is driven by light of 700 nm or less and its primary function is to transfer electrons to the final electron acceptor, $NADP^+$. PSII is driven by light of 680 nm or less, and its primary function is to split water to yield electrons, as well as protons and oxygen. During linear electron flow, electrons move as follows: PSII (water split to produce electrons) -> plastoquinone (Q) -> cytochrome *bf* complex -> Plastocyanin -> PSI -> $NADP^+$. The energy stored as NADPH is used to fix CO_2 and ultimately synthesize carbohydrates.

15. The Calvin cycle reactions are inactivated in the dark to conserve ATP for the synthesis of other cell molecules. The mechanism of inactivation depends on the enzyme; examples include pH-dependent and Mg^{2+}-dependent enzyme regulation, as well as reversible reduction-oxidation of disulfide bonds within certain Calvin cycle enzymes.

Analyze the Data

a. The electron transport system normally pumps protons out of the mitochondrial matrix, increasing the pH of the matrix; thus the fluorescence of matrix-trapped BCECF would increase in intensity. The observed decrease in intensity of BCECF trapped inside the vesicles suggests that the vesicles have an inverted (inside-out) orientation, so that protons were pumped from the outside to the inside of the vesicles.

b. The concentrations of ADP, P_i, and oxygen should decrease over time as the process of oxidative phosphorylation utilizes oxygen as an electron acceptor and uses ADP and P_i to synthesize ATP.

c. Dinitrophenol compromises the pH gradient and the resulting equilibration of protons leads to an increase in the intravesicular pH and corresponding increase in emission intensity. Valinomycin, a potassium ionophore, affects the electric potential more than the pH gradient. Since BCECF fluorescence reflects the pH of the milieu, it is largely unaffected by valinomycin-induced changes in the transmembrane electric potential.

d. The fluorescence intensity inside the vesicles should remain constant over time since inner mitochondrial membranes from brown fat tissue would likely contain thermogenin, a protein that functions as an uncoupler of oxidative phosphorylation. Since thermogenin is a proton transporter, its presence would prevent the generation of a proton gradient and thus no fluorescence change would be expected.

9

Molecular Genetic Techniques and Genomics

Review the Concepts

1. A recessive mutation is defined as a mutation that must be present in both alleles of a diploid organism in order for the mutant phenotype to be observed. That is, the individual must be homozygous for the mutation. In contrast, a dominant mutation produces a mutant phenotype even in the presence of one mutant and one wild type allele, such as in a heterozygous individual. A temperature-sensitive mutation is a mutation that is expressed conditionally, i.e., only at certain temperatures. For example, the wild type phenotype would be expressed at the permissive temperature, whereas the mutant phenotype would be expressed at the non-permissive temperature. Presumably, a conformational change occurs in the protein at the non-permissive temperature that results in the mutant phenotype. A temperature-sensitive mutation is particularly useful for identifying and studying genes that are essential for survival.

2. Complementation analysis can be used to determine whether two recessive mutations are present in the same or different genes. If a heterozygous organism containing both mutations shows the mutant phenotype, then the two mutations are in the same gene because neither allele provides a functional copy of the gene. In contrast, if a heterozygous organism shows a wild type phenotype, then the two mutations are in different genes because a wild type allele of each gene is present. A suppressor mutation is a compensatory mutation in another or the same gene, which leads to suppression of the original mutant phenotype. A synthetic lethal mutation is a mutation that enhances rather than suppresses the deleterious effect of another mutation.

3. Bacteria that synthesize restriction enzymes also synthesize a DNA modifying enzyme to protect its own DNA. The modifying enzyme is a methylase, which methylates the host DNA. Methylated DNA is no longer a substrate for the encoded restriction enzyme. Restriction enzyme sites commonly consist of 4-8 base pair, palindromic sequences. After being cut with a restriction enzyme, the ends of the cut DNA molecule can exist as a single-stranded tail (sticky end) or as a blunt (flush) end. DNA ligase is an enzyme that catalyzes the reformation of the phosphodiester bond between nucleotides in the presence of ATP.

4. A plasmid is a circular, extrachromosomal DNA molecule that contains an origin of replication, a marker gene that permits selection, and a region in which foreign DNA can be inserted (cloning site). A plasmid is useful for cloning DNA fragments up to approximately 20 kb. A lambda phage vector contains genes required for head and tail formation and genes required for the lytic cycle. Non essential genes are removed and can be replaced with foreign DNA. Lambda phage are useful for cloning fragments up to about 25 kb. Cloning DNA into lambda phage is about 1000 times more efficient than cloning into plasmid vectors.

5. DNA libraries are a collection of randomly cloned DNA fragments. A cDNA (complementary DNA) library is a collection of DNA molecules that are copied from messenger RNA molecules using the enzyme reverse transcriptase. A cDNA library does not contain every gene, only those that are expressed as mRNA at the time of RNA isolation. In contrast, a genomic DNA library consists of random fragments of the total genome. This would include not only genes, but also areas of the genome that do not encode for genes. For hybridization screening, the lambda phage infected E. coli are plated in petri dishes and allowed to grow. The resultant phage are transferred to nylon filters where the phage are lysed and the DNA fixed to the filter. A radioactive probe is then hybridized to the filters and the location of lambda phage DNA that hybridizes to the radioactive probe is determined

by autoradiography. For expression screening, the protein encoded in the cloned DNA is expressed in the bacteria, which is then fixed to the nylon membrane. The presence of the protein can be detected by the presence of an antibody or an activity stain. Because of the degeneracy of the genetic code there is some ambiguity in the codons used to direct the amino acid sequence. For the peptide Met-Pro-Glu-Phe-Tyr, the nucleotide sequence could be 5'-ATG, CC(A, T, G or C), GA(A or G), TT(T or C), and TA(T or C)-3' for Met, Pro, Glu, Phe, and Tyr, respectively. Thus there would be 32 possible 15 nucleotide sequences that could encode for the peptide Met-Pro-Glu-Phe-Tyr.

6. The PCR reaction is performed as multiple cycles of a three-step process. The first step involves heat denaturation of a target DNA molecule. The second step involves cooling the DNA solution to allow annealing of short single-stranded oligonucleotide primers that are complementary to the target DNA molecule. In the final step, the hybridized oligonucleotides serve as primers for DNA synthesis. The resultant double-stranded DNA molecules are then subjected to further rounds of denaturation, annealing, and DNA synthesis (extension). A thermostable DNA polymerase was essential for automation of the PCR process. A non-thermostable DNA polymerase would be inactivated by heat denaturation during each cycle of the PCR process and would necessitate the addition of new enzyme prior to each DNA synthesis step.

7. Southern blotting is a technique in which DNA fragments are separated by size in a gel and then transferred to a solid support such as a nitrocellulose or nylon membrane. The DNA is fixed to the nylon membrane and hybridized to a labeled DNA or RNA probe. The hybridized probe is then detected by some technique such as autoradiography. Northern blotting is similar to Southern blotting, except that RNA is denatured and then separated on the gel instead of DNA. Southern blotting can be used to identify a DNA fragment that contains a DNA sequence of interest. Northern blotting can be used to determine the steady-state levels of a specific RNA.

8. In order to express a foreign gene, a recombinant plasmid would require a promoter for efficient transcription of the foreign gene. A promoter that is inducible would provide even higher expression levels of the foreign gene product. To facilitate purification of the foreign protein, a molecular tag can be added to the recombinant protein. An example of this type of molecular tag includes a short sequence of histidine residues (a polyhistidine sequence). The resultant His-tagged protein will bind specifically to a bead that has bound nickel atoms. Other proteins can be washed out and the His-tagged protein can be released from the nickel atoms by lowering the pH of the solution. Bacterial cells are limited in their capacity to synthesize complex proteins because of their inability to perform many post-translational modifications, such as glycosylation, that mammalian cells can perform. These post-translational modifications are essential for the biological activity of the recombinant protein.

9. Genomics is defined as the genome-wide analysis of the organization, structure and expression of genes. Proteomics is the global analysis of the function and expression of proteins. Computer searches for open reading frames are more useful for bacterial genomes because bacterial genes are present as an uninterrupted stretch of nucleotides. In contrast, in eukaryotes many of the genes are divided into exons and introns, which makes the search for genes more complex. Paralogous genes are genes that have diverged as a result of a gene duplication, i.e., two genes in an organism that have different functions but very similar nucleotide sequences. Orthologous genes are genes that arose because of speciation, i.e., genes found in different species that have very similar nucleotide sequences. Because of alternative splicing, a gene can give rise to numerous protein products. Thus a small increase in gene number could result in a very large increase in protein number. Thus the number of proteins and protein-protein interactions could be much greater in the organism with the larger genome.

10. A DNA microarray consists of hundreds or thousands of individual, closely packed gene-specific sequences attached to the surface of a glass microscope slide. The expression of each of

these genes can be analyzed globally following hybridization of the array with labeled cDNA prepared from RNA. Microarrays allow the simultaneous analysis of the expression of thousands of genes, whereas Northern blotting allows the analysis of gene expression a single gene at a time. Because Northern blotting involves the separation of RNA on a gel, the different sizes of a given mRNA can be observed, whereas microarray analysis does not allow the examination of different mRNA sizes.

11. To generate a knockout mouse, mouse embryonic stem cells are first transfected with a disrupted allele of the target gene. Through a process known as homologous recombination the disrupted allele replaces the functional homologous gene in the chromosome, resulting in a non-functional chromosomal gene. The ES cells, which now contain a mutant gene, are injected into a blastocyst. The blastocyst is transferred into a recipient mouse. Pups that are born will be chimeras. The loxP-Cre system can be used to conditionally knockout a gene. Using the above technology, loxP sites can be engineered to flank the gene of interest. Expression of the recombinase, Cre, in a specific tissue will result in loss of the flanked gene in that tissue. Knockout mice serve as models for human diseases. For example, if a human disease is known to result from a mutation in gene X, a knockout mouse can be generated that lacks gene X.

12. A dominant negative mutation is a mutation that produces a mutant phenotype even in cells carrying a wild type copy of the gene. This type of mutation produces a loss of function phenotype. RNA interference (RNAi) is a method of inactivating gene expression by selectively destroying RNA. In this method, a short double stranded RNA molecule is introduced into cells. This double stranded RNA base pairs with its target mRNA, promoting degradation of the mRNA by specific nucleases.

13. Restriction fragment length polymorphisms (RFLP) result from mutations that create or destroy restriction enzyme sites. As a result, DNA molecules with or without the restriction enzyme sites are cleaved into different sized fragments.

Single nucleotide polymorphisms (SNPs) are changes in a single nucleotide between two individuals. Simple Sequence Repeats (SSRs), also known as microsatellites, consist of a variable number of repeating one-, two-, or three-base sequences. The number of these repeat units at a specific genetic locus varies between individuals. All of these types of polymorphisms can be used as a molecular marker for mapping studies. The recombination frequency between two polymorphisms can be determined and serve as the basis for development of a genetic map. In general, the further two markers are separated on a chromosome, the greater the recombination frequency between those two markers and vice versa.

14. Once a gene is roughly located along a chromosome by genetic linkage studies, further analysis is required to identify the disease gene. One strategy for identifying a disease gene involves gene expression analysis. Comparison of gene expression in tissues from normal and affected individuals by Northern blot analysis may indicate a gene that is involved in the disease process. Northern blot analysis allows a comparison of both the level of expression and the size of the transcripts between normal and disease tissues. Sometimes expression levels and/or size of the transcripts are not altered between the normal and the disease states. In this case, DNA sequencing of a potential disease gene from tissues of a normal and disease state could reveal a single nucleotide change that results in the disease phenotype.

Analyze the Data

a. The Northern blot reveals the steady state mRNA levels for p24 (in the center) and p25 (on the right). For p24, only the transfection of siRNA-p24 reduces the amount of p24 mRNA; whereas the control transfection without siRNA and transfection with siRNA-p25 did not affect p24 mRNA levels. Similarly, for p25, only siRNA-p25 reduced p25 RNA levels. These results demonstrate that the siRNAs can cause specific degradation of their target RNAs.

b. In the cultured cells, transfection of either siRNA-p24 or siRNA-p25 yielded a viral titer that was slightly lower than the control transfection. This result indicates that reduction of either p24 mRNA or p25 mRNA and presumably p24 and p25 protein only minimally affects the ability of the virus to infect the cells. Transfection, however, of both siRNA-p24 and siRNA-p25 did result in a significant reduction in viral titer. In combination with the previous results, these results indicate that either p24 or p25 can be used as a viral receptor. You could hypothesize that loss of p24 and p25 mRNA would lead to a decrease in p24 and p25 protein, resulting in inhibition of viral infection and replication. Transfection of siRNA to a viral protein resulted in an even greater reduction of virus titer compared to transfection with siRNA to both p24 and p25. One possible explanation is that transfection of siRNA to a viral protein directly inhibited viral replication/assembly. A second possibility has to do with the half-life of p24 and p25 proteins. A reduction of p24 and p25 mRNA may not lead to an immediate reduction in p24/p25 protein due to the stability of the protein. The protein level is determined by the half-life of the protein and the length of time after transfection of siRNA. For example, if the half lives of p24 and p25 proteins are greater than 24 hours, then during the 20 hour incubation after transfection with siRNA there should still be more than half of the protein still present. In this problem no information about the half-life of the protein was given, so a definitive conclusion cannot be made. But based on the results, it is reasonable to conclude that the half-life of the p24 and p215 protein is relatively long, such that 20 hours after transfection with the siRNA, sufficient protein remains to act as a receptor for the virus.

c. Using the loxP-Cre system, the p24 or p25 gene can be selectively knocked-out in the liver or lung of transgenic mice. Using this approach, the receptor required for viral infection in specific tissues can be examined in tissue sections by immunohistochemistry. In the control, infection of a wild type mouse resulted in viral infection in both liver and lung tissue. In the transgenic mice lacking the p24 gene in lung, viral infection was seen in both the liver and lung. These results indicate that the other receptor protein, p25, can be used as a viral receptor in liver and lung. The mice, however, which have p24 knocked out in the liver had no viral infection in the liver, but did have viral infection in the lung, indicating that p24 is required for viral infection in the liver but not the lung. This result was unexpected because the cell culture experiment in part b indicated that either p24 or p25 could be used as a receptor for viral infection. Similarly, mice with a knockout of the p25 gene in liver demonstrate that p24 can be used as a viral receptor in liver and lung. Mice with a knockout of the p25 gene in lung had no viral infection in the lung, indicating that p25 was required for viral infection in lung but not liver. Again, this result was unexpected due to the results of the cell culture experiment in part b.

d. With the added information that p24 is expressed only in the liver and p25 is expressed only in the lung, the apparent discrepancy between the results from the knockout mouse study and the cell culture studies can be resolved. Both the p24 and p25 protein can be used as a viral receptor in cells, liver, and lung tissue. Because p24 is expressed only in the liver of mice, knockout of p24 in liver blocks viral infection because p25 is not present. Thus in liver, p24 is the protein used as the viral receptor. Similarly, because p25 is expressed only in the lung of mice, knockout of p25 in lung blocks viral infection because p24 is not present. Thus in lung, p25 is the protein used as the viral receptor. The Northern blot in part a shows that in the cultured cells both p24 and p25 are present. Thus in the cultured cells both p24 and p25 can act as the viral receptor protein.

10

Molecular Structure of Genes and Chromosomes

Review the Concepts

1. A gene is commonly defined as the entire nucleic acid sequence that is necessary for the synthesis of a functional gene product (RNA or polypeptide). This includes introns and the regulatory regions, e.g., promoters and enhancers of the gene. A complex transcription unit consists of multiple exons. These multiple exons can be spliced together in a number of different ways to generate different mRNAs. For example, assume that a hypothetical gene contains four exons, 1, 2, 3, and 4. Assuming that all mRNA from this gene must start with exon 1 and end with exon 4, which is not true for all genes, the possible combinations of alternatively spliced exons are exons 1-2-3-4, exons 1-2-4, exons 1-3-4, and exons 1-4. Because the sequence of each of these alternatively spliced mRNAs differ, the amino acid sequence of the encoded protein would differ.

2. Single or solitary genes are present once in the haploid genome. In multicellular organisms roughly 25-50% of the protein-coding genes are solitary genes. Gene families consist of a set of duplicated genes that encode proteins with similar but nonidentical amino acid sequences. An example of a gene family is the β-like globin genes. Pseudogenes are copies of genes that are nonfunctional even though they seem to have the same exon-intron structure as a functional gene. Pseudogenes probably arise from a gene duplication event followed by the accumulation of mutations that render the gene nonfunctional. Tandemly repeated genes are present in a head-to-tail array of exact or almost exact copies of genes. Examples of tandemly repeated genes include ribosomal RNAs, transfer RNAs, and histones.

3. Satellite, or simple sequence DNA, can be categorized as microsatellite or minisatellite DNA depending upon the size of the DNA fragment. Microsatellites contain repeats that contain 1-13 base pairs. Minisatellites consist of repeating units of 15-100 base pairs, present in relatively short regions of 1 to 5 kb made up of 20-50 repeat units. The number of copies of the tandemly repeated DNA varies widely between individuals. DNA fingerprinting is a technique that examines the number of repetitive units at a specific genetic locus. This technique can distinguish and identify individuals based on differences in their amount of repetitious DNA.

4. A bacterial insertion sequence, or IS element, is a member of the class of mobile DNA elements. The IS element usually contains inverted repeats at the end of the insertion sequence. Between the inverted repeats is a region that encodes the enzyme transposase. Transposition of the IS element involves a three step process. First, transposase excises the IS element in the donor DNA; second, it makes staggered cuts in a short sequence in the target DNA; and, third, it ligates the 3' termini of the IS element to the 5' ends of the cut donor DNA. The final step involves a host cell DNA polymerase, which fills in the single-stranded gaps, generating 5-11 base pair short direct repeats that flank the IS element before DNA ligase joins the free ends.

5. Retrotransposons transpose through an RNA intermediate. One class of retrotransposons, named LTR retrotransposons, contain long terminal repeats (LTRs) at their ends and a central protein-coding region that encodes the enzymes reverse transcriptase and integrase. The retrotransposon is first transcribed into RNA using host RNA polymerase. This RNA intermediate is then converted into DNA by the action of reverse transcriptase. The DNA copy is then inserted into the target site by the action of integrase. Retrotransposons that lack LTRs transpose by a different mechanism. Non LTR retrotransposons, of which LINES are an example, consist of direct repeats that flank a region that encodes for two proteins: ORF1, a RNA-binding

protein, and ORF2, which is similar to reverse transcriptase. The LINE element is first transcribed by host RNA polymerase, resulting in the translation of ORF1 and ORF2. ORF2 makes staggered nicks in the target DNA and reverse transcribes the LINE RNA. The RNA strand of the resulting RNA/DNA hybrid is hydrolyzed. A double stranded DNA is synthesized and then ligated into the nick introduced into the target DNA by ORF2.

6. Insertion of mobile DNA elements can generate spontaneous mutations that may influence evolution. In addition, recombination between mobile DNA elements may result in gene duplications and DNA rearrangements. Unequal crossing over between mobile elements could lead to gene duplications. Subsequent divergence of these duplicated genes leads to members of gene families with distinct functions. The inclusion of flanking DNA during transposition results in the movement of genomic DNA to another region of the genome. This could result in new combinations of transcriptional control regions. Recombination between mobile elements present in introns can generate novel combinations of exons, a process known as exon shuffling.

7. A nucleosome comprises a protein core of histones with DNA wound around its surface. The protein core consists of an octamer, containing two copies of histones H2A, H2B, H3 and H4. Approximately 150 base pairs of DNA are wrapped less than two times around the octameric histone core. The histone H1 binds to the linker region, which varies in length from 15 to 55 base pairs and is located between nucleosomes. The nucleosomes are packaged into an irregular spiral or solenoid arrangement to form a 30 nm fiber.

8. A eukaryotic chromosome consists of a long, linear DNA molecule. The DNA is wrapped around octameric histone cores to form chromatin, which is then condensed into a 30 nm fiber. Long loops of the 30 nm fiber are attached to a chromosome scaffold composed of non histone proteins. Scaffold associated regions (SARs) or matrix attachment regions (MARs) are regions in the DNA that bind to the chromosome

scaffold. Genes are primarily located within chromatin loops, i.e., between SARs or MARs.

9. Metaphase chromosomes can be identified by size, shape, banding patterns or hybridization to fluorescent probes (chromosome painting). G banding involves treating metaphase chromosomes with mild heat or proteolysis and staining them with Giemsa reagent. G band staining corresponds to regions that have low G+C content. Treating chromosome spreads with a hot alkali treatment prior to Giemsa staining results in a banding pattern that is almost the inverse of G bands, known as R bands. Chromosome painting is a technique for visualizing chromosomes using fluoresecent probes. In this method, DNA probes labeled with specific fluorescent tags are hybridized to metaphase chromosomes. After unbound probes are washed off, the chromosomes are visualized with fluorescence microscopy. Each chromosome then fluoresces with a different combination of fluorescent tags. Then a computer analyzes the tags and assigns a false color image. This way each chromosome can be identified by its false color image.

10. Replication origins are the points at which DNA synthesis is initiated. The centromere is the region to which the mitotic spindle attaches. The telomeres are specialized structures located at the ends of linear chromosomes. Telomerase is essential for maintaining the ends of chromosomes. Because of the mechanism of DNA replication, the lagging strand at the ends of a linear DNA molecule would shorten after each round of replication. Telomerase is a RNA-dependent DNA polymerase, which regenerates the ends of linear chromosomes. A yeast artificial chromosome is a synthetic "chromosome" that contains a centromere, telomeres, and origins of replication or autonomously replicating sequences. Because YACs act like mini chromosomes, large DNA fragments up to 1 megabase can be cloned into YAC vectors.

11. Mitochondrial genomes encode their own mitochondrial rRNAs, tRNAs and some essential mitochondrial proteins. Plant mitochondrial genomes are generally larger and more variable in

size than the mitochondrial genomes of other organisms. For example, watermelon mitochondrial DNA is about 330 kilobases, while human mitochondrial DNA is about 16.6 kilobases, and that of yeast is about 78 kilobases. Furthermore, plant mitochondrial genomes contain multiple mitochondrial DNA molecules that recombine with each other, whereas mammals contain a single mitochondrial DNA molecule. In addition, plant mitochondrial genomes encode a unique plant 5S mitochondrial rRNA and the α subunit of the F_1 ATPase. Plant mitochondria use the standard genetic code for protein translation, whereas the mitochondria from mammals and fungi (yeast) use a modified genetic code.

12. Similarities between bacteria, mitochondria, and chloroplasts reflect the proposed endosymbiotic origin of mitochondria and chloroplasts. All three have circular genomes. Mitochondrial and bacterial ribosomes differ from eukaryotic ribosomes in their RNA and protein compositions, their size, and their sensitivity to antibiotics. Bacterial and mitochondrial ribosomes are sensitive to chloramphenicol but resistant to cycloheximide. Eukaryotic ribosomes are sensitive to cycloheximide and resistant to chloramphenicol. Chloroplast DNA encodes four subunits of RNA polymerase and eight proteins that are highly homologous to bacterial RNA polymerase and ribosomal proteins, respectively.

Analyze the Data

a. In the telomerase-positive cell A, telomerase activity, as indicated by the presence of the ladder of bands, is observed in the control transfection. As expected, an increase in telomerase activity is observed when a wild type hTERT is transfected. Transfection of the mutant hTERT suppresses the endogenous telomerase activity. In telomerase-negative cell B, no telomerase activity is seen in the control transfection and telomerase activity is seen after transfection with wild type hTERT, as expected. Again, no telomerase activity is observed after transfection with the mutant hTERT. The telomerase mutant acts as a dominant-negative mutation, i.e., a mutant which can suppress wild type activity.

b. In the case of cell A, transfection of wild type hTERT increases the length of the telomeres, as indicated by the increase in size of the hybridizing smear on the gel. In this assay, a single band is not observed, indicating that the telomeres are of varying lengths. Transfection of mutant hTERT into cell A results in a shortening of the telomeres, which is consistent with the suppression of endogenous telomerase activity by the mutant hTERT. In contrast, in cell B, the control transfection or transfection with wild type or mutant hTERT resulted in the same size telomere bands. These results indicate that in cell A, telomere length is maintained by telomerase; whereas in cell B, telomere length is telomerase-independent.

c. The transfection of mutant hTERT into cell A results in inhibition of cell growth compared to control transfection or transfection with wild type hTERT. In contrast, transfection of mutant hTERT had no effect on cell growth in cell B. these results suggest that for cell A, telomerase is required for cell growth, whereas for cell B, telomerase is not required.

d. For cell A, the results support the role of telomerase in long-term survival of cells in culture. The suppression of wild type hTERT by the mutant hTERT resulted in inhibition of cell proliferation. For cell B, telomerase does not seem to play a role in long-term survival of the cells. These cells grow in the absence of telomerase activity. Cell B serves as an important control for cell A because it is possible that the effects of transfection of mutant hTERT on cell survival could be due to a pathway independent of suppression of telomerase. If this was the case then transfection of mutant hTERT into cell B, which does not require telomerase, should also show growth inhibition.

e. Cell A, transfected with wild type hTERT, expresses telomerase and grows continuously in culture and therefore would be expected to form a tumor in nude mice. In contrast, cell A, transfected with mutant hTERT, does not express telomerase and does not grow continuously in culture and therefore would not be expected to form tumors in nude mice.

11

Transcriptional Control of Gene Expression

Review the Concepts

1. Nuclear run-on transcription analysis of many different genes in several different cell types has shown that the transcription rate is usually proportional to the amount of encoded protein found in that same cell. This means that protein levels are most often correlated with transcription initiation, or that the limiting factor in most cases is whether transcription initiation occurs.

2. RNA polymerase I is responsible for transcribing rRNA genes. RNA polymerase II is responsible for mRNA transcription. RNA polymerase II transcribes tRNAs, 5S RNAs and several other small RNAs. Since RNA polymerase II is uniquely inhibited by α-amanitin, one can determine whether any gene requires this polymerase by measuring gene transcription in the presence and absence of this compound. If RNA polymerase II is responsible for transcribing a gene, then transcription should only occur in the absence of α-amanitin.

3. Phosphorylation of the CTD occurs during transcriptional initiation (when the initiation complex becomes an elongation complex). The CTD remains phosphorylated during transcription. For some genes, transcription beyond 50 bases requires hyperphosphorylation of the CTD.

4. TATA boxes, initiators, and CpG islands are all promoter elements. The TATA box was the first to be identified, and it was found in viral genes that are very actively transcribed.

5. To identify DNA control elements within promoters, investigators utilize 5' deletion mutants and linker scanning mutants. These mutants contain a loss of sequence in the promoter and are compared to the normal promoter to identify regions required for transcription. Once a region has been identified, it can then be added back to a minimal promoter to determine if it can enhance transcription of this promoter.

6. Proximal promoter elements are usually somewhat constrained in their function if moved very far from the promoter. In contrast, distal enhancers will function at a variable distance and can even be located 3' of the gene.

7. Once a putative control region is identified, DNA Footprinting with a nuclear extract can identify the precise DNA sequence that is bound by a protein in the extract. This assay depends on the ability of protein factors to "protect" DNA from DNaseI digestion. The electrophorectic mobility shift assay can also be used to determine whether specific extracts or proteins bind to DNA. In this assay, an extract is incubated with a labeled piece of DNA. If protein binds to the DNA, then it "shifts" in terms of its migration on a polyacrylamide gel. This technique is often used as an assay to purify DNA-binding proteins.

8. Transcriptional activators and repressors contain a modular structure in which one or more transcriptional activation or repression domains are connected to a sequence-specific DNA-binding domain through a flexible "hinge-like" domain.

9. In patients with Wilm's tumor the WT1 gene is non-functional, so no WT1 transcriptional repressor is made. Since this repressor normally binds to the control region of the EGR-1 gene and represses its transcription, patients with Wilm's tumor have abnormally high levels of EGR-1. This high EGR-1 level allows for aberrant transcription of growth control genes in kidney cells resulting in kidney tumors.

10. CREB binding to its co-activator (CBP) is regulated by cAMP which stimulates phosphorylation of CREB. The phosphorylated acidic activation domain within CREB is a random coil that undergoes a conformational change to form two α helices that wrap around CBP. In contrast, nuclear receptors contain ligand binding domains that when bound by ligand, undergo a large conformational change. This change allows the structured globular ligand-binding domain to interact with a short α helix in the nuclear co-activator.

11. The first protein to bind to a polymerase II promoter is the TATA box binding protein (TBP). This protein folds into a saddle-like structure which binds to the minor groove of DNA near the TATA box and bends the DNA. TFIIB then binds and makes contact with both TBP and DNA on either side of the TATA box. TFIIF and Pol II bind, and Pol II is positioned over the start site. TFIIE and TFIIH bind, and a helicase activity unwinds the DNA.

12. Integration of gene X near the telomere is not ideal for good expression of gene X. Telomeres are usually contained in heterochromatin, which is tightly packed and less accessible for the transcriptional machinery. If the yeast line used for expression contained mutations in the H3 and H4 histone genes, the outcome could be different, depending on the specific mutations. For example, if the DNA sequence encoding lysines in the histone N-termini were mutated so that glutamine residues were substituted in their place, then repression of gene X would not take place. Repression would not take place in such a mutant because the glutamine residues simulate acetylated lysine residues which causes the DNA to adopt a more open chromatin structure. This open chromatin structure would facilitate transcription initiation formation and inhibit binding of proteins like SIR3 and SIR4, which are involved in formation of silenced genes.

13. A good prediction is that STICKY functions as a transcriptional repressor. Repressors contain two domains, one that binds DNA and a second domain that represses transcription. The bHLH domain is a DNA-binding domain that has been found in many different transcription factors. The Sin3- interacting domain is likely to attract a Sin3 containing histone deacetylase complex. This complex can repress transcription because deacetylation of histones promotes a more closed conformation of DNA.

14. Using the alpha-globin gene as bait in a yeast two-hybrid experiment, one expects to clone a gene encoding beta-globin because these two proteins are known to interact. Similarly, using the catalytic subunit of protein kinase A will yield its own gene plus the gene for encoding the regulatory subunit of protein kinase A. Recall from Chapter 3 that this enzyme is a tetramer of two catalytic subunits and two regulatory subunits. The catalytic subunit of aspartate transcarbamylase, if used as bait in a two-hybrid experiment, will also bind to itself and to its regulatory subunit.

Analyze the Data

a. Bone and skin extracts contain "shifted" bands that differ in size from the free probe found at the bottom of the gel. Thus, these two tissues contain a protein that binds to the DNA probe and hence to the promoter region of gene X. Because the binding activities are shifted to different positions in the gel, it is unlikely that the DNA-binding factors in bone and skin are identical.

b. An in vivo transfection assay (Figure 11-16 in the text) would be an ideal test for this. To perform this assay, the gene encoding the putative transcription factor must be cloned first. Once the transcription factor has been purified, the amino acid sequence of the protein can be determined easily. This amino acid sequence can be used to produce a degenerate oligonucleotide probe used for cloning the transcription factor gene. Once the transcription factor gene is cloned, it can be transfected into a donor cell along with a separate plasmid containing the promoter of gene X fused to a reporter gene. If the putative transcription factor gene encodes a factor that binds to the gene X promoter and activates transcription, then an assay for the reporter will yield positive results. If the putative transcription factor gene does not encode a factor that activates transcription, the assay for the reporter will yield negative results.

c.   A DNA Footprinting experiment can be used to determine the specific DNA sequence to which the transcription factor binds.

d.   Since the DNA-binding activity is present in cell types that do not have transcription of gene X, the DNA-binding activity must be associated with gene repression. Repressor complexes can contain histone deacetylases.

12

Post-transcriptional Gene Control and Nuclear Transport

Review the Concepts

1. In the case of a protein-coding gene, gene control beyond regulation of transcriptional initiation can be regulated by controlling the stability of the corresponding mRNA in the cytoplasm, by controlling the rate of translation, and by controlling the cellular location so that newly synthesized protein is concentrated where it is needed.

2. Biochemical and chromatin IP experiments suggest that the polyadenylation complex is associated with CTD of RNA polymerase II following initiation.

3. A mutation in the TAR sequence that abolishes Tat binding would result in shorter HIV transcripts (i.e. no antitermination) because Cdk9 would not be positioned correctly to phosphorylate the CTD of RNAP, which would terminate transcription at the transcriptional block. Loss of Cdk9 activity would also result in no antitermination and shorter HIV transcripts.

4. The heat shock genes are induced by harsh conditions such "heat shock." Paused RNAP molecules stay associated with ≈25 nucleotide-long nascent heat shock gene transcripts and the template DNA until they are stimulated to continue chain elongation. This mechanism shortens the time for RNA expression because no time is required to assemble a transcription pre-initiation complex.

5. You should be surprised! A 5' cap is a well-conserved structure amongst all eukaryotes and it contributes to several processes. For example, during transcription, the CTD of RNA polymerase II interacts with capping components. During translation, the cap promotes engagement of ribosomes and mRNA by binding to translation initiation factors (eIF-4E). During splicing, the cap promotes splicing via the cap binding complex which enhances U1 snRNP and U6 snRNA /5'site interactions. As well, polyadenylation requires the cap binding complex.

6. The gene product encoded by the fragile-X gene (Fmr1) contains a known 45 residue RNA-binding motif called the KH motif. Proteins containing this motif can be predicted to bind RNA. Further X-ray crystallographic studies have shown how RNA binds to the KH motif.

7. The term hnRNA describes heterogeneous nuclear RNAs that consist of several different types of RNA molecules that are found in the nucleus. Small nuclear RNAs (snRNAs) bind to splice sites and participate in splicing reactions. Small nucleolar RNAs play a similar role in tRNA processing and can help to position methyltransferases near methylation sites. Micro RNAs (miRNAs) and short, interfering RNAs (siRNAs) are involved in gene silencing. Both are derived from longer precursor molecules and become part of the RISC complex.

8. Electron microscopy studies with adenoviral DNA complexed with the mRNA encoding the a viral capsid protein revealed that such RNA-DNA hybrids contained "loops" of DNA that had no counterparts in the RNA (Figure 12-5 in the text). The locations of exon-intron junctions are predicted by comparing the genomic and cDNA sequences of a gene.

9. Both types of splicing involve two transesterification reactions and similar intermediates and products. In Group II intron self-splicing events, the introns alone form a complex secondary structure involving numerous stem loops, whereas spliceosomal splicing utilizes snRNAs interacting with the 5' and 3' splice sites of pre-mRNAs, which forms a three-dimensional

RNA structure functionally analogous to the group II intron. Evidence that supports the idea that introns in pre-mRNAs evolved from group II self-splicing introns comes from experiments with mutated group II introns. In these experiments, domains I and V are deleted and this yields a group II intron incapable of self-splicing. When RNA molecules equivalent to the deleted portions are added back "in trans" in the in vitro reaction, self-splicing is restored. This shows that portions of group II introns can be trans-acting like snRNAs.

10. Researchers believe these events occur mostly within discrete foci in the nucleus. Supporting evidence for discrete nuclear regions comes from imaging studies utilizing rhodamine labeled poly (dT) to detect polyadenylated RNA. In the nucleus this probe binds to about 100 discrete foci.

11. In muscle cells, the internal polyadenylation site could be spliced out of the mature RNA when the fifth intron is removed. This would leave the site in the 3' UTR as the sole polyadenylation site. In other cells, the fifth intron may not be spliced out. This would result in earlier polyadenylation and a shorter mRNA transcript. In this scenario, a muscle-specific splicing factor could facilitate removal of the fifth intron.

12. In yeast, mutants have been identified that do not export polyadenylated RNA, which accumulates in the nucleus. The cloning of the normal copy of the defective gene revealed that this exporter contains two subunits, and together they form a domain that interacts with the nucleoporins in the nuclear pore complex. Highly conserved gene homologues of the exporter are found in all eukaryotes, suggesting that this exporter functions in most vertebrates as well.

13. Ran–Guanine nucleotide exchange factor (Ran-GEF or RCC1) must be present in the nucleus and Ran-GAP must be in the cytoplasm for unidirectional transport of cargo proteins across the nuclear pore complex. When the Ran is bound to GTP it has high affinity for cargo proteins. During nuclear export, Ran-GTP picks up cargo proteins in the nucleus and carries them to the cytoplasm. To be able to release the cargo on the cytoplasmic side, GTP must be hydrolyzed to GDP, and this is stimulated by Ran-GAP. Once translocated back into the nucleus, Ran needs to be in the GTP bound state to pick up more cargo. Ran-Gef in the nucleus stimulates the exchange of GDP to GTP, and this process of export can start again.

14. A plant deficient in Dicer activity shows increased susceptibility to RNA viruses because Dicer is not present to degrade a portion of the viral double-stranded intermediates that viruses synthesize during replication. Without Dicer, all of these viral mRNAs are available for further viral infection.

15. β-actin mRNAs can be seen to accumulate on the leading edges of myoblasts during in situ hybridization experiments. Further, specific portions of the actin genes (the 3' UTRs) can be linked to a reporter gene and transfected into cultured cells. When the cells are assayed for activity of the reporter gene product, the results show that the 3'UTR directs localization of the reporter gene.

Analyze the Data

a. In fibroblasts, the three exons from construct 1 are spliced and joined together and translation results in a 38.5 kD protein as shown by the immunoprecipitation data (in b). In contrast, in muscle cells this same construct gives rise to a smaller protein, suggesting that differential splicing has occurred. From analysis of constructs 2 and 3, it appears that the smaller muscle protein results from the joining of exons 1 and 3. Analysis of transcripts (mRNA) could confirm this hypothesis. An alternative, but less likely explanation for the smaller muscle protein is that proteolytic processing gives rise to this smaller protein.

b. Most likely the intron sequences surrounding exon 2 contribute to the regulation of this alternative splicing event.

c. When exon 3 is present (as in constructs 1 and 3), an active protein kinase protein is synthesized. This can be concluded by comparing the activity levels of each construct. Exon 2 also influences the activity of the encoded protein. When exons 2 separates exons 1 and 3, the kinase acts upon substrate A, but not substrate B. However, if exon 2 is omitted, this substrate preference is reversed.

13

Signaling at the Cell Surface

Review the Concepts

1. Endocrine, paracrine and autocrine signaling differ according to the distance over which the signaling molecule acts. In endocrine signaling, signaling molecules are released by a cell and act on target cells at a distance. In animals, the signaling molecule is carried to target cells by the blood or other extracellular fluids. In paracrine signaling, the signaling molecules are released and affect only target cells in close proximity. In autocrine signaling, the cell that releases the signaling molecule is also the target cell. Growth hormone is an example of endocrine signaling because the growth hormone is synthesized in the pituitary, located at the base of the brain, and travels to the liver via the blood.

2. The ligand-receptor complex that shows the lower K_d value has the higher affinity. Because the K_d for receptor 2 (10^{-9} M) is lower than that for receptor 1 (10^{-7} M), the ligand shows greater affinity for receptor 2 than for receptor 1. To calculate the fraction of receptors with bound ligand, $[RL]/R_T$, use formula 13-2 $[RL]/R_T = 1/(1 + K_d/[L])$. For receptor 1, the K_d is 10^{-7} M and the concentration of free ligand $[L]$ is 10^{-8} M. Thus the $[RL]/R_T$ for receptor 1 is 0.091 or only 9% of the receptors have bound ligand at a free ligand concentration of 10^{-8} M. In contrast, the $[RL]/R_T$ for receptor 2 is 0.91 or 91% of the receptors have bound ligand.

3. To purify a receptor by affinity chromatography, a ligand for the receptor must first be chemically linked to a bead used to form a column. A detergent solubilized cell membrane extract containing the receptor is then passed through the column. The receptor will bind to the ligand attached to the bead and other proteins will wash out. The receptor can then be eluted from the column with an excess of ligand. A cell surface receptor can be cloned using a functional expression assay. In this approach, cDNAs are synthesized using mRNA extracted from cells expressing the receptor and are inserted into expression vectors. These recombinant plasmids are transfected into cells that do not express the receptor. The rare cells that express the receptor of interest can then be identified by biochemical assays.

4. For trimeric G proteins in the inactive state, $G_{s\alpha}$ is bound to GDP and complexed with $G_{\beta\gamma}$. Upon ligand binding to its receptor, the receptor undergoes a conformational change that affects the associated trimeric G protein. GTP is exchanged for GDP and the $G_{s\alpha}$-GTP complex dissociates from the $G_{\beta\gamma}$ complex. The released $G_{s\alpha}$-GTP or $G_{\beta\gamma}$ complexes then activate downstream effector proteins. Hydrolysis of GTP to GDP by the intrinsic GTPase activity of $G_{s\alpha}$ returns $G_{s\alpha}$ to the inactive state bound to GDP and the $G_{\beta\gamma}$ complex. A mutant G_α subunit with increased GTPase activity would be expected to hydrolyze GTP to GDP at a faster rate and thus reduce the time that the G_α subunit remains in the active state. This, in turn, would lead to reduced activation of the effector protein.

5. Some cytosolic adapter proteins have protein-binding domains known as PDZ domains. The PDZ domain is an approximately 90 amino acid domain that binds to three amino acid sequences at the carboxy terminus of target proteins, such as cell surface receptors and transporters. An adapter protein with multiple PDZ domains is capable of binding to multiple target proteins and effectively clustering them in one location in the membrane. Other target proteins are found clustered in lipid rafts termed caveolae. These rafts are marked by the presence of the protein caveolin. The precise method, however, of how signaling proteins are anchored to caveolin is unclear. The advantage to clustering receptors and signal transduction proteins would be to facilitate the interaction between these proteins and thus enhance the signal transduction pathway.

6. Epinephrine binding to the β-adrenergic receptor causes an activation of adenylyl cyclase through the activation of $G_{s\alpha}$, a stimulatory G protein. In contrast, epinephrine binding to the α-adrenergic receptor causes an inhibition of adenylyl cyclase through the activation of $G_{i\alpha}$ an inhibitory G protein. An agonist acts like the normal hormone, which in this case would be epinephrine. Thus agonist binding to a β-adrenergic receptor would result in activation of adenylyl cyclase. In contrast, an antagonist binds to the receptor but does not activate the receptor. Thus antagonist binding to a β-adrenergic receptor would have no effect on adenylyl cyclase activity. In fact, it would reduce a normal epinephrine stimulated response because it would prevent epinephrine from binding to the receptor.

7. Epinephrine binds to its receptor and activates the $G_{s\alpha}$ subunit of the trimeric G protein. The activated $G_{s\alpha}$-GTP complex then binds to and activates adenylyl cyclase, leading to an increase in the levels of cAMP. The resultant rise in cAMP activates protein kinase A. To attenuate the signal, cAMP is hydrolyzed to 5'-AMP by the action of cAMP phosphodiesterase. In addition, the intrinsic GTPase activity of the $G_{s\alpha}$ subunit hydrolyzes GTP to GDP, which converts the $G_{s\alpha}$ subunit to its inactive form bound to $G_{\beta\gamma}$. Thus, adenylyl cyclase is inactivated blocking the further synthesis of cAMP.

8. Receptor desensitization can involve phosphorylation of the receptor itself. The increase in cAMP levels as a result of ligand binding to the receptor leads to an activation of protein kinase A. Protein kinase A can phosphorylate target proteins as well as cytosolic serine and threonine residues in the receptor itself. Phosphorylated receptor can bind ligand but is reduced in its ability to activate adenylyl cyclase. Thus the receptor is desensitized to the effect of ligand binding. Phosphorylated receptors are resensitized by the removal of phosphates by phosphatases. A mutant receptor that lacked serine or threonine phosphorylation sites could be resistant to desensitization by phosphorylation and thus would continuously activate adenylyl cyclase in the presence of ligand.

9. Rhodopsin is a G protein-coupled receptor that is activated by light. Rhodopsin contains a light absorbing pigment, 11-cis-retinal, that is covalently linked to opsin. In the presence of light, 11-cis-retinal is converted to all-trans-retinal. This activated opsin then interacts and activates an associated G protein named transducin. The activated G_{α}-GTP complex binds to the inhibitory subunit of a phosphodiesterase. The released catalytic subunits of the phosphodiesterase hydrolyzes cGMP to 5'-GMP. As a result, the cGMP level declines, leading to the closing of a nucleotide-gated ion channel.

10. During receptor desensitization, the β-adrenergic receptor is phosphorylated by protein kinase A and also by the β-adrenergic receptor kinase (BARK). BARK phosphorylates only activated receptor, i.e., receptor bound to its ligand. The phosphorylated receptor is reduced in its ability to activate downstream effector proteins. During visual adaptation, activated opsin is phosphorylated by rhodopsin kinase. Phosphorylated opsin is reduced in its ability to activate its associated G protein, $G_{t\alpha}$. In both cases, receptor resensitization is mediated by dephosphorylation of the receptor by a phosphatase.

11. Cleavage of PIP_2 by phospholipase C generates IP_3 and DAG. IP_3 interacts with and opens Ca^{2+} channels in the endoplasmic reticulum (ER) membrane resulting in release of Ca^{2+} from the ER. Ca^{2+}-ATPase pumps located in the ER membrane pump cytosolic Ca^{2+} back into the ER lumen. Because Ca^{2+} is also pumped from the cytosol to the exterior of the cell by Ca^{2+}-ATPase pumps in the plasma membrane, Ca^{2+} must be transported back into the cell. The IP_3-gated Ca^{2+} channels bind to and open store-operated Ca^{2+} channels in the plasma membrane, allowing an influx of Ca^{2+}. The principal function of DAG is to activate protein kinase C, which then phosphorylates specific target proteins.

12. Nitric oxide is synthesized by nitric oxide synthase from O_2 and arginine. Nitric oxide binds to the nitric oxide receptor and causes relaxation of smooth muscle cells. Nitric oxide binding to its receptor activates an intrinisic guanylyl cyclase

activity, leading to a rise in cGMP levels. The rise in cGMP activates protein kinase G, resulting in an inhibition of the actin-mysoin complex, relaxation of the cell, and dilation of the blood vessel.

13. The transcription regulator, Tubby, is tightly bound to PIP$_2$ in the plasma membrane. Receptor stimulation activates phospholipase C, which cleaves PIP$_2$ into DAG and IP$_3$. Tubby is released from the plasma membrane and translocates to the nucleus where it binds to DNA. G protein-coupled receptor activation of adenylyl cyclase leads to an increase in cAMP and subsequent activation of protein kinase A. Active protein kinase A catalytic subunits translocate to the nucleus and phosphorylate the transcription factor CREB. Phosphorylated CREB can interact with the coactivator CBP/P300 forming a complex, which can bind to DNA and stimulate transcription.

Analyze the Data

a. For the wild type G protein, the activity of adenylyl cyclase is what you would expect. In the presence of GTP there is a basal level of adenylyl cyclase activity, which can be greatly stimulated by the addition of isoproterenol. Isoproterenol binds to the β$_2$-adrenergic receptor and causes activation of adenylyl cyclase. In comparing adenylyl cyclase activity in the presence of GTP or GTPγS, again the expected result is seen. The addition of GTPγS leads to an increase in adenylyl cyclase activity because GTPγS is non hydrolyzable. Thus the G$_{s\alpha}$ subunit remains active, leading to prolonged activation of adenylyl cyclase. In the case of the mutant, again the addition of isoproterenol results in an increase in adenylyl cyclase activity as expected. The adenylyl cyclase activity, however, is not different in the presence of GTP or GTPγS. Thus the mutation causes an increase in the basal activity of adenylyl cyclase, likely due to a change in the GTPase activity.

b. In cells transfected with the mutant G protein, higher levels of adenylyl cyclase would be present relative to cells transfected with wild type G protein. Thus mutant-transfected cells would have higher cAMP levels, which would result in higher levels of active protein kinase A. The higher protein kinase A levels would result in more extensive phosphorylation of target proteins, which would affect normal cell development and proliferation.

c. From the GTPase results, it is clear that the mutation affects the intrinsic GTPase activity of the G$_{s\alpha}$ subunit. These results are consistent with the adenylyl cyclase results. For the mutant G protein, binding GTP or GTPγS to the G$_{s\alpha}$ subunit leads to the same level of adenylyl cyclase activation because the G$_{s\alpha}$ subunit has greatly reduced ability to cleave GTP.

14

Signaling Pathways That Control Gene Activity

Review the Concepts

1. In multiple cell types, TGFβ activates a conserved signaling pathway that results in translocation of Smad2 or Smad3 to the nucleus in complexes with co-Smad4. Once in the nucleus, the Smads interact with other transcription factors to regulate the expression of target genes. The complement of these other transcription factors is cell-type specific, and thus, the TGFβ signaling pathway will induce the transcription of different genes in different cell types.

2. TGFβ binds to its type II receptor either directly or when presented by the type III receptor. The type II receptor is a constitutively active serine/threonine kinase. When the type II receptor binds TGFβ, it forms a tetrameric complex consisting of two molecules of the type II receptor and two molecules of the type I receptor. The type II receptor then phosphorylates the type I receptor, activating the type I receptor as a serine/threonine kinase. The type I receptor then phosphorylates R-Smad2 or 3, inducing a conformational change that exposes a nuclear localization signal on the R-Smad. The R-Smad then forms a complex consisting of two molecules of R-Smad and one molecule each of co-Smad4 and importin-β. This complex translocates to the nucleus where it interacts with other transcription factors to elicit changes in gene expression. The TGFβ signaling pathway is diagrammed in Figure 14-2 of *Molecular Cell Biology*, Fifth Edition.

3. Both cytokine receptors and receptor tyrosine kinases (RTKs) form functional dimers upon binding their ligands. Upon dimerization, one of the poorly active cytosolic kinases phosphorylates the other on a particular tyrosine residue in the activation lip. This phosphorylation activates the kinase, which phosphorylates the second kinase in the dimer as well as other tyrosine residues in the receptor. Both cytokine receptors and RTKs then serve as docking sites for signaling molecules that bind to these phosphotyrosine sites. A major distinction between cytokine receptors and RTKs is that the RTK is itself the tyrosine kinase, whereas the cytokine receptor has no catalytic activity but rather is associated with a JAK kinase.

4a. When JAK phosphorylates and activates STAT5, STAT5 itself translocates to the nucleus as a homodimer and functions as a transcription factor.

4b. When GRB2 binds the Epo receptor, the Ras/MAP kinase signaling pathway is activated, resulting in the translocation of MAP kinase to the nucleus, where it phosphorylates and regulates the activity of transcription factors.

5a. TGFβ signaling induces the expression of SnoN and Ski, two proteins that bind to Smads and inhibit their ability to regulate transcription. TGFβ signaling also induces the expression of I-Smads, which prevent the phosphorylation of R-Smads by the TGFβ receptor. When a signaling pathway induces expression or activation of its own inhibitor, it is called negative feedback.

5b. When erythropoietin binds to its receptor, EpoR, an SH2 domain on the phosphatase SHP1 binds to a phosphotyrosine on the receptor. Binding induces a conformation change that activates SHP1, which is in close proximity to JAK (associated with EpoR). SHP1 dephosphorylates and inactivates JAK, thus inhibiting signal transduction. The erythropoietin signaling pathway also possesses negative feedback that triggers long-term downregulation in which STAT proteins induce the expression of SOCS proteins. SOCS proteins contain SH2 domains that bind EpoR, preventing the binding of signaling molecules. One SOCS protein also binds the activation lip of JAK2 and inhibits its kinase activity. SOCS proteins also recruit E3 ubiquitin ligases, which ubiquitinate JAKs and target them for degradation by the proteasome.

6. GRB2 serves as an adapter molecule that contains both SH2 and SH3 domains. GRB2 binds to phosphotyrosine residues on activated receptor tyrosine kinases via its SH2 domain and binds to the guanine nucleotide exchange factor Sos via two SH3 domains. Sos then binds to and activates the Ras protein. Although all SH2 domains bind to phosphotyrosine residues, specificity is determined by the conformation of the binding pocket, which interacts with amino acids on the carboxy-terminal side of the phosphotyrosine.

7. Constitutive activation is the alteration of a protein or signaling pathway such that it is functional or engaged even in the absence of an upstream activating event. For example, Ras^D is constitutively active because it cannot bind GAP, and therefore remains in the GTP-bound, active state even when cells are not stimulated by growth factor to activate a receptor tyrosine kinase. Constitutively active Ras is cancer promoting because cells will proliferate in the absence of growth factors, and thus normal regulatory mechanisms for cell proliferation are bypassed.

7a. A mutation that resulted in Smad3 binding Smad4, entering the nucleus, and activating transcription independent of phosphorylation by the TGFβ receptor would render Smad3 constitutively active.

7b. A mutation that made MAPK active as a kinase and able to enter the nucleus without being phosphorylated by MEK would render MAPK constitutively active.

7c. A mutation that prevented NFκ-B from binding to Iκ-B or that allowed NF-κB to enter the nucleus and regulate transcription even when bound to Iκ-B would render NF-κB constitutively active.

8. In the mating factor signaling pathway, Ste7, another serine/threonine kinase is the substrate for Ste11. When the mating factor signaling pathway is activated, Ste7, Ste11, and the other relevant kinases in the cascade form a complex with the scaffold protein Ste5. Binding to Ste5 ensures that Ste7 is the only substrate to which Ste11 has access.

9. Maximal activation of protein kinase B requires: 1) release of inhibition by its own PH domain, which is achieved when PI 3-phosphates bind the PH domain; 2) phosphorylation of a serine in the activation lip of protein kinase B by PDK1, which occurs when both protein kinase B and PDK1 are recruited to the cytosolic surface of the plasma membrane by binding PI 3-phosphates; and, 3) phosphorylation of an additional serine residue located outside of the activation lip. In muscle cells, insulin-stimulated activation of protein kinase B causes the GLUT4 glucose transporter to locate to the cell surface, resulting in increased influx of glucose. Insulin-stimulated activation of protein kinase B also promotes glycogen synthesis because protein kinase B phosphorylates and inhibits glycogen synthase kinase 3 (GSK3), an inhibitor of glycogen synthase.

10. PTEN phosphatase removes the 3-phosphate from PI 3, 4, 5 triphosphate, thus reversing the reaction catalyzed by PI-3 kinase and rendering the PI phosphate unable to bind protein kinase B and PDK1. Loss-of-function mutations are cancer-promoting because constitutive activation of protein kinase B results in constitutive phosphorylation and inactivation of proapoptotic proteins such as Bad and Forkhead-1. Cancers are typically characterized by cells that are resistant to apoptosis. A gain-of-function mutation in PTEN phosphatase would promote cell death by causing the apoptotic pathway to be active even in the presence of survival factors that signal through protein kinase B.

11. The signaling pathway that activates NF-κB is considered irreversible because I-κB, the inhibitor of NF-κB, is degraded when the pathway is activated. Thus, the signal cannot be switched off rapidly as with a kinase-induced signal that can be reversed by the action of an opposing phosphatase. The NF-κB pathway is eventually disengaged by negative feedback in which NF-κB stimulates the transcription of I-κB. However,

synthesis of the I-κB inhibitor *de novo* is a relatively slow process, and thus, the NF-κB pathway can remain active for some time after the original stimulus, such as TNF-α, is removed.

12. Presenilin I is a protease that is responsible for cleaving the integral membrane Notch receptor to release a cytosolic fragment of Notch that diffuses to the nucleus to activate transcription factors. In neurons, presenilin I is a component of γ-secretase, which cleaves the integral membrane protein APP after the extracellular domain of APP is cleaved by β-secretase. The 42-amino acid peptide that remains in the membrane has been implicated in the formation of amyloid plaques in Alzheimer's patients.

13. A decoy receptor is a receptor that binds to a ligand but does not transduce a signal. Decoy receptors compete with functional signaling receptors for ligand binding and thereby downmodulate a signal. A decoy receptor for PDGF could bind this ligand and prevent PDGF from binding its receptor tyrosine kinase, which transduces a signal to induce cell proliferation through the Ras/MAP kinase pathway.

Analyze the Data

a. This experiment reveals that MEK5 and MEKK2 co-localize within a complex since immunoprecipitation with a MEK5 antibody also precipitates MEKK2. However, this data gives no evidence as to whether MEKK2 activates MEK5 or MEK5 activates MEKK2.

b. MEKK2 is required for the activation of ERK. However, MEKK2 cannot activate ERK unless MEK5 is present. MEKK2 alone partially activates ERK5 because there is some endogenous MEK5, but in the presence of MEK5AA, which inhibits endogenous MEK5, MEKK2 cannot activate ERK5. These experiments clearly place ERK5 downstream of MEKK2 and MEK5 in the signaling pathway; however, they do not unambiguously order MEKK2 and MEK5. To do so would require co-expression of ERK5, MEK5AA and a constitutively active form of MEKK2. If ERK5 were phosphorylated, then MEKK2 would be downstream of MEK5. If not, then MEKK2 would be upstream of MEK5.

15

Integration of Signals and Gene Control

Review the Concepts

1. Several examples are given in this chapter. Notch functions in the human immune system to prevent formation of T cells. The worm Notch protein functions in the determination of cell fate in vulval development. EGF receptors control the formation of adipocytes from PC12 cells. Gurken binds to the *Drosophila* EGF receptor stimulating the production of Spatzle in dorsal/ventral specification.

2. The glycogen phosphorylase kinase (GPK) enzyme catalyzes this reaction. GPK has a multimeric subunit structure $(\alpha\beta\gamma\delta)_4$. The α and β subunits are phosphorylated by protein kinase A, while the δ subunit is a calmodulin-like protein which is activated by calcium. When the α and β subunits are phosphorylated, this increases the affinity of the δ subunit for calcium. Thus, an increase in cAMP can activate GPK, and an increase in calcium can also stimulate activation of GPK.

3. After a meal, the blood glucose level rises in Type I diabetics, however, no insulin is released into the bloodstream. Since no insulin is available for binding to the insulin receptor on muscle cells and adipocytes, no new GLUT4 glucose transporters are transported to the membrane. This results in a lack of glucose transport into tissues, and an abnormally high blood glucose level. Type II diabetics synthesize insulin and have insulin receptors present on muscle cells and adipocytes, yet very little to no GLUT4 transporters get transported to the cell membrane. This suggests that the defect in Type II Diabetes is in the insulin signal transduction pathway downstream of the insulin receptor.

4. In *Drosophila*, both Spatzle and Dpp proteins play critical roles in determining the dorsal/ventral axis. Spatzle sets off a signal transduction pathway that eventually leads to Dorsal protein accumulation in the nuclei of ventral cells. In ventral nuclei, Dorsal turns on expression of genes that induce ventral cell fates. Dorsal also represses Dpp. In the dorsal side of the embryo, no Dorsal protein is present, so Dpp is expressed and dorsal cell fates are induced. In *Xenopus*, the TGFβ family members BMP2 and BMP4 have similar effects to those of the *Drosophila* Dpp protein. One difference in *Xenopus* dorsal/ventral axis specification is that the axis is flipped in vertebrates as compared with invertebrates.

5. *Dorsal* mRNA would be expressed in all areas of the syncytial embryo. Dorsal protein, is also expressed in all regions of the embryo, but only accumulates in the ventral side nuclei where this protein acts to repress genes whose expression is needed for dorsal specification (i.e. *dpp*).

6. Microarray analysis reveals patterns of gene expression for many genes. If wildtype and *dorsal* mutant embryos were compared, the results would yield a wildtype gene expression pattern that could be compared to the pattern when the Dorsal protein is not expressed. The only difference in the types of embryos is the presence/absence of Dorsal, so differences in the two expression patterns would reveal genes regulated by Dorsal, either directly or indirectly. The genes we would expect to see increased when *dorsal* gene function is lost are: *dpp, tolloid, short gastrulation,* and *zerknullt*, since Dorsal normally represses these genes. The genes we would expect to see decreased when dorsal gene function is lost are: *twist, snail* and *rhomboid*, since Dorsal normally activates expression of these genes.

7. *Bicoid* RNA localization at the anterior end produces a Bicoid protein gradient. This localization is dependent on the *bicoid* 3' UTR. If this 3' UTR regulatory sequence was removed, *bicoid* RNA and protein would NOT be localized at the anterior end, so anterior structures would not be exclusive to this region. Instead, formation of anterior structures would be expanded to occupy a greater proportion of the embryo.

8. Kinesin functions to move oskar mRNA along microtubules to the posterior pole where it is translated. In the posterior pole, Oskar binds to Smaug, inactivating it. Since Smaug functions to turn off translation of *nanos* RNA, the posterior pole effectively translates *nanos* RNA. In contrast, at the anterior pole Smaug is free to inhibit Nanos protein production, resulting in no Nanos production at the anterior end. Finally, Nanos functions to repress hunchback RNA translation, so at the anterior end, Hunchback protein is synthesized, whereas it is not in the posterior pole region. If Kinesin only functioned during embryo development, then loss of Kinesin function would result in a lack of Oskar localization at the posterior pole. This would allow Oskar to inactivate Smaug in both ends of the embryo, and this would allow for Nanos production at both ends. So the final phenotype would be the development of two posterior poles instead of the normal anterior/posterior gradient.

9. The fact that invertebrates, vertebrates and also plants contain Toll-like receptors that stimulate similar signal transduction pathways suggests that a common ancestor from a billion years ago or more had a similar pathway.

10. The five gap genes are expressed in specific spatial domains (i.e. broad overlapping stripes) and function to regulate one another. This complexity creates combinations of transcription factors that can lead to the development of more than five cell types.

11. Homeosis is the development of an organ or body part that has the characteristics of another organ or body part. There are three main flower homeotic mutants. Mutants in A function contain carpels instead of sepals, and stamens in place of petals. A genes thus function to specify sepals and part of petal identity. Mutants in B function contain sepals instead of petals, and carpels in place of stamens. B genes thus function to specify part of petal identity and part of stamen identity. Mutants in C function contain petals instead of stamens, and sepals in place of carpels. C genes thus function to specify part of stamen identity and all of carpel identity.

12. Hedgehog becomes associated with the plasma membrane where it stimulates the 7 membrane-spanning domain Smoothened receptor. When Hedgehog binds to Smoothened, it relieves the repression on Smoothened caused by the Patched protein. Wnt protein is also tethered to the membrane and acts through two cellular receptors, Frizzled and Lrp. In contrast to Patched, Lrp is not an antagonist of Frizzled as seen by the fact that mutations in the *wnt, frizzled* and *lrp* genes give rise to the same phenotypes.

13. Antibodies to Sonic hedgehog (Shh) protein block the formation of different ventral neural-tube cells in the chick. This shows that Shh is required for ventral neural-tube cell development. In addition, different concentrations of Shh were added to chick neural-tube explants. These different concentrations induced development of different types of cells.

14. Because ephrin B2 is only expressed on arteries one would expect that an ephrin B2 knockout would only affect arterial development. Since EphB4 is expressed on veins and it is the receptor for ephrin B2, it suggest that interaction between an arterial cell expressing ephrin B2 and a venous cell expressing EphB4 causes the induction of both cell types.

15. Several examples are described. Argos is a *Drosophila* protein that is secreted and competes with (or antagonizes) signals that activate the EGF receptor. This is accomplished by blocking dimerization of the stimulated receptor which is necessary for subsequent signal transduction. Transcription of the *argos* gene is induced by EGF receptor ligands. In this way, different amounts of Argos protein is synthesized that functions to buffer the EGF signal.

Analyze the Data

a. The putative rice *bicoid* gene is expressed in a specific region of the embryo that roughly corresponds to the apical half of the embryo.

b. Since this gene has a specific, restricted gene expression pattern and is similar to bicoid, one possibility is that this gene encodes a transcription factor that activates expression of genes required for apical development.

c. If the hypothesis in b. is correct, then we expect microinjection of rice bicoid mRNA in the basal end to result in the acquisition of apical structures at the basal end of the embryo.

d.   If plant embryo development utilizes similar regulatory strategies as Drosophila anterior/posterior patterning, then one would expect that a rice *nanos* gene would be expressed preferentially in the basal end of the embryo.

16

Moving Proteins into Membranes and Organelles

Review the Concepts

1. In eukaryotes, protein translocation across the endoplasmic reticulum (ER) membrane is most commonly cotranslational, but can also occur post- translationally, particularly in yeast. a) The energy source for cotranslational translocation comes from the translation process itself. In other words, the nascent chain is pushed through the translocon channel. Please note, however, that as translation is completed a portion of the newly synthesized protein still resides within the translocon. This portion is drawn into the ER lumen rather than being pushed. In post-translational translocation, the newly synthesized polypeptide chain is drawn through the translocon by an energy input from ATP hydrolysis by BiP. BiP is luminal protein of the ER and is a member of the Hsc70 family of molecular chaperones. BiP-ATP activates by binding to the Sec63 complex that in turn binds to the Sec61 translocon complex. Activated BiP is enymatically active and cleaves ATP to ADP plus Pi. It is BiP-ADP that binds to the entering, unfolded nascent chain. Sequential binding of BiP-ADP to the nascent chain serves to block any sliding of the chain back and forth in the translocon and serves to rachet the nascent chain through the translocon. b) Translocation across the bacterial cytoplasmic membrane also occurs post-translationally. In gram-negative bacteria, SecA, a cytosolic protein, acts to drive the incompletely folded, newly synthesized polypeptide through the translocon (SecY, SecE, SecG complex). SecA associates with the tranlocon and newly synthesized polypeptide. SecA uses ATP hydrolysis to push the polypeptide through the translocon. c) Translocation into the mitochondrial matrix occurs through a bipartite Tom/Tim complex in which Tom is the outer membrane translocon and Tim is the inner membrane translocon. Three energy inputs are required. First, ATP hydrolysis by a cytosolic Hsc70 chaperone keeps the newly synthesized mitochondrial precursor protein unfolded in the cytosol. Second, ATP hydrolysis by multiple ATP-driven matrix Hsc70 chaperones may serve to pull the translocating protein into the matrix. Matrix Hsc70s interact with Tim44 and hence may be analogous to the BiP/Sec63 interaction at the ER membrane. Third, energy input from the H^+ electrochemical gradient or proton-motive force is required. The inside-negative membrane electric potential may serve to electrophorese the amphipathic matrix-targeting sequence towards the matrix

2. SRP (signal recognition particle) acts as a cycling cytosolic factor for the translocation of ER targeted proteins. It binds to both the signal sequence and SRP receptor, a heterodimer associated with the ER membrane. In doing this, SRP initiates ribosome binding to ER membranes and positions the nascent chain proximal to the translocon. Both SRP and the SRP receptor are GTPases. The unfolded nascent chain then translocates. Cytosolic Hsc70 functions as a cytosolic factor required for protein translocation into mitochondria. It acts as a molecular chaperone to keep the post-translationally targeted mitochondrial precursor protein in an open, extended conformation. At least two different cytosolic proteins are required for translocation of peroxisomal matrix proteins. These are Pex5, the soluble receptor protein for matrix proteins containing a C-terminal PTS1 targeting sequence, and Pex7, the soluble receptor protein for matrix proteins containing a N-terminal PTS2 targeting sequence. A different receptor, Pex19, is required for peroxisomal membrane proteins.

3. Many membrane proteins are embedded in the membrane by virtue of transmembrane α-helical segment(s). Such segments can be referred to as topogenic sequences. These segments share general principles or properties. They tend to be about 20 amino acids long, a length sufficient to span the membrane, and hydrophobic, an

appropriate property for a sequence embedded in the hydrophobic lipid bilayer. Application of these principles through computer algorithms is predictive. In brief, amino acid sequences of polypeptides may be scanned for hydrophobic segments of about 20 amino acids in length. Each amino acid may be assigned a hydrophobic index value based on relative solubility in hydropbobic media versus water and these values then can be summed by a computer for all 20 amino acid segments of a protein. Segments exceeding a threshold value are expected to be topogenic transmembrane segments. Similar reasoning with, in practice, less predictive value can be applied based on the properties of signal sequences for targeting to ER, mitochondria, chloroplasts, peroxisomes, etc. In general, such an algorithm must examine the full sequence of a precursor protein. In general, the computer identification of topogenic sequences can be predictive. To illustrate, let us consider an ER targeted, multipass membrane protein with an N-terminal, cleavable signal sequence. Based on principles, such an ER targeting N-terminal, cleavable signal sequence can be computer-identified by sequence scanning. Subsequent internal signal anchor and stop-transfer anchor segments similarly can be identified. Such sequences alternate within a multipass membrane protein. Because of this, the overall arrangement of the protein can predicted as described in detail within Chapter 16.

4. Improperly folded or oligomerized proteins fail to exit the ER. In other words, exit from the ER is protein quality dependent. A fairly wide array of accessory proteins are involved in the processing of newly synthesized proteins within the ER lumen. These include proteins necessary for N-glycosylation of appropriate proteins, protein folding chaperones such as BiP, disulfide bond processing proteins such as protein disulfide isomerase, enzymes that accelerate rotation about peptidyl-prolyl bonds such as peptidyl-prolyl isomerases, and proteins such as calnexin and calreticulin that bind to glucose and act to prevent aggregation of unfolded or misfolded proteins. Proteins unable to exit the ER are generally degraded. To the surprise of most investigators, this degradation occurs in the cytosol in an ubiquitin-dependent, proteasome-dependent manner. The Sec61p translocon serves to transport these misfolded proteins into the cytosol for degradation.

5. The 7-sugar intermediate is synthesized by sugar addition to cytosolic facing dolichol phosphate. The intermediate is flipped from the cytosolic face of the ER membrane to the luminal face. Further sugar additions then occur within the lumen of the ER. Short forms of the intermediate are on the wrong side of the membrane to add to nascent polypeptides within the ER lumen. Incomplete adductants within the ER lumen are located appropriately to N-glycosylate nascent polypeptide.

6. Several proteins facilitate the modfication or folding of secretory proteins within the ER. These include signal peptidase, BiP, oligosaccharyl transferase, various glycosidases, calnexin and calreticulun, protein disulfide isomerase, peptidyl-propyl isomerase and others. Of these, BiP and peptidyl-propyl isomerase act to facilitate conformation changes. Protein disulfide isomerase facilitates the making/breaking of disulfide bonds to ensure correct protein folding. Calreticulin and calnexin are lectins that bind to glycoproteins during folding. The others all directly support the covalent modification of proteins within the ER lumen.

7. It is BiP-ADP that binds to nascent chains. Replacement of ADP with ATP releases BiP from nascent chains. The result observed is consistent with this mechanism for BiP binding.

8. In essence, the type III secretion apparatus of Yersenia acts like a microinjection needle to transfer proteins into animal cells. Designer Yersinia cells can be engineered that do not secrete animal harmful proteins. These cells, if engineered to express desirable protein with the proper type III signal sequence, can deliver the engineered protein via the type III mechanism to macrophages, for example.

9. Each mutation has a different effect. a) Tom22 together with Tom20 act as outer mitochondrial membrane receptor proteins for N-terminal matrix targeting sequences. A defective Tom22

receptor protein would result in accumulation of mitochondrial matrix targeted proteins in the cytosol, possibly followed by their turnover within the cytosol. b) Tom70 signal receptor is an outer mitochondrial membrane protein recognizing multipass mitochondrial membrane proteins that have internal signal sequences. Mutation in Tom70 will have no immediate effect on mitochondrial matrix protein import as Tom70 does not recognize this class of protein. c) Matrix Hsc70 has a role in the folding of matrix proteins. Also, it is one source of energy for powering translocation. Defective matrix Hsc70 should result in clogging the Tom/Tim translocon complex with incompletely translocated proteins. d) Retention of the matrix targeting N-terminal signal sequence because of a defective matrix signal peptidase might well result in defective folding of the imported protein. The sequence normally is removed.

10. On the whole, protein import into the mitochondrial matrix and the chloroplast stroma, topologically equivalent locations, is by functionally equivalent mechanisms. Functionally analogous proteins mediate each process. However, the proteins are not homlogous indicating a separate evolutionary origin of mitochondria and chloroplasts. Energetically, unlike the situation for mitochondria, there is no need for a membrane electrochemical gradient for import into chloroplasts. Presumably stromal Hsc70 pulls proteins into the stroma.

11. This is basically a molecular ruler question. How many amino acids must span the Tom/Tim complex to expose the matrix-targeting sequence to the matrix-processing protease? DHFR in the presence of the drug, methotrexate, is locked into a folded state. A chimeric mitochondrial protein with folded DHFR fails to translocate fully into the mitochondria matrix. Instead, it is stuck in the Tom/Tim complex. The number of amino acids between the matrix targeting sequence and the folded DHFR sequence could be varied to provide a molecular ruler. Any unfolded N-terminal DHFR sequence must be included within the ruler. With respect to channel length, an overestimate will result from this approach as the

matrix targeting sequence must be spaced out from Tom/Tim to be accessible for cleavage.

12. Mitochondrial targeting sites include: matrix, inner membrane, intermembrane space, and outer membrane. Chloroplasts targeting sites include: stroma, inner membrane, intermembrane space, outer membrane, and thylakoid membrane. Despite some commonalities, significant differences exist between the mechanisms by which mitochondrial ADP/ATP antiporter and cytochrome b_2 are targeted in a site-specific manner. ADP/ATP antiporter has internal signal sequences recognized by Tom70 receptor and by Tim22 complex. Cytochrome b_2 has a N-terminal matrix targeting sequence that is followed by an intermembrane space-targeting sequence. Both use Tom40. However, the Tim components differ. Both are released from the Tim complex into the inner mitochondrial membrane. The ADP/ATP antiporter stays there. Cytochrome b_2 is released into the inner membrane space by proteolytic cleavage.

13. Separate mechanisms are used to import peroxisomal matrix and membrane proteins. Hence, mutations can selectively affect one or the other. Either can result in the loss of functional peroxisomes. One approach to asking whether the mutant is primarily defective in insertion/assembly of peroxisomal membrane proteins or matrix proteins is to use antibodies to ask by microscopy if either class of proteins localize to "peroxisomal" structures, e.g., peroxisome ghots. An alternate approach is cell fractionation, in which the assay is for whether the appropriate proteins are present in a membrane organelle fraction.

Analyze the Data

a. Messenger RNA lengths vary by steps of 20 codons or 20 amino acids each in corresponding synthesized product. When translated in the absence or presence of microsomes, only mRNAs of 130 and 150 codons produce a product that displays any difference in size with the addition of microsomes. The 130 codon mRNA gives a product that is either full length—showing no difference in size when compared to the minus microsome product—or a smaller product that is the presumed result of signal peptidase cleavage. This suggests variable or incomplete accessibility of the product to signal peptidase. In contrast, the next step size longer mRNA codes for product that is fully sensitive to signal peptidase cleavage when synthesized in the presence of microsomes. The key datum is the smaller size (faster mobility) of the product plus microsomes versus the product minus microsomes. Hence the conclusion is that the prolactin chain must be somewhere between 130 and 150 amino acids in length for the signal sequence to be fully accessible for cleavage.

b. The polypeptide must be mostly α-helical. A 100-amino acid polypeptide as a α-helix spans 150 Å, the length of the ribosome channel. Thirty amino acids span about 50 Å, the membrane thickness. In sum, a total length of about 160 amino acids (130 amino acid spacer) is required to space the signal peptide out by 150 Å, the necessary minimum length. Only a 60 amino acid spacer is required to give a minimally accessible signal peptide if the polypeptide were in extended conformation.

c. The fact that there is no prolactin cleavage if microsomal membranes are added after prolactin translation is complete indicates that prolactin must translocate cotranslationally.

d. A series of parallel reactions were done and a microsomal membrane fraction was prepared by centrifugation. No prolactin labeled polypeptide is seen in the membrane fraction for mRNA shorter than 90 codons. Hence, it is only at this nascent chain length that any engagement of ribosomes with SRP and hence binding to membrane occurs. If the mRNA is 110 codons or longer, the membrane fraction contains all labeled product found in the total reaction as indicated by the equal intensity of the gel bands. So, roughly a total chain length of 100 amino acids is required to expose the 30 amino acid signal peptide for SRP binding. The two bands seen with the 130-codon reaction is due to partial cleavage of the product by signal peptidase. The single band seen with the 150-codon reaction reflects full cleavage of the signal peptide by signal peptidase.

17

Vesicular Traffic, Secretion, and Endocytosis

Review the Concepts

1. The pancreatic acinar cell is a specialized secretory cell. It packages trypsinogen, chymotrypsinogen, and other digestive enzymes into secretory granules that then are released in response to signaling. Most protein synthesis in these cells is devoted to secretion. The vast majority of ribosomes are found in association with the rough endoplasmic reticulum. The cells are also decidedly polarized with a definite gradient in organelle distribution. A radioactive amino acid label is mostly incorporated into molecules associated with membrane bound ribosomes and, subsequently, with secretory organelles. Autoradiographic grains formed in the photographic emulsion coated, thin sections can be associated with specific organelles over time. The actual organelles can be identified by morphology under electron microscopy. In comparison, HeLa cells are a relatively non-differentiated cancer cell line. There is little to no polarity to organelle distribution. Amino acid label is incorporated into a wide range of proteins, few of which associate with ribosomes of the rough ER. The autoradiographic technique results in exposed silver grains scattered about the entire cell. The HeLa cell simply does not lend itself to the autoradiographic, pulse-chase amino acid label approach of Palade and colleagues.

2. NSF, through its ATPase activity, catalyzes the dissociation of v-SNARE/t-SNARE complexes. Such complexes are essential in specific membrane fusion at several stages of the secretory and endocytic pathways. Why then does the *Sec18* NSF mutation produce a class C phenotypeaccumulation of ER-to-Golgi transport vesicles? This can be explained readily if one considers the need for NSF to generate free v-SNAREs and t-SNAREs to support multiple rounds of vesicle membrane fusion. In the absence of NSF activity, vesicles bud from the ER but are unable to fuse with downstream membranes because of the lack of v-SNAREs. Vesicles accumulate at what is the first vesicle organelle fusion step within the secretory pathway.

3. Coat proteins play two roles in vesicle budding: 1) they provide a scaffold that establishes membrane curvature, and 2) they interact with cargo proteins or cargo protein receptors to provide enrichment of certain proteins in the bud. Small GTPase of either the Sar or Arf family recruit coat proteins to membranes. It is the GTP bound form of Sar or ARF that is active. The exact mechanisms by which each acts are unknown. The mechanism is particularly unclear in the case of ARF, which recruits clathrin and different adapter proteins at different sites within the cell. As cited above, vesicles are enriched in cargo proteins. Moreover, newly formed vesicles are programmed for subsequent fusion events by the selective inclusion of v-SNAREs and rabs in their membrane. Dynamin is the one protein well known to have a role in pinching off vesicles at the cell surface and at the *trans* Golgi network. In all known cases, these are clathrin-coated vesicles.

4. The observation that "decoating" Golgi membranes by treatment of cells with the drug brefeldin A (BFA) results in the redistribution of Golgi proteins to the ER suggests that COPI, the major Golgi associated coat, has a role in stabilizing Golgi structure. To some extent, COPI may be equivalent to the dam that holds back the water in a reservoir. Arf1 is the small GTPase that recruits COPI to Golgi membranes. Mutation of Arf1 to give a GDP restricted form of the protein would result in a GTPase that is now unable to recruit COPI to membranes. Since the COPI association with membranes is dynamic, this mutation would shortly lead to uncoating of Golgi membranes. Note that mutation of Arf1 to the GTP restricted form would have the opposite effect; the Arf would now be permanently in the on-state and the association of COPI with Golgi

membranes would be permanently stabilized. Since GDP-restricted Arf1 produces the same phenotype as BFA, this suggests that the drug evokes a normal physiological possibility.

5. Coat proteins must be able to come on and off membranes to produce a cycle of vesicle production and consumption. An antibody since as EAGE that binds to a "hinge" region of βCOPI is likely to be a blocking or function inhibitory antibody. Molecular flexibility is likely to be required for either uncoating or coating. In reality, EAGE blocks uncoating of COPI from Golgi membranes and vesicles. Inhibition of COPI function should inhibit anterograde transport from the ER to the plasma membrane. COPI is required both for cisternal progression and for retrieval of proteins from the *cis* Golgi to the ER. Cisternal progression stops because Golgi enzymes, for example, are not retrieved from *trans* to *medial* to *cis* Golgi. Similarly v-SNAREs would not be retrieved from *cis* Golgi to ER. Note here that the effect on anterograde transport is indirect. In reality, the direct effect is on retrograde transport. One possible experiment to test whether the effect of EAGE microinjection is initially on anterograde or retrograde transport is to characterize the properties of COPI derived vesicles. If these are retrograde carriers, they should be enriched in Golgi enzymes, for example, rather than anterograde cargo.

6. Vesicle fusion involves two stages, first a docking stage mediated by long coiled-coil proteins such as p115 or EEA1 and then a specific membrane fusion step mediated by SNAREs. The docking or tethering proteins, p115 and EEA1, are recruited to vesicles by rab proteins in their GTP-activated state. Rab5 plays a role in prompting vesicle fusion with early endosomes. Overexpression of the GTP-restricted form, the activated form, of rab5 would have the effect of prompting such fusions and lead to enlarged early endosomes.

7. In essence, SNARE mediated membrane fusion is a process of specifically forcing two membranes very close together because of paired, four helix bundled SNARE complexes. How this then leads to membrane fusion is not known. HA is the hemaglutinin protein of influenza virus. It

mediates an acid pH activated membrane fusion process between the viral envelope and a cellular membrane. The viral envelope itself is a membrane. At acid pH, HA changes in conformation to expose fusion peptides. The fusion peptides are hydrophobic and insert into endosomal membranes. This has the effect of drawing the viral envelope and endosomal membrane very close together. This then leads to fusion. In summary, SNAREs and HA are very different proteins in amino acid sequence and in structure. Both have the effect of drawing membranes very close together. How this then produces membrane fusion is not fully understood.

8. Lys-Asp-Glu-Leu (KDEL) and Lys-Lys-X-X (KKXX) are both retrieval sequences for ER proteins. KDEL is a sequence feature of soluble ER luminal proteins while KKXX is found on the cytosolic domain of ER membrane proteins. Retrieval of a normally ER luminal protein from the *cis* Golgi is a COPI-dependent process. COPI is found on the cytosolic face of the *cis* Golgi membrane. The KDEL containing protein is within the lumen of the *cis* Golgi. The two interact through a bridging membrane protein, the KDEL receptor. It is the KDEL receptor/KDEL containing protein complex that is retrieved to the ER. In the cisternal progression model, *trans* Golgi proteins, for example, must be retrieved to the *medial* Golgi to generate a new *trans* Golgi cisterna. This is a COPI mediated process. There must be interactions between COPI and Golgi proteins to promote such retrieval.

9. There are four known clathrin adapter protein complexes: AP1 (*trans*-Golgi to endosome), GCA (*trans*-Golgi to endosome), AP2 (plasma membrane to endosomes), and AP3 (Golgi to lysosome, melanosome, or platelet vesicles). Each contains one copy of four different adapter protein subunits proteins. The clathrin coat, unlike the COPI or COPII coat is a double layered coat with a core coat of adapter proteins and an external clathrin coat. Each adapter complex is different but related. Presently, it is not known if the coat of AP3 vesicles contains clathrin. This is consistent with the possibility that evolutionarily

the adapter complex may well be the core coat with clathrin being an accessory layer added later.

10. I cell disease is a particularly severe form of lysosomal storage disease. Multiple enzymes are lacking in the lysosome and the organelle becomes stuffed with non-degraded material and therefore generates a so-called inclusion body. I cell disease is inherited and caused by a single gene defect in the N-acetylgucosamine phosphotransferase that is required for the formation of mannose 6-phosphate (M6P) residues on lysosomal enzymes in the *cis* Golgi. This enzyme recognizes soluble lysosomal enzymes as a class and hence a defect in this protein affects the targeting of a large number of proteins. A defect in the phosphodiesterase that removes the GlcNAc group that initially covers the phosphate group on mannose 6-phosphate would also produce an I cell disease phenotype. Similarly, defects in mannose 6-phosphate receptors would affect the targeting of lysosomal enzymes as a class.

11. The trans-Golgi network (TGN) is the site of multiple sorting processes as proteins exit the Golgi complex. The sorting of soluble lysosomal enzymes occurs via binding to mannose 6-phosphate (M6P) receptors. Binding is pH dependent and occurs at the TGN pH of 6.5 but dissociates at the late endosomal pH of 5.0-5.5. Hence, lysosomal enzymes reversibly associate with M6P receptors. Clathrin/AP1 vesicles containing M6P receptors and bound lysosomal enzymes bud from the TGN, lose their coats and subsequently fuse with late endosomes. Vesicles budding from late endosomes recycle the M6P receptors back to the TGN. Packaging of proteins such as insulin into regulated secretory granules is a very different process. This sorting is thought to be due to selective aggregation followed by budding. The TGN also may be the site of protein sorting to the apical and basolaterial cell surfaces in polarized cells. This is the case in MDCK cells, a line of cultured epithelial cells, where there is direct basolateral-apical sorting at the TGN cells. In contrast, hepatocytes use different mechanisms for sorting to basolateral versus apical surfaces. Here newly made apical and basolateral proteins are first transported in vesicles from the TGN to the basolaterial surface and incorporated into the plasma membrane by exocytosis. From there, both basolateral and apical proteins are endocytosed in the same vesicles, but then their paths diverge. The endocytosed basolateral proteins are recycled back to the basolateral membrane. In contrast, the apically destined endocytosed proteins are sorted into transport vesicles that move across the cell and fuse with the apical membrane in a process termed transcytosis.

12. Antibody proteins, members of the IgG class, for example, are organized in domains such that the antigen combining portion is variable in sequence and portions distal from this are much more constant in sequence from one antibody molecule to another. The Fc portion of an IgG is constant in sequence from IgG to IgG, and it is this portion of the antibody protein to which the macrophage phagocytosis receptor is directed. These receptors are termed Fc receptors. Antibodies may be cleaved by gentle proteolysis into Fc and antigen binding domains. These antibody fragments can be separated and then purified. In a competition experiment, one can test the ability of an added antibody fragment to block the enhanced uptake of opsonized bacteria by macrophages.

13. Within the endocytic pathway, there is a progressive acidification (increased hydrogen ion concentration) in compartments going from early to late endosomes to lysosomes. The pH drops from almost neutral to pH 4.5. The binding of LDL to LDL receptor is pH sensitive. At the cell surface, neutral pH, LDL binds to LDL receptor. At an acidic pH, pH equals 5.5, LDL dissociates from receptor. LDL is then transported to lysosomes and LDL receptor is sorted and recycled back to the cell surface. Mannose 6-phosphate bearing lysosomal enzymes dissociate from mannose 6-phophate receptors in the acidic pH late endosomal compartment. Elevating pH prevents this. Receptors become saturated with lysosomal enzymes and the cell no longer has the capacity to direct newly synthesized lysosomal enzymes to lysosomes. Instead, the enzymes are secreted from the *trans* Golgi network by constitutive secretion.

14. In terms of membrane topology, both the formation of multivesicular endosomes by

budding into the interior of the endosome and the outward budding of HIV virus at the cell surface are equivalent. Important mechanistic features are shared. Both processes involve an ubiquitination step. In multivesicular endosome formation, cargo proteins to be included in the budding endosome and the Hrs protein are ubiquitinated. For HIV budding, it is the HIV Gag protein that is ubiquitinated. In closing off the budding endosome or the budding HIV a cellular ESCRT protein complex recognizes the ubiquitin and cellular Vps4 is used later to dissociate the ESCRT complex. The viral Gag protein mimics the function of cellular Hrs, redirecting ESCRT complexes to the plasma membrane. ESCRT binds to the C-terminal portion of HIV Gag protein. One logical peptide inhibitor/competitor of HIV budding is a synthetic peptide corresponding to the portion of Gag protein that binds to ESCRT. Such a peptide might well compete or interfere with normal cellular proteins such as ESCRT binding to ubiquitnated Hrs.

15. Calcium influx at axon terminus is sensed by synaptotagmin. Synaptotagmin is a transmembrane protein of the synaptic vesicle that binds Ca^{++}, interacts with phospholipids, and binds SNARE proteins. At low Ca++ concentrations, it binds to a complex of the plasma membrane proteins neurexin and syntaxin. The presence of synaptotagmin in this complex blocks the binding of other essential fusion proteins to the neurexin-syntaxin complex, preventing vesicle fusion. When synaptotagmin binds Ca^{++}, it is displaced from the complex, allowing docking/fusion to proceed further. *Shi* or *shibere* mutants are defects in dynamin. Dynamin is a GTPase involved in vesicle pinching off. A nonhydrolyzable analog of GTP will have the effect of promoting the recruitment of dynamin to vesicle necks but as a nonhydrolyzable analog the overall pinching off process will be blocked. Hence, an analog induced equivalent to the mutant state can result.

Analyze the Data

a. *Pseudomonas* toxin B subunit has a KDEL sequence at its C-terminus while Shiga toxin B subunit does not. Hence, Pseudomonas toxin B subunit is expected to bind to KDEL receptors while Shiga toxin B subunit should not.

b. Microinjected antibodies directed against either COPI or KDEL-receptor elevate protein synthesis in cells treated with *Pseudomonas* toxin but not in cells treated with Shiga toxin. This is the expected result if Pseudomonas toxin binds to the KDEL-receptor that in turn binds to COPI coat protein. Toxicity appears to be dependent on both. COPI coat proteins and vesicles are known to mediate retrograde traffic from the *cis* Golgi to ER.

c. Anti-EAGE (anti-COPI) has no effect on Shiga B toxin transport from the Golgi to the ER. Only when the Shiga B toxin subunit has a KDEL sequence added is there effect. The net conclusion, therefore, must be that Shiga B toxin is transported normally from the Golgi to the ER by means other than COPI-coated vesicles.

18

Metabolism and Movement of Lipids

Review the Concepts

1a. Membrane phospholipids are synthesized at the interface between the cytosolic leaflet of the endoplasmic reticulum (ER) and the cytosol. Water-soluble, small molecules are synthesized and activated in the cytosol. Membrane bound enzymes of the ER then link these small molecules together to create larger, hydrophobic membrane phospholipids.

1b. Membrane phospholipids can be flipped from the cytosolic leaflet of the ER membrane to the exoplasmic leaflet. This process, mediated by flippases, results in the incorporation of newly synthesized phospholipids into both leaflets.

1c. Phospholipids can be moved from their site of synthesis to other membrane, e.g., to the plasma membrane. Some of this transport is by vesicles. Some is due to direct contact between membranes. Small, soluble lipid-transfer proteins also mediate transfer. On the whole, the mechanism of phospholipids transfer between membranes is not well understood.

2. The common fatty acid chains in glycerophospholipids are tabulated in Table 18-1 and include myristate, palmidtate, stearate, oleate, linoleate, and arachidonate. These fatty acids differ in carbon atom number by multiples of 2 because they are elongated by the addition of 2 carbon units. For example, the acetyl group of acetyl CoA is a 2-carbon moiety.

3. Sterols, i.e., cholesterol-like lipids, are found in the membranes of plants and fungi as well as animals. Plant sterols are not absorbed efficiently from the intestine. Their chemistry is slightly different than cholesterol and they are not metabolically useful to mammals. Rather, they are excreted. In fact, the ABC protein, ABCG5/8, located in the apical membrane of intestinal epithelial cells appears to pump plant sterols out of the cells and back into the intestinal lumen. Less than 1% of dietary plant sterols enter the bloodstream.

4. The key regulated enzyme in cholesterol biosynthesis is HMG (β-hydroxy-β-methylglutaryl)-CoA reductase. This enzyme catalyzes the rate-controlling step in cholesterol biosynthesis. The enzyme is subject to negative feedback regulation by cholesterol. In fact, the cholesterol biosynthetic pathway was the first biosynthetic pathway shown to exhibit this type of end-product regulation. As the cellular cholesterol level rises, the need to synthesize additional cholesterol goes down. The expression and enzymatic activity of HMG-CoA reductase is suppressed. HMG-CoA reductase has eight transmembrane segments and, of these, five compose the sterol-sensing domain. Sterol sensing by this domain triggers the rapid, ubiquitin-dependent proteasomal degradation of HMG-CoA reductase. Homologous domains are found in other proteins such as SCAP (SREBP cleavage-activating protein) and Niemann-Pick C1 (NPC1) protein, which take part in cholesterol transport and regulation.

5. Diet is indeed a source of cholesterol. However, cells also synthesize cholesterol. This is the source of much of the cholesterol in the human body. In fact, for the vegetarian, this is the total cholesterol source. Production of cholesterol is a normal part of human metabolism.

6. Several aspects of cholesterol metabolism lead to the conclusion that cholesterol is a multifunctional lipid. Cholesterol is a metabolic precursor to steroid hormones critical for intercellular signaling, vitamin D, bile acids (which help emulsify dietary fats for digestion and absorption in the intestine), stored cholesterol esters, and sterol modified proteins such as Hedgehog.

7. Most phospholipids and cholesterol membrane-to-membrane transport in cells is not by Golgi-mediated vesicular transport. One line of evidence for this is the effect of chemical and mutational inhibition of the classical secretory pathway. Either fails to prevent cholesterol or phospholipids transport between membranes, although they do disrupt the transport of proteins and Golgi-derived sphingolipids. Membrane lipids produced in the ER can not move to the mitochondria by classic secretory transport vesicles. No vesicles budding from the ER have been found to fuse with mitochondria. Other mechanisms are thought to exist. However, presently these are poorly defined (see Figure 18-8). They include direct membrane-membrane contact and small, soluble lipid-transfer proteins.

8. The two major mechanisms by which lipids are exported from cells are ABC superfamily proteins and secretion of lipoprotein complexes via the classic secretory pathway. Often the precise mechanism of overall transport is not known. For example, the ABCB4 flippase flips phosphatidylcholine from the cytosolic leaflet to the exoplasmic leaflet of the apical membrane in hepatocytes. However, the mechanism by which the excess phospholipids desorb from the exoplasmic leaflet into the extracellular space is unknown. The three major methods by which lipids are imported into cells are transport proteins such as sodium-coupled symporters that mediate import, CD36 and SR-BI superfamily proteins that can mediate unidirectional and bidirectional transport from HDL and other lipoprotein complexes, and receptor mediated endocytosis.

9. Bile acid cycling plays a major role in the maintenance of lipid homeostasis. Bile acids are derived from cholesterol. In enterohepatic circulation, bile that contains micelles phospholipids, bile acids, and cholesterol is exported from hepatocytes to bile ducts and delivered to the gallbladder where it is a stored and concentrated. In response to a fat containing meal, bile is released into the small intestine where bile acids help emulsify dietary fats. Cholesterol present in bile may be excreted from the intestine. Bile acids are imported by intestinal epithelial cells and exported by them into the blood, from which they can be subsequently cycled back to hepatocytes for import. High liver bile acid levels due to effective recycling suppress the conversion of cholesterol to bile. The sequestrant clearance of bile acids from the liver can be an effective approach to lowering cholesterol levels. Lowering bile acid levels promotes the conversion of cholesterol to bile acids that may then be cleared from the body through intestinal excretion.

10. The four major classes of lipoproteins in the mammalian circulatory system are chylomicron, very low density lipoprotein (VLDL), low density lipoprotein (LDL), and high density lipoprotein (HDL). Chylomicrons mainly carry triglcerides. VLDL is intermediate, carrying some cholesterol esters (25% of core lipids) but mostly carrying triglycerides, while LDL and HDL mostly carry cholesterol esters ($\approx$90% of core lipids). LDL associated cholesterol is considered "bad" cholesterol and HDL associated cholesterol is considered "good" cholesterol. LDL associated cholesterol can be oxidized in a region of inflammation and lead to accumulation of excess cholesterol in macrophages (foam cells). In comparison, HDL can protect against the development of atheroschlerosis through reverse cholesterol transport, suppression of LDL oxidation, and stimulation of nitric oxide production. Nitric oxide stimulates artery dilation (widening) by relaxing smooth muscle around the artery.

11. In reality, circulating lipids are beneficial and necessary. Lipid movements help maintain appropriate intracellular and whole-body lipid levels. Lipids absorbed from the diet can be circulated throughout the body. Moveover, lipids stored in adipose tissue can be distributed to cells throughout the body. In this way, essential dietary lipids, linoleate for example, can be distributed to cells and the body benefits from the dietary buffering capacity of lipid storage. Humans are evolved to withstand short periods of starvation. The ability of some cells to export lipids permits the excretion of excess lipids from the body or their secretion into body fluids such as milk. Circulation permits the coordination of lipid and energy metabolism throughout the

organism and allows for lipid based hormone signaling.

12. Atherosclerosis affects blood circulation, a property of the organism, not an individual cell. It is debatable whether the lipid droplets in the foam cell of an atherosclerotic plaque are of any disadvantage to the individual cell. The liver is key in cholesterol regulation because it is the site of about 70 percent of the body's LDL receptors. It is also the site where unesterified cholesterol and its bile acid derivatives are secreted into bile. Familial hypercholesterolemia is typically due to a defect in LDL receptor expression. Liver transplantation is an approach to LDL receptor replacement therapy. Evolution is selection based on the success or non-success of passing genes along. Since atherosclerosis is a late onset disease, typically post reproduction, it has little obvious effect on passing genes along and hence is of little consideration from the standpoint of evolutionary selection.

13. SREBPs (sterol regulatory element binding proteins) are transmembrane proteins of the ER. The C-terminal cytosolic domain contains the SRE binding domain. When cells have adequate concentrations of cholesterol. SREBP is complexed with SCAP, Insig-1 and perhaps other proteins. SCAP has a sterol-sensing domain and binds to insig-1 at high membrane cholesterol concentrations. This blocks the inclusion of these proteins in COPII vesicles. When cholesterol concentrations drop SCAP no longer binds to Insig-1. The SREBP-SCAP complex is now transported to the *cis* Golgi and the SRE binding domain (nSREBP) is released by regulated intramembrane proteolysis (RIP). It is nSREBP that activates genes containing SRE in their promoters.

14. FXR senses the total level of bile acids in a cell irrespective of whether they are synthesized in the cell or imported. It is activated when on binding of bile acids. Both hepatocytes and intestinal epithelial cells are part of enterohepatic circulation. In the hepatocyte, bile acids are exported to the bile duct by transporters and imported from the blood by another transporter. At high bile levels, expression of both the exporter and the importer should be activated. However, expression of enzymes catalyzing the production of bile acids from cholesterol should be suppressed. The intestinal epithelial does not synthesize bile acids, but similar considerations do apply to transporters in these cells and to the intestinal epithelial cell bile acid binding protein or carrier.

15. Inflammation, cholesterol, and foam cells are key elements in atherosclerosis. Inflammation is a response to infection or tissue damage. In arteries or blood vessels, leukocytes that differentiate into macrophages are attracted. They are thought to internalize inflammation damaged LDL in an unregulated manner via scavenger receptors. This produces macrophage foam cells filled with lipid droplets containing cholesteryl esters. The localized production of macrophage foam cells is the first step in the production of an atherosclerotic plaque. The higher the LDL level relative to HDL, the higher the chance that foam cells will be produced in the coarse of a normal response to infection or tissue damage.

16. Statins and bile acid sequestrants are two classes of cholesterol lowering drugs discussed in Chapter 18. Of these, statins act by binding to HMG-CoA reductase and thereby lowering cholesterol biosynthesis. This causes a SREBP mediated increase in hepatocyte LDL receptor levels and increased clearance of LDL cholesterol from circulation. Bile acid sequestrants are insoluble resins that bind bile acids in the intestine. The bile acid-sequestrant complexes are excreted. This leads to a compensating conversion of intracellular cholesterol to bile acids and, because of the lowered cellular cholesterol levels, a SREBP mediated increase in hepatic LDL receptor synthesis. Increased receptor number means decreased circulating LDL cholesterol levels. In the end, both classes of drugs promote the production of LDL receptors and hence a receptor mediated decrease in circulatory LDL cholesterol.

Analyze the Data

Based on the SDS-PAGE analysis, there are clearly significant differences in relative mobility or molecular weight between the various molecular forms. I SREBP has an apparent mobility of ≈190 kDa, and there are likely to be 2-3 slightly different mobility species. M SREBP has an apparent mobility of ≈75 kDa and 3 slightly different mobility species, and N SREBP has a mobility of ≈50-60 kDa with 3-4 slightly different mobility species.

Product formation without apparent decrease in the relative amount of I SREBP could be due to a number of factors. A trivial or technical possibility is that the probe, an antibody for SREBP, is being used at saturating amounts. In other words, the probe results in a constant, saturating amount of I SREBP specific product. This is certainly a real possibility. However, it is not sufficient to explain the differences between gel lane 1 and 2. In lane 1, there is N SREBP and M SREBP. However, in lane 2, there is little, if any. The experimental difference is the presence of sterols during the final 5 h incubation in 2 but not 1. A plausible and simple explanation here is that N and M SREBP are metabolically labile and almost all N and M SREBP degraded during the 5 h + sterol, condition 2 incubation. In other words, product build up is limited and transient due to protein turnover. Hence one would not expect much product build up relative to I SREBP.

Added sterols as indicated by the comparison between lanes 1 and 2 suppress the production of nuclear SREBP. The normal half-life of N SREBP based on a comparison of lanes 1 and 2, and assuming that the addition of sterols results in a rapid suppression of N SREBP, production must be short, say 1 h or so. During the 5 h + sterol incubation, N SREBP has gone to almost nothing.

N SREBP is a transcription factor activating genes with sterol regulatory elements. Metabolic turnover here of N SREBP means that as sterols increase in amount, N SREBP turns over, and the genes are no longer activated. In other words, the system can be both turned on and off by variation in sterol levels. This makes for an effective regulatory system.

BFA addition, as shown by comparison of lanes 3 and 4, results in the continued presence of N and M SREBP even when sterol is added. Assuming no unexpected drug side effects here, a simple and plausible interpretation of these results is that BFA treatment results in the presence of Golgi proteins in the ER and that this then results in the continued production of N and M SREBP even in the presence of added sterols.

Let us look first at lanes 1, 2, and 5 in analyzing the second figure. These show first of all that there is no pelletable complex in the absence of a specific precipitating antibody (lane 1). Lanes 2 and 5 indicate that a 3 component complex (SREBP, SCAP and INSIG-1) coimmunoprecipitates whether anti-SCAP or anti-MycINSIG-1 antibody is used. Finally, let us consider lanes 3 and 4. The result in lane 3 indicates that INSIG interacts with SCAP; SCAP is found in the pellet with anti-MycINSIG-1 antibody. Lane 4 indicates, however, that INSIG does not directly interact with SREBP; there is no SREBP in the pellet in the absence of amplified SCAP production. Overall, the second figure indicates a complex in which INSIG interacts with SCAP which then interacts with SREBP.

19

Microfilaments and Intermediate Filaments

Review the Concepts

1. For actin filaments, polarity refers to the fact that one end is different from the other end. This difference is generated because all subunits in an actin filament point toward the same end of a filament. By convention, the end at which the ATP-binding cleft of the terminal actin subunit is exposed is designated the (-) end, while the opposite end, at which the cleft contacts the next internal subunit, is termed the (+) end. Polarity may be detected by electron microscopy in "decoration" experiments in which myosin S1 fragments (essentially myosin head domains) are incubated with actin filaments. The S1 fragments bind along the actin filament with a slight tilt, leaving the actin filament decorated with arrowheads that all point toward one end of the filament. The pointed end corresponds to the (-) end and the barbed end corresponds to the (+) end.

2. Cells utilize various actin cross-linking proteins to assemble actin filaments into organized bundles or networks. Whether the actin filaments form a bundle or network depends on the specific actin cross-linking proteins involved and the structure of the actin cross-linking protein. Actin cross-linking proteins that generate bundles typically contain a pair of tandem (i.e. closely spaced) actin-binding domains, while actin cross-linking proteins that generate networks typically contain actin-binding sites that are spaced far apart at the ends of flexible arms.

3. Once actin has been purified, its ability to assemble into filaments can be monitored by viscometry, sedimentation, fluorescence spectroscopy, or fluorescence microscopy. The viscometry method measures the viscosity of an actin solution, which is low for unassembled actin subunits but increases as actin filaments form, grow longer and become tangled. The sedimentation assay utilizes ultracentrifugation to pellet (sediment) actin filaments but not actin subunits, and thereby separates assembled actin from unassembled actin. Fluorescence spectroscopy measures a change in the fluorescence spectrum of fluorescent-tagged actin subunits as they assemble into actin filaments. Lastly, fluorescence microscopy can be used to visualize the assembly of fluorescent-tagged actin subunits into actin filaments.

4. ATP-G-actin assembles onto the ends of actin filaments and the ATP bound to the subunit is subsequently hydrolyzed to ADP and phosphate. As a result, most of the filament consists of ADP-F-actin with the exception of a small amount of ATP-F-actin at the (+) end. At the (-) end, exposed ADP-F-actin dissociates to become ADP-G-actin. If a mutation in actin prevents the protein from binding ATP, actin filaments would fail to assemble (or may assemble at very low levels if the actin could still bind ADP). If a mutation prevents actin from hydrolyzing ATP, the subunits would assemble into a filament that would be unable to disassemble normally.

5. Treadmilling is a form of actin filament assembly that occurs when the rate of subunit addition at one end equals the rate of subunit loss at the other end, so that filament length remains constant as subunits flow, or treadmill, through the filament. For treadmilling actin filaments, ATP-actin subunits add to the (+) end and ADP-actin subunits dissociate from the (-) end. Treadmilling occurs when the actin subunit concentration is greater than the C_c value for the (+) end but less than the C_c value for the (-) end.

6. Inactivation of profilin should reduce the amount of actin polymer and may disrupt

signaling pathways that involve this protein. Inactivation of thymosin β4 should increase the cell's free actin concentration and consequently actin filaments. Inactivation of gelsolin should inhibit processes such as cell locomotion and cytokinesis that depend on actin filament turnover. Inactivation of tropomodulin should reduce actin filament stability in certain muscle cells, where tropomodulin serves to cap the (-) ends of actin filaments in the sarcomere. Inactivation of Arp2/3 will inhibit assembly of actin filaments and formation of branched networks.

7. All myosins use energy derived from ATP hydrolysis to "walk" along actin filaments. Depending on the specific type of myosin, this movement is used to generate contraction or to transport specific cellular components relative to actin filaments. All myosins are composed of one or two heavy chains (the motor subunit) and several light chains. The heavy chains of all types of myosin have similar head domains (which interact with actin, bind and hydrolyze ATP, and generate force to move) but have different tail domains, which specify the particular cellular component that a given myosin recognizes and hence the function of each myosin. For each myosin type, different light chains may be present, but all associate with the neck region just adjacent to the head domain.

8. Myosin motility may be observed in the sliding-filament assay. In this approach, myosin motors are adsorbed onto the surface of a glass coverslip and the coverslip is placed onto a slide to create a chamber. Fluorescent-labeled actin filaments and ATP are then delivered to the chamber and the myosin motors will walk along the actin filaments. Since the motors are attached to the coverslip, the movement of myosin causes the actin filaments to slide across the coverslip surface. ATP must be added to these assays to provide the energy for myosin movement. The sliding-filament assay can reveal the direction of myosin movement if the polarity of the moving actin filaments is known. The force generated by myosin can be measured with an "optical trap," in which optical forces are used to determine the force just needed to hold a sliding actin filament still.

9. The principal contractile bundles of nonmuscle cells are the circumferential belt of epithelial cells, the stress fibers present on cells cultured on artificial substrates, and the contractile ring. The circumferential belt and stress fibers appear to function in cell adhesion rather than cell movement. The contractile ring generates the cleavage furrow during cytokinesis that eventually leads to division of a single cytoplasm into two.

10. The mechanism by which a rise in Ca^{2+} triggers contractions differs in skeletal and smooth muscle. In skeletal muscle, binding of Ca^{2+} to troponin C leads to muscle contraction. Contraction of smooth muscle is triggered by activation of myosin light-chain kinase by Ca^{2+}-calmodulin.

11. Keratinocytes and fibroblasts have been used for experiments on cell locomotion. The movement of these cells is pictured as a series of four steps. First, the leading edge of the cells is extended by actin polymerization. Second, the newly extended membrane forms an attachment (containing embedded actin filaments) to the substrate, which anchors this portion of the cell to the substrate and prevents its retraction. Third, the bulk of the cell body is translocated forward, perhaps by myosin-dependent contraction of actin filaments. Finally, the focal adhesions at the rear of the cell are broken, perhaps by stress fiber contraction or elastic tension, so that the tail end of the cell is brought forward.

12. Ras-related G proteins are involved in the signal-transduction pathways that are activated in fibroblasts as part of the wound healing response. Depending on the specific G protein, activation of this type of signal pathway may lead to formation of filopodia, lamellipodia, focal adhesions, and/or stress fibers. Ca^{2+} is probably involved in activation of gelsolin, cofilin, profilin, and contraction of myosin II, and may be found in intracellular gradients important for steering in chemotactic cells.

13. The fundamental subunit of intermediate filaments is the tetramer. Each tetramer is formed by the antiparallel association of two dimers, which in turn are formed by the parallel association of two monomers. Each dimer is a polar molecule with both heads at one end and both tails at the other, but the antiparallel association of two dimers produces a nonpolar tetramer with two heads and two tails at each end. Since the tetramer is nonpolar, i.e., both ends are the same, the end-to-end association of tetramers into a filament will produce a nonpolar filament.

14. Intermediate filament subunits do not bind ATP like actin, but instead assembly appears to be regulated by phosphorylation (by kinases) and dephosphorylation (by phosphatases). Disassembly of several types of intermediate filament can be induced by phosphorylation of a serine residue in the N-terminal domain. This typically occurs early in mitosis and the phosphorylation reaction may be catalyzed by cdc2 kinase. Disassembly of the cytoplasmic intermediate filament arrays probably facilitates the cytoplasmic rearrangement involved in mitosis and cytokinesis, while disassembly of the nuclear lamina contributes to the disassembly of the nuclear envelope. Phosphatase-dependent removal of the phosphate group permits filament reassembly as cells leave mitosis.

15. Intermediate filament-associated proteins cross-link intermediate filaments, organize the intermediate filament cytoskeleton and integrate it with the microfilament and microtubule cytoskeletons, and attach intermediate filaments to the plasma membrane and the nuclear membrane. Thus different intermediate filament-associated proteins may bind to microfilaments, microtubules, or various membrane proteins.

Analyze the Data

a. At concentration A, there would be no growth at the (+) end and shortening at the (−) end. At concentration B, there would be growth at the (+) end and shortening at the (−) end but less than at concentration A. At concentration C, there would be growth at both ends, but more at the (+) end.

b. In the presence of protein X, the concentration of actin needed to promote polymerization is shifted to a much higher concentration. This is the expected finding if protein X binds to the (+) end of microfilaments. If the protein bound to the (−) end, one would observe a shift of much lesser magnitude. Myosin-decorated microfilaments, in the presence or absence of protein X, could be incubated with high concentrations of actin. If protein X specifically binds to and inhibits addition at the (+) end, filament growth should be visible at the nonbarbed (−) end, but not at the barbed (+) end.

c. Extract A causes a net depolymerization of actin filaments, as evidenced by the steady increase in the amount of soluble labeled actin over time. In the presence of extract B, the actin filaments maintain a constant length; however, they must be losing and adding actin filaments at identical rates, otherwise there would be no soluble actin monomers in solution. The filaments in extract C are likewise maintaining a constant length; the low basal level of labeled monomeric actin in this case suggests that C contains an unidentified capping entity that binds to the actin filaments and inhibits their assembly and disassembly.

20 Microtubules

Review the Concepts

1. The basis of microtubule polarity is the head-to-tail assembly of αβ-tubulin heterodimers, which results in a crown of α-tubulin at the (−) end and a crown of β-tubulin at the (+) end. In nonpolarized animal cells, (−) ends are typically associated with MTOCs, and (+) ends may extend toward the cell periphery. Other arrangements occur in different types of cells, but the (−) ends are associated with a MTOC in most cases. Microtubule motors can "read" the polarity of microtubule, and a specific motor protein will transport its cargo toward either the (+) or the (−) end of the microtubule.

2. During dynamic instability, microtubules alternate between growth and shortening. The current model to account for dynamic instability is the GTP cap model. According to this model, GTP-tubulin (subunits with GTP bound to β-tubulin) can add to the end of a growing microtubule, but sometime after assembly, the GTP will be hydrolyzed to GDP, leaving GDP-tubulin, which makes up the bulk of the microtubule. Thus GTP-tubulin is present only at the microtubule end, and as long as this situation holds, the microtubule will continue to grow since the cap stabilizes the entire microtubule. However, if GDP-tubulin becomes exposed at the end then the stabilizing cap is lost and the microtubule will begin to shorten. The microtubule will continue to shorten until it disappears or until GTP-tubulin returns to the end and a new GTP cap is formed.

3. The best understood proteins involved in regulating microtubule assembly are the stabilizing MAPs. These proteins bind to microtubules and promote assembly, increase microtubule stability, and, in some cases, cross-link microtubules into bundles. The other main group of MAPs functions to destabilize microtubules. This group includes proteins such as katinin, which severs microtubules, and Op 18, which promotes the frequency of microtubule catastrophe.

4. Microtubule organizing centers (MTOCs), also known as centrosomes, are responsible for determining the arrangement of microtubules within a cell. The typical cell contains a single MTOC, although mitotic cells contain two (the spindle poles), and certain types of cells may contain several hundred. Microtubules are nucleated by γ-tubulin ring complexes, which are located in the pericentriolar material of an MTOC. The γ-tubulin provides a binding site for αβ-tubulin dimers, and the ring complex structure appears to provide a template for nucleating microtubule formation.

5. Microtubule/tubulin-binding drugs are used to treat a variety of diseases, including gout, certain skin and joint diseases, and cancer. Such drugs either prevent microtubule assembly (e.g., colchicine) or prevent microtubule disassembly (e.g., taxol). Although the effects are opposite, the results are the same: inhibition of cellular processes that depend on microtubules and the dynamic rearrangement of these polymers.

6. Kinesin I was first isolated from squid axons, which were manipulated to produce extruded cytoplasm, a cell-free system to study synaptic vesicle motility. Video microscopy was used to follow the ATP-dependent movement of synaptic vesicles along individual microtubules in the extruded cytoplasm, but when purified synaptic vesicles were added to purified microtubules, no movement was observed even in the presence of ATP. Subsequent addition of a squid cytosolic extract to the purified components restored ATP-dependent vesicle movement along microtubules, indicating that a soluble protein in the cytosol was responsible for driving vesicle movement. To identify the soluble "motor" protein, researchers took advantage of previous experiments with

extruded cytoplasm that demonstrated that a nonhydrolyzable ATP analog (AMPPNP) caused vesicles to bind so tightly to microtubules that movement was stopped. To purify the motor, AMPPNP was added to a mixture of cytosolic extract and microtubules with the goal of forcing the motor to bind microtubules. The microtubules and any bound proteins were then recovered by centrifugation and treated with ATP to release proteins that bound microtubules in an ATP-dependent manner (a fundamental property of microtubule-dependent motor proteins). The predominant protein released in this approach was kinesin I.

7. Although microtubule orientation is fixed by the MTOC (and any given motor only moves in one direction), some cargoes are able to move in both directions along a microtubule because they are able to interact with both (+) end- and (-) end-directed motor proteins. The direction that a given cargo moves along a microtubule appears to be controlled by swapping one motor protein for the other (it may also be possible to activate one motor and inactivate the other). Certain cargoes may move on both microtubules and actin filaments if the cargo contains binding sites for both microtubule and actin motor proteins.

8. The kinesin family motor (or head) domain contains the ATP binding site and the microtubule-binding site of the motor, while the neck is a flexible region connecting the motor domain to the central stalk domain. The kinesin motor domain is required to generate movement but does not appear to determine the direction a kinesin motor will move on a microtubule. Instead, the neck determines the direction of movement. These conclusions are based on experiments in which the motor domains of (+) and (-) end-directed motors were swapped with no effect on direction of motor movement, and on experiments in which mutations of the neck region caused a change from a (-) to a (+) end-directed motor.

9. The appendages (cilia and flagella) used for cell swimming contain a highly organized core of microtubules and associated proteins. This core, termed the axoneme, is typically made of nine

outer doublet microtubules and two central pair microtubules (termed the 9 + 2 arrangement). Each outer doublet consists of a 13-protofilament microtubule and a 10-protofilament microtubule, while the central pair contains 13 protofilaments each. Cell movement depends on axoneme bending, which, in turn, depends on force generated by axonemal dyneins. These motor proteins act to slide outer doublet microtubules past each other, but this sliding motion is converted into bending due to restrictions imposed by cross-linking proteins in the axoneme, and perhaps by the action of inner arm dyneins.

10. The three types of microtubules that comprise the spindle are kinetochore microtubules, polar microtubules and astral microtubules. The (-) ends of all three types associate with spindle poles. Kinetochore microtubules connect chromosomes, via the kinetochore attachment site, to the spindle poles. Polar microtubules from each pole overlap and are involved in holding the poles together and regulating pole-to-pole distance. Astral microtubules radiate from each spindle pole toward the cortex of the cell, where they help position the spindle and determine the plane of cytokinesis.

11. Inhibition of Kin-C would likely produce an elongated spindle or cause the spindle poles to completely separate into two asters. Inhibition of BimC would prevent spindle pole separation and hence formation of a bipolar spindle. Inhibition of cytosolic dynein would eliminate or reduce poleward kinetochore movement during prometaphase, metaphase, and anaphase A, and would block or reduce spindle pole separation during anaphase B. Inhibition of CENP-E may inhibit the ability of kinetochores to remain attached to microtubules during prometaphase, metaphase, and anaphase A. For each motor, inhibition of the motor's function would lead to problems with the accurate segregation of chromosomes to the daughter cells.

12. Motor proteins such as CENP-E at the kinetochore may allow this structure to hold onto shortening microtubules. It is not clear whether this activity requires ATP hydrolysis and subsequent force generation by the motor, since

kinetochores have been shown to hold onto depolymerizing microtubules in vitro in the absence of ATP. The MCAK motor, which is also present at kinetochores, may promote microtubule depolymerization.

13. The separation of spindle poles during anaphase B is thought to depend on BimC motors present on microtubules in the overlap zone between the poles, which act to push the spindles apart, as well as on cytosolic dynein motors on the inner surface of the cell membrane, which act to pull astral microtubules and hence the poles apart (different organisms may utilize pushing and pulling forces to differing degrees during anaphase B). In addition, elongation of microtubules in the overlap zone appears to increase the extent of pole separation.

14. Depending on the type of cell, the plane of cytokinesis is determined either by the spindle or the asters. In animal cells, the spindle determines the cleavage plane, while in sand dollar eggs, the asters determine the cleavage plane. In either case, it is believed that microtubules (spindle or astral) send a signal to the region of the cell cortex midway between the asters that promotes the assembly of actin and myosin II into the contractile ring and subsequent activation of contractile ring contraction. Microtubules are therefore involved with determining the plane of cytokinesis while actin filaments, as components of the contractile ring, carry out the process of cytokinesis.

b. In the control (Ctrl) and kinesin heavy chain (KHC) samples, all tubulin remained assembled in microtubules as indicated by the presence of tubulin in the pellet fractions. However, Kin I motor in the presence of ATP must have depolymerized the microtubules since the tubulin was found completely in the supernatant fraction (indicating it was no longer assembled). In addition, the Kin I motor must require energy from ATP hydrolysis to produce disassembly since there was no disassembly in the presence of the nonhydrolyzable ATP analog, AMPPNP. These results suggest that ATP was present in the experiments in part (b) since the Kin I produced microtubule disassembly.

c. Since Kin I can disassemble GMPCPP microtubules, the mechanism by which Kin I causes disassembly is not as simple as promoting hydrolysis of the GTP bound to the subunits that form the GTP cap. Perhaps the motor physically separates tubulin subunits from the microtubule end.

d The Kin I motor is probably present at microtubule ends, which are lower in concentration than tubulin subunits, and which are the sites of tubulin addition and loss.

Analyze the Data

a. Substitution with the tubulin solution has no effect on microtubule growth while replacement with buffer alone causes microtubule shortening because there is no free tubulin to support elongation. In comparison, the plot shown in part (b) indicates that even when free tubulin is available, the Kin I motor acts to promote microtubule shortening (although at a rate slower than results from dilution of the tubulin in the chamber).

21 Regulating the Eukaryotic Cell Cycle

Review the Concepts

1. The unidirectional and irreversible passage through the cell cycle is brought about by the degradation of critical protein molecules at specific points in the cycle. Examples are the proteolysis of securin at the beginning of anaphase, proteolysis of cyclin B in late anaphase, and proteolysis of the S-phase Cdk inhibitor at the start of S phase. The proteins are degraded by a multiprotein complex called a proteasome. Proteins are marked for proteolysis by the proteasome by the addition of multiple molecules of ubiquitin to one or more lysine residues in the target protein. Securin and cyclin B are both polyubiquitinated by the APC complex. The S phase Cdk inhibitor is polyubiquitinated by the Cdc34 pathway.

2. When fused with a cell in S phase, a cell in G_1 will immediately enter S phase and begin DNA replication because S-phase promoting factors (SPFs), which are the S-phase Cdk complexes, can activate prereplication complexes on replication origins in G_1 nuclei. However, prereplication complexes have already been destroyed in G_2 and M phase nuclei, ensuring that DNA is replicated once and only once per cell cycle. Therefore, G_2 and M phase cells will not initiate DNA replication when fused to an S-phase cell. Instead, the M phase cell will induce the S phase cell to prematurely begin chromatin condensation, since the S phase cell is susceptible to the mitotic Cdk complexes, which function as mitosis-promoting factor (MPF). When G_1 and G_2 phase cells are fused, each will enter S phase according to its own timetable since the G_1 nucleus is licensed for replication and the G_2 nucleus will not be licensed until that cell progressed through M phase and forms pre-replication complexes early in G_1.

3. The wee phenotype in *S. pombe* displays smaller than usual cells. Premature entry into mitosis, before the cell has grown to the size that normally signals cell division, is the cause of this phenotype. Wee cells result from the excess activity of Cdc2, the cyclin-dependent kinase of *S. pombe* MPF. The wee phenotype can result from a mutation in the *wee1* gene, which encodes the Wee1 protein kinase responsible for catalyzing the addition of phosphate to tyrosine 15 of Cdc2, which inhibits Cdc2 function and prevents premature entry into mitosis. In addition to a loss-of-function mutation in *wee1*, the wee phenotype results when a mutation renders Cdc2 insensitive to Wee1 or a mutation in which Cdc25, the phosphatase that opposes Wee1, is overexpressed. Discovery of the wee phenotype and the characterization of the *wee1* gene revealed the intimate link between cell size and cell cycle progression as well as the important role that tyrosine phosphorylation plays in regulating the activity of Cdks.

4. Murray and Kirschner performed a classic set of experiments in frog egg extracts to reveal the essential role of cyclin B synthesis and degradation in cell cycle progression. In one experiment, extracts were treated with RNase to destroy all endogenous mRNAs. These extracts arrested in interphase, suggesting that an essential a protein needed to be translated to drive the cell cycle into mitosis. When RNase-treated extracts were supplemented with a single exogenous mRNA, encoding wild-type cyclin B, the extract progressed into mitosis, indicating that cyclin B was the essential protein that had to be synthesized to drive entry into mitosis. When RNase-treated extracts were supplemented with mRNA encoding a nondegradable form of cyclin B, the extract entered mitosis and arrested there with high MPF activity, instead of eventually destroying MPF and exiting mitosis. This experiment revealed that degradation of cyclin B was necessary for mitotic exit.

5. A cell cycle checkpoint is a place in the cell cycle where a cell's progress through the cycle is

monitored, and, if a signaling pathway is engaged, the activity of the relevant Cdks will be inhibited. The cell cycle will arrest at this checkpoint. Checkpoints exist at G_1 and S phases to assess DNA damage, at G_2 to assess DNA damage and to determine whether DNA replication is complete, and at M phase to identify any problems with assembly of the mitotic spindle or chromosome segregation. Because these checkpoints identify problems with the genome (unreplicated, damaged, or improperly segregated DNA), and arrest the cell cycle so that these problems can be fixed, checkpoints can prevent the propagation of mutations into the next cell generation and thereby preserve the fidelity of the genome.

6. MPF (cyclin B, Cdc2) is the substrate of Cdc25. Cdc25 removes inhibitory phosphate groups from Cdc2. MPF activates Cdc25 (by phosphorylating it) and Cdc25, in turn, activates MPF (by dephosphorylating it). Therefore, when a small amount of active MPF is injected into an immature egg, it will phosphorylate and activate the endogenous Cdc25, which can then dephosphorylate and activate large stores of inactive MPF. This explains the autocatalytic nature of MPF.

7a. CDKs are active as kinases only when bound to a cyclin. Cyclin-binding exposes the active site of the CDK and also helps to form the substrate-binding pocket.

7b. CAK is a kinase that phosphorylates cyclin-CDKs on a threonine residue in the T loop. This phosphorylation event induces a conformational change that increases affinity of CDKs for their substrates, thereby greatly enhancing the catalytic activity of the CDK.

7c. Wee1 is a kinase that phosphorylates CDKs on tyrosine 15 in the ATP-binding region. This phosphorylation interferes with ATP binding and thereby inhibits the catalytic activity of the CDK.

7d. p21 is a stoichiometric inhibitor that binds and inhibits the activity of cyclin-CDKs, usually in response to damaged DNA.

8. Phosphorylation of the nuclear lamins by MPF causes their depolymerization, leading to the breakdown of the nuclear envelope during mitosis. Phosphorylation of condensin by MPF or a kinase regulated by MPF promotes chromatin condensation. Phosphorylation of myosin light chain by MPF prevents premature cytokinesis until late anaphase, when MPF is destroyed by the destruction of mitotic cyclins.

9. To initiate sister chromatid segregation at anaphase, the APC polyubiquitinates securin, targeting it for degradation by the proteasome. Degradation of securin releases the enzyme separase, which cleaves Scc, a protein that cross-links sister chromatids.

10. Cdc28-Clns phosphorylate and target Sic1 for degradation. Cdc28-Clns turn off the APC. Cdc28-Clns promote the synthesis of Clbs by activating their transcription factor, MBF.

11. In *S. cerevisiae*, S-phase CDKs become active at the beginning of S, when the CDK inhibitor Sic1 is degraded, and remain active until telophase. Pre-replication complexes can assemble on origins of replication only during G_1 when S-phase CDK activity is low. However, initiation of replication requires the phosphorylation of components of the pre-replication complex by S-phase CDKs. Once an origin has "fired" (i.e. replication has been initiated), the persistence of S-phase CDK activity during S, G_2 and M prevents reassembly of pre-replication complexes on that origin until the cell has completed mitosis and entered G_1 of the next cell cycle. Therefore, each origin initiates replication once and only once per cell cycle due to the oscillating activity of S-phase CDKs.

12. The restriction point is the place in the cell cycle beyond which cells are committed to completing DNA replication and mitosis even if growth factors, or mitogens, are removed. To enter the cell cycle, quiescent cells in G_0 require growth factors, which bind to cell surface receptors and trigger a signaling cascade that leads to the transcription of early-response genes and then delayed-response genes. Among the delayed-response genes is the cyclin D gene, which partners with Cdks 4 and 6, and the cyclin-Cdk

complex phosphorylates the Rb protein. When Rb is phosphorylated by cyclin D-Cdk4/6, it can no longer bind the transcription factor E2F. When E2F is released from Rb, then it induces transcription of the genes that promote entry into S phase.

12a. High levels of cyclin D bypass the requirement for growth factors, which normally induce synthesis of cyclin D.

12b. If Rb is not functional, then growth-factor induced synthesis of cyclin D is not required to promote phosphorylation and inactivation of Rb by Cdk 4/6.

12c. Virally-encoded Fos and Jun bypass the requirement for growth factors to induce expression of cellular *fos* and *jun*, which are early-response genes.

13. ATM phosphorylates p53. Phosphorylation by ATM stabilizes p53, which can then function as a transcription factor to induce the expression of the CDK inhibitor, p21. p21 binds to and inhibits cyclin-CDKs, thereby arresting the cell cycle. ATM also phosphorylates Chk1 and Chk2 kinases. Chk1 and Chk2 phosphorylate the phosphatases Cdc25A and Cdc25C, targeting Cdc25A for degradation and inactivating Cdc25C. In the absence of Cdc25 phosphatases, CDKs are maintained with inhibitory phosphorylations, thereby arresting the cell cycle.

14a. Ime2 replaces the G_1-CDK function of phosphorylating Sic1, allowing the cell to enter S phase. Since Ime2 is expressed during meiosis I but not meiosis II, DNA replication is prevented during meiosis II, allowing for reduction to 1n chromosome content in the resulting gametes.

14b. Rec8, a homolog of the mitotic cohesin subunit Scc1, maintains centromeric cohesion of sister chromatids during meiosis I. Centromeric Rec8 is protected from degradation by separase during meiosis I, so that sister chromatids remain attached. Rec8 is degraded during meiosis II when sister chromatids must separate.

14c. Monopolin is required for the formation of specialized kinetochores during meiosis I that co-orient sister chromatids of synapsed homologous chromosomes.

Analyze the Data

The boxed regions are areas where the amino acid sequence is highly conserved, in fact identical, among *cdc2* in human, *cdc2* in *S. pombe*, and *cdc28* in *S. cerevisiae*. The assumption made is that these regions would also be conserved in plant CDKs, and therefore degenerate primers based on these sequences would be the most likely to amplify a plant CDK gene from a cDNA library.

The level of amino acid sequence similarity is high given the evolutionary distances among plants, yeast and mammals. The high degree of sequence similarity suggests that the CDK genes perform essential, conserved functions that have not diverged during the course of evolution.

Cdc28^ts cells possess a mutation in the *cdc28* gene product such that it is functional at low temperatures (25°C) and non-functional at higher temperatures (37°C). Typically, temperature-sensitive mutant proteins are partially denatured at the higher, restrictive temperature. Because Cdc28 is essential for progression through the cell cycle, only *cdc28*^ts cells with functional Cdc28, that is, those grown at 25°C, will proliferate and form colonies.

When *cdc28*^ts cells containing maize *cdc2* are grown at the restrictive temperature (37°C), they form colonies. This indicates that the maize *cdc2* gene can replace or **complement** the non-functional yeast *cdc28* gene.

This experiment, like many others that have been performed, indicates that CDK genes from distant eukaryotic species not only exhibit a high degree of similarity in amino acid sequence, but that they are also functionally very similar.

22

Cell Birth, Lineage, and Death

Review the Concepts

1. By definition, a stem cell divides to give rise to a copy of itself and to a differentiated cell or a cell capable of differentiating into multiple cell types, such as a mutipotential progenitor cell. Totipotential stem cells can give rise to every tissue in an organism. Pluripotent stem cells give rise to multiple, but not necessarily all, cell types. Progenitor cells give rise to more than one cell type but, unlike a stem cell, do not self-renew.

2. In plants, stem cells are located in meristems, such as shoot apical meristems (SAMs) and floral meristems. In adult animal cells, stem cell populations are thought to exist in low numbers in many organs including skin, intestine, and bone marrow. SAMs in plants are embryonic-like in their concentration of totipotent stem cells. However, stem cells are difficult to purify from adult animals, and the only totipotent stem cells found in animals are in very early stage embryos.

3. Because Dolly was derived from an egg containing a nucleus from an adult, differentiated cell, we know that differentiated nuclei (at least adult mammary cells) have the potential to dedifferentiate and become totipotent. Since Dolly was derived from a differentiated nucleus placed into an egg and not from an intact, differentiated cell, we can conclude nothing about the ability of a differentiated cell to become totipotent. Other than the nucleus, the other organelles, including mitochondria, which also contain genetic material, were derived from a germ cell. Differentiation of cells is maintained by cytoskeletal structures, organelles that confer cell properties, particular modifications of key regulatory proteins, and accessibility of regulatory genes in the chromatin. The Dolly experiment best indicates that the chromatin of differentiated nuclei can be remodeled from a differentiated to a totipotent state.

4. Because *C. elegans* consist of a small, invariant number of cells, it has been possible to generate a fate map of every cell from the fertilized egg to adulthood. *C. elegans* is also very amenable to genetic manipulation. Therefore, it is possible to alter the expression of specific genes and then determine the effect of this manipulation on cell division, cell differentiation and cell death. Because many differentiation pathways are highly conserved between *C. elegans* and mammals (e.g. the apoptotic pathway) much of the information derived from studies in *C. elegans* can be applied by analogy to mammalian systems and homologous genes can be discovered.

5a. MCM1 binds efficiently to the P site of **a**-specific genes in **a** cells to promote transcription.

5b. MCM1 alone does not bind efficiently to the P site of α-specific genes in **a** cells.

5c. MCM1 binds efficiently to the P site of α-specific genes when α1 is bound to the adjacent Q site in α cells.

5d. MCM1 bound to the P site complexed with α2 dimers bound to the adjacent α2 sites inhibit transcription of **a**-specific genes in α cells.

6. **a** cells secrete only **a**-type mating phermone and express only α-type phermone receptor. α cells secrete only α-type mating phermone and express only **a**-type phermone receptor. Therefore, each haploid cell is able to attract and respond only to cells of the opposite mating type.

7. Exposure of C3H 10T1/2 cells to 5-azacytidine is thought to induce muscle differentiation by incorporation of this compound, which cannot be methylated, into DNA. As a result, genes that previously had been inactivated by methylation are re-activated, and a different phenotype is

expressed. The first step in isolating the genes involved in muscle differentiation was to demonstrate that DNA isolated from cells treated with 5-azacytidine, called azamyoblasts, could transform untreated C3H 10T1/2 cells into muscle. The mRNAs isolated from azamyoblasts were then converted to cDNAs and subjected to subtractive hybridization with mRNAs extracted from untreated cells. The azamyoblast-specific cDNAs then were used as probes to screen an azamyoblast cDNA library. The genes isolated by this procedure were tested for their ability to promote muscle differentiation by transfecting them into C3H 10T1/2 cells and assaying for the muscle protein myosin with an immunofluorescence assay.

8. MyoD is a member of the helix-loop-helix family of transcription factors.

8a. MyoD binds the E-box of target genes with 10-fold higher affinity when dimerized with E2A.

8b. MEF homodimers synergize with MyoD-E2A dimers to induce transcription of target genes.

8c. Id contains a dimerization region but lacks a DNA-binding region. Id forms heterodimers with MyoD, preventing its association with DNA and thereby blocking its activity as a transcription factor.

9. MyoD = Achaete and Scute; myogenin = Asense; Id = Emc; E2A = Da. Injection of MyoD mRNA into *Xenopus* embryos should result in an increase in the amount of muscle tissue as measured by the expression of muscle-specific genes.

10a. Since HO endonuclease is required to catalyze allele switching, neither cell should be able to undergo mating-type switching.

10b. Constitutive expression of HO endonuclease should render both mother and daughter cell capable of mating-type switching.

10c. Since SWI/SWF is the transcription factor that induces expression of the HO gene, both mother

and daughter cells should be capable of mating-type switching.

11. In *S. cerevisiae*, the myosin motor protein, Myo4p, localizes Ash1 mRNA to the bud that will form the daughter cell. In *Drosophila* neuroblasts, microtubules are required for assembly of the Baz/Par6/PKC3 protein complex at the apical end.

12. In mutant mice in which either neurotrophins or their receptors are knocked out, specific classes of neurons die by apoptosis. These results indicate that apoptosis occurs by default unless a specific extracellular signal is transduced to block the apoptotic program.

13. Apoptosis is characterized by cell shrinkage and DNA fragmentation. During necrosis, cells swell and lyse, inducing damage and inflammation in surrounding tissue. Although external signals such as TNF and Fas ligand induce apoptosis, the responding cell must still transduce the death signal through an intracellular pathway and induce its own death by activation of the caspase enzymes. The morphologic events of this death are indistinguishable from apoptosis triggered by an intrinsic pathway.

14a. The cell should undergo apoptosis even in the presence of trophic factors.

14b. The cell should not undergo apoptosis even in the absence of trophic factors.

14c. The cell should not undergo apoptosis even in the absence of trophic factors.

Both b and c could be found in cancer cells since either mutation would block apoptosis even in the absence of trophic factors.

Analyze the Data

a. Because male mouse cells contain a Y chromosome and female cells do not, the presence of a Y chromosome serves as a marker for cells derived from the donor mouse. Although the converse experiment is possible, it would be much harder to detect the infrequent cells that lacked a Y chromosome (or contained two rather than one X chromosome) than it is to screen for the presence of rare Y chromosome-positive cells among a Y-negative background of female tissue.

b. Since irradiation destroyed all the bone marrow stem cells in the recipient mouse, all blood cells subsequently generated had to be derived from the donor cells. Transplanting bone marrow into a non-irradiated mouse may also have resulted in a percentage of blood cells derived from donor stem cells, but the frequency would have been much lower.

c. Using cell surface markers, FACS (fluorescence activated cell sorting) could have been used to purify a population of stem cells from bone marrow. A pure population may have resulted in higher levels of engraftment or a higher survival rate.

d. The detection of Y-chromosome positive cells in various organs indicates that the bone marrow cells or their progeny are capable of migrating throughout the organism. Since these cells bear some markers of the tissue of localization, it is possible that bone marrow stem cells differentiated into non-blood lineages appropriate for the site of their localization. Because these cells were rare and no functional tests could be performed, it is impossible to draw conclusions about the functionality of donor-derived cells beyond their ability to reconstitute a functional hematopoetic system.

23

Cancer

Review the Concepts

1. Benign tumors remain localized to the tissue of origin, often maintaining normal morphology and function, and are pathological only if their sheer mass interferes with tissue function or if they overproduce a hormone or other factor that disrupts normal body homeostasis. Malignant tumors possess cells that divide more rapidly than normal, fail to die by apoptosis, invade surrounding tissues and may metastasize to other parts of the body. The genetic difference between benign colon polyps and malignant colon carcinoma is in the number of cancer-promoting mutations. The polyp possesses a loss-of-function mutation in the *APC* gene whereas the malignant carcinoma possesses the *APC* mutation as well as other cancer-promoting mutations in the *K-ras* and *p53* genes.

2. Metastasis is the process by which cancer cells escape their tissue of origin, travel through the circulation and invade and proliferate within another tissue or organ.

2a. By inhibiting enzymes that degrade the extracellular matrix, cancer cells will be unable to digest the basal lamina and escape the tissue of origin.

2b. Inhibition of integrin function will prevent attachment of cancer cells to the basal lamina, an early step in metastasis.

2c. By inhibiting osteoclasts, which are recruited and activated by many cancer cells, particularly cancers that originate in bone marrow, the cells will not be able to escape from surrounding bone tissue to enter the circulation and metastasize or populate other bone marrow tissues.

3. The growth factors bFGF, TGFα, and VEGF all promote angiogenesis, the proliferation of blood vessels. If cancer cells acquire the ability to induce angiogenesis, then the tumor can develop its own vasculature and grow to a virtually unlimited size.

4. The increased incidence of cancer with age is explained by a "multi-hit" model; successive mutations or alterations in gene expression correspond to the discrete stages leading to a lethal tumor. For example, many colon cancers contain mutations in *APC*, *DCC* and *p53*, tumor suppressor genes, and in *ras*. The *APC* mutation is found in polyps, an early stage of colon cancer, while *p53* mutation is required for malignancy. In mice, overexpression of *myc* or expression of *ras*^{D} causes cancer only after a long lag. However, these two genes act synergistically to cause cancer in at least one-third the time of either alone.

5. Proto-oncogenes are genes that become oncogenes by mutations that render them constitutively or excessively active. They promote cell growth, inhibit cell death or promote some other aspect of the cancer phenotype such as metastasis. Tumor-suppressor genes restrain growth, promote apoptosis, or inhibit some other aspect of the cancer phenotype. Gain-of-function mutations convert proto-oncogenes to oncogenes, and thus only a single copy of the proto-oncogene needs to be mutated to an oncogene to be cancer promoting. Loss-of-function mutations in tumor-suppressor genes are cancer promoting, and thus, both copies of the gene usually need to be inactivated unless mutation of a single copy functions in a dominant-negative manner as is the case with some mutations in the *p53* gene. The *ras*, *bcl-2*, *telomerase* and *jun* genes are proto-oncogenes. The *p53* and *p16* genes are tumor-suppressor genes.

6. In hereditary retinoblastoma, individuals have inherited one mutated copy of the *RB* gene, and

therefore require only a spontaneous mutation in the other copy to lack functional Rb protein. The relative frequency of a single spontaneous mutation is high enough that these individuals develop retinoblastoma early in life in both of their eyes. However, in spontaneous retinoblastoma, individuals have inherited two normal copies of the *RB* gene. Therefore, spontaneous mutations in each copy of *RB* must occur within a single cell for it to lack functional Rb. The likelihood of a cell possessing both mutations is extremely low, and thus these mutations rarely occur until adulthood and then usually in a single eye. Because the chance of an individual with hereditary retinoblastoma receiving an inactivating mutation in the other copy of the *RB* gene in any one of the susceptible cells is quite high, the disease is inherited in a dominant manner.

7.   Many individuals are genetically predisposed to cancer due to the loss or inactivation of one copy of a tumor-suppressor gene. Loss-of-heterozygosity (LOH) describes the loss or inactivation of the second, normal copy in a somatic cell, a prerequisite for the development of at tumor since one functional copy of a tumor-suppressor gene is usually sufficient for normal function. Since the development of cancer requires loss-of-function in one or more tumor-suppressor genes (e.g. *RB*, *p53*), LOH of at least one allele is found in virtually all malignant tumors. One mechanism by which loss-of-heterozygosity develops is the mis-segregation of chromosomes during mitosis. The spindle assembly checkpoint normally arrest cells in mitosis until chromosomes are properly aligned on the mitotic spindle. If this checkpoint is not functional, mis-segregation events leading to LOH are more frequent.

8.   Transmembrane growth factor receptors such as the EGF receptor are protein tyrosine kinases. Cytokine receptors such as the erythropoietin receptor activate associated JAK kinases.

8a.   gp55 binds to the erythropoietin receptor, causing the receptor to dimerize and become activated, leading to the constitutive activation of associated JAK kinases, even in the absence of erythropoietin.

8b.   The chimeric protein generated when the extracellular domain of the Trk receptor fuses with the N-terminal region of nonmuscle tropomyosin can dimerize due to the coiled-coil structure of the tropomyosin region. The Trk receptor tyrosine kinase is thus constitutively active even in the absence of ligand.

8c.   A point mutation in the Her2 receptor likewise causes receptor tyrosine kinase dimerization and activation even in the absence of EGF ligand. The resulting constitutively active receptor is called the Neu oncoprotein.

9.   Gain-of-function (GOF) mutations in the *ras* gene (i.e. *ras*^D), renders Ras constitutively active in the GTP-bound form. Constitutively active Ras activates the growth-promoting MAPK signaling pathway, even in the absence of upstream signals from growth factor-bound receptor tyrosine kinases. Loss-of-function mutations (LOF) in *NF-1* have the same effect as GOF mutations in *ras* since *NF-1* encodes a protein that hydrolyses GTP bound to Ras, converting Ras to the inactive, GDP-bound form. Since GOF mutations (such as the formation of Ras^D) require only a single allele to be mutated, whereas in LOF mutations (such as the inactivation of NF-1), usually both alleles must be mutated, cancer-promoting mutations in Ras are more common than cancer-promoting mutations in NF-1.

10.   The v-Src protein lacks the carboxy terminal 18 amino acids, including tyrosine 527. Phosphorylation of tyrosine 527 on c-Src by Csk causes a conformational change that inactivates Src. Because v-Src lacks this phosphorylation site, it is insensitive to Csk and therefore constitutively active.

11.   In Burkitt's lymphoma, a translocation event places the *c-myc* gene under the influence of the antibody heavy chain gene enhancers. Thus, *myc* is expressed at high levels, but only in cells in which antibodies are produced, e.g. B-lymphocytes. Thus, this mutation is found in lymphomas rather than in other types of cancers. *Myc* can also be rendered oncogenic by amplification of a DNA segment containing the

myc gene. This type of mutation is not restricted to lymphomas.

12. Smad4 is a transcription factor that transduces the signal generated when TGFβ binds to its receptor on the plasma membrane. Smad4 promotes expression of the p15 gene, which, like p16, inhibits cyclin D-Cdk function, promoting cell cycle arrest in G_1. Smad4 also promotes expression of extracellular matrix genes and plasminogen activator inhibitor 1 (PAI-1), both of which inhibit the metastasis of tumor cells. Thus a loss of Smad4 abrogates both the proliferation and metastasis inhibiting effects of TGFβ signaling.

13. p53 inhibits malignancy in multiple ways: When cells are exposed to ionizing radiation, p53 becomes stabilized and functions as a transcription factor to promote expression of p21^{CIP}—leading to cell cycle arrest in G_1—and to repress expression of cyclin B and topoisomerase II, leading to cell cycle arrest in G_2. Thus, p53 functions in DNA damage checkpoints during both G_1 and G_2 of the cell cycle. p53 can also promote apoptosis, in part by promoting transcription of Bax. A loss of cell cycle checkpoints and apoptosis are both characteristics of cancer cells.

13a. Human papillomavirus encodes the E6 protein, which binds to and inhibits p53.

13b. The carcinogen benzo(*a*)pyrene is activated by enzymes in the liver to become a mutagen that coverts guanine to thymine bases, including several guanines in *p53*, rendering the gene nonfunctional.

14a. UV irradiation causes thymine-thymine dimers. These are usually repaired by the nucleotide excision repair system, which utilizes XP complexes and the transcriptional helicase TFIIH to unwind and excise the damaged DNA. The gap is then filled in by DNA polymerase.

14b. Ionizing radiation causes double-stranded breaks in DNA. Double-stranded breaks are repaired either by homologous recombination or DNA end-joining. Homologous recombination requires the BRCA1, BRCA2 and Rad51 proteins to use the sister chromatid as template for error-free repair. DNA end-joining is error-prone since nonhomologous ends are joined together.

Since formation of a malignant tumor requires multiple mutations, cells that have lost DNA repair function are more likely to sustain cancer-promoting mutations. Examples of this are individuals who suffer from xeroderma pigmentosum due to mutations in XP genes that prevent repair of thymine dimers and individuals with a genetic predisposition to breast cancer due to germ-line mutations in the BRCA1 or BRCA2 genes.

15. In humans, normally only germ cells and stem cells possess telomerase activity. Telomerase maintains telomere ends and promotes immortality of cells, one characteristic of cancer. Since stem cells express telomerase, they may have a greater likelihood of becoming malignant, a concern that needs to be addressed if stem cells are to be used therapeutically to treat human disease.

Analyze the Data

1. Breast tissue is induced to proliferate during pregnancy and lactation. Since cancer cells are characterized by enhanced proliferation, comparison of the gene expression profiles in mammary tissue of pregnant and lactating rats to malignant carcinomas should reveal those genes with altered expression specific to malignancy and not normal proliferation.

2. The growth factor PDGF, the cyclin-dependent kinase Cdk4, and cyclin D all promote self-sufficiency in growth by participating in the signaling cascade that inactivates Rb and allows cells to pass the restriction point.

3. If extracellular matrix production is decreased, cancer cells may have an easier time escaping the tissue of origin to become metastatic.

4. Since STAT5a regulates the expression of cyclin D, a gain-of-function mutation in STAT5a could constitutively activate a normal, unmutated cyclin D gene. Since transcription factors coordinate the expression of many genes, a mutation in a transcription factor can have multiple consequences. In this case, cancer promoting effects in the expression of genes that promote self-sufficient growth (e.g. cyclin D) and regulate apoptosis (e.g. Bcl-X$_L$) is a possbility.

5. At least with respect to these two carcinogens, PhIP and DMBA, a similar pattern of gene expression characterizes the resulting carcinomas, suggesting that many of the same signaling pathways have been altered to generate the characteristics of cancer described throughout Chapter 23.

6. Most likely, the gene expression profile from a tumor of different origin, even if treated with the same carcinogen, would demonstrate greater differences because each tumor would retain some of the properties of its tissue of origin. Different mutations would be cancer-promoting for each tumor based on the signal and signaling pathways characteristic of that tissue.